The Anthropology of
IRAQ

Informed by the author's extensive fieldwork in Iraq, this work is an invaluable resource for all those interested in the anthropology of Iraq. Providing the reader first with important background information about the geography and climate of Iraq, the author goes on to give a detailed account of its peoples, presenting information on their physical characteristics and health in clear prose as well as in numerous readable tables. The work is supplemented by appendices which describe Iraq's mammals, insects and plants.

Henry Field was Curator of Physical Anthropology at the Field Museum of Natural History, Chicago.

THE KEGAN PAUL
ARABIA LIBRARY

ARABIA AND THE ISLES
Harold Ingrams

STUDIES IN ISLAMIC MYSTICISM
Reynold A. Nicholson

LORD OF ARABIA: IBN SAUD
H. C. Armstrong

AVARICE AND THE AVARICIOUS
Abu 'Uthman 'Amr ibn Bahr al-Jahiz

TWO ANDALUSIAN PHILOSOPHERS
*Abu Bakr Muhammad ibn Tufayl
& Abu'l Walid Muhammad Ibn Rushd*

THE PERFUMED GARDEN OF SENSUAL DELIGHT
Muhammad ibn Muhammad al-Nafzawi

THE WELLS OF IBN SAUD
D. van der Meulen

ADVENTURES IN ARABIA
W. B. Seabrook

THE SAND KINGS OF OMAN
Raymond O'Shea

THE BLACK TENTS OF ARABIA
Carl S. Raswan

BEDOUIN JUSTICE
Austin Kennett

THE ARAB AWAKENING
George Antonius

ARABIA PHOENIX
Gerald De Gaury

HOW GREEK SCIENCE PASSED TO THE ARABS
DeLacy O'Leary

SOBRIETY AND MIRTH
Jim Colville

IN THE HIGH YEMEN
Hugh Scott

ARABIC CULTURE THROUGH ITS LANGUAGE AND LITERATURE
M. H. Bakalla

IBN SA'OUD OF ARABIA
Ameen Rihani

IRAQ
Philip Willard Ireland

IRAQ FROM MANDATE TO INDEPENDENCE
Ernest Main

SOUTHERN ARABIA
J. Theodore Bent

THE SYRIAN DESERT
Christina Phelps Grant

THE TRAGEDY OF THE ASSYRIAN MINORITY IN IRAQ
R. S. Stafford

THE TRIBES OF THE MARSH ARABS OF IRAQ
Fulanain

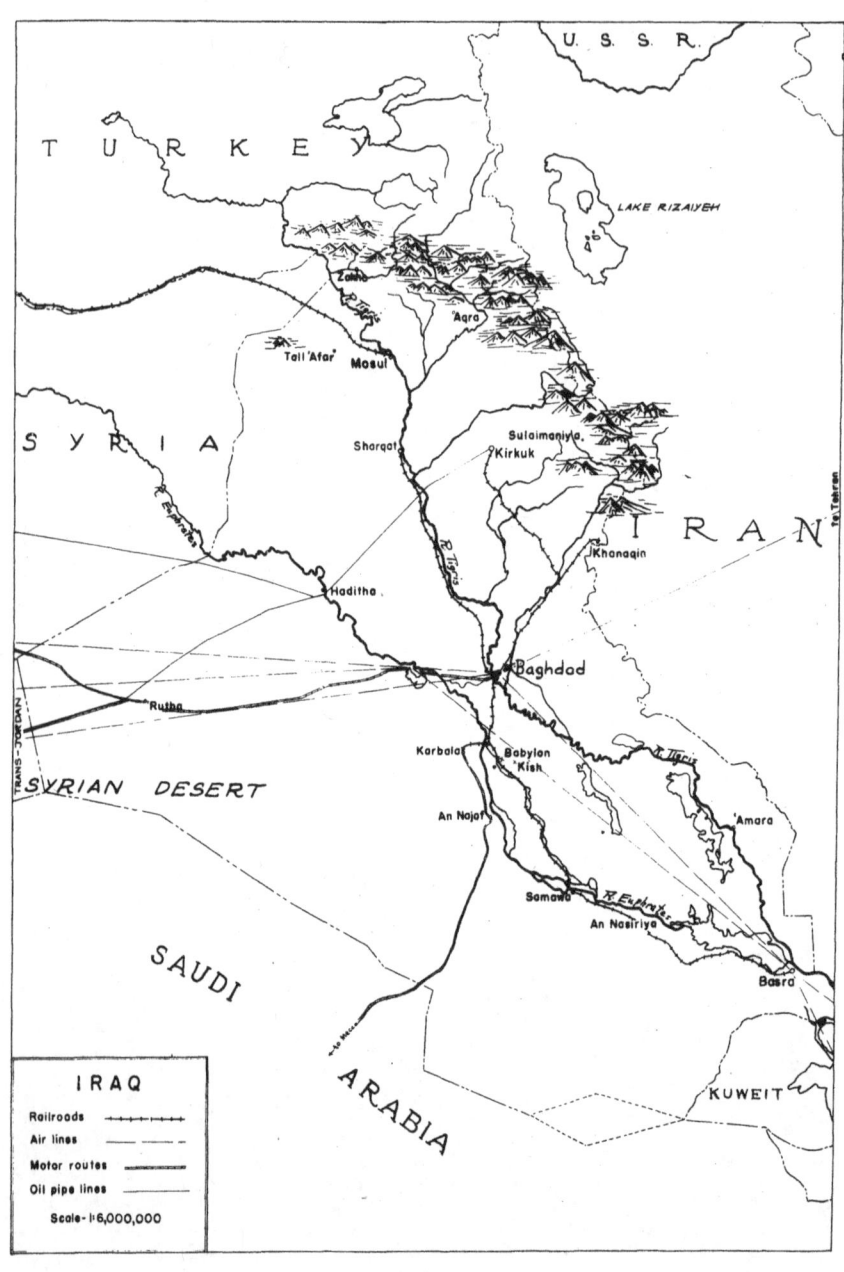

The Anthropology of
IRAQ

The Upper Euphrates

HENRY FIELD

Routledge
Taylor & Francis Group
LONDON AND NEW YORK

First published in 2004 by
Kegan Paul Limited

Published 2014 by Routledge
2 Park Square, Milton Park, Abingdon, Oxfordshire OX14 4RN
711 Third Avenue, New York, NY 10017

First issued in paperback 2014

Routledge is an imprint of the Taylor & Francis Group, an informa business

© Kegan Paul, 2004

All Rights reserved. No part of this book may be reprinted or
reproduced or utilised in any form or by any electric, mechanical
or other means, now known or hereafter invented, including
photocopying or recording, or in any information storage or retrieval
system, without permission in writing from the publishers.

ISBN 978-0-710-30996-9 (hbk)
ISBN 978-1-138-87003-1 (pbk)

British Library Cataloguing in Publication Data
A catalogue record for this book is available from the British Library.

Library of Congress Cataloging-in-Publication Data
Applied for.

CONTENTS

		PAGE
	List of Illustrations	5
	Preface	7
I.	Introduction	13
II.	The Land and the People	17
III.	The Physical Anthropology of the Dulaim and the Anaiza	32
	Anthropometric Methods and Technique	32
	List of Anthropometric Abbreviations	33
	The Dulaim	33
	The Anaiza	54
	Ram-faced Types among the Dulaim and the Anaiza	73
IV.	Additional Anthropometric Data from Iraq	75
	Arabs of the Kish Area	76
	Iraq Army Soldiers	83
	Ba'ij Beduins	86
	Summary *by Sir Arthur Keith*	89
V.	The Tribes and Sub-Tribes of the Upper Euphrates	91
	Appendices	103
	A. The Population of Iraq *by Major C. J. Edmonds*	103
	B. Land Tenure in Iraq *by Sir Ernest Dowson*	106
	C. Notes on General Health of the Kish Arabs	110
	D. Anthropometric Data from Royal Hospital, Baghdad, *by Dr. B. H. Rassam*	122
	E. Individuals Measured in Royal Hospital, Baghdad, *by Winifred Smeaton*	131
	F. Mammals from Iraq *by Colin C. Sanborn*	156
	G. Notes on Insects from Iraq	163
	H. Plants Collected by the Expedition *by Paul C. Standley*	165
	Glossary	198
	Bibliography	199
	Indexes	204
	Tribes Referred to in Chapter V	204
	Dulaimis Illustrated in Plates	207
	Anaiza Tribesmen Illustrated in Plates	207
	Tribal Names Appearing on Map of Iraq (A)	208
	Tribal Names Appearing on Map of Iran (B)	212
	General	214

LIST OF ILLUSTRATIONS

PLATES

1. General view of Haditha.
2, 3. Classic Mediterranean type.
4. Fine and Coarse Mediterranean types.
5. Iraqo-Mediterranean types.
6, 7. Dolichocephals.
8. Brachycephals.
9, 10. Facial types.
11, 12. Mixed-eyed individuals.
13–17. Variations in nasal profile.
18, 19. Variations in hair form.
20–35. Dulaimis measured at Haditha.
36. Hairless Dulaimi.
37–47. Anaiza tribesmen.
48. Water-wheel at Haditha.

TEXT FIGURES

	PAGE
1. Geographical position of Iraq	14
2. Communications with Iraq	19
3. The Upper Euphrates region	21
4. Environs of Lake Habbaniya	29
5–10. Tribes and sub-tribes of the Anaiza Beduins	56–61

MAP

General Map of Iraq . . . . Frontispiece

SUPPLEMENTS

Map A. Distribution of tribes in Iraq
Map B. Distribution of tribes in western Iran

PREFACE

In December, 1925, Dr. L. H. Dudley Buxton, Reader in Physical Anthropology at Oxford, accompanied me to Iraq, where the Field Museum–Oxford University Joint Expedition was excavating the ancient city of Kish, which lies eight miles due east of Babylon. Our trip was financed by my great-uncle, Mr. Barbour Lathrop, a firm believer in the benefits of practical experience. During our brief visit to the Expedition we were enrolled by Professor Stephen Langdon as volunteer physical anthropologists.

At that time excavations were in progress in the Babylonian levels of mound "W" and on the southern flank of the great temple complex dedicated to Harsagkalemma. Dr. Buxton instructed me in the technique of excavating human skeletal remains. Several questions arose in relation to the physical appearance of these ancient dwellers in Mesopotamia. Were they similar to, or different from, the modern Arabs of the Kish area? Had the basic population of Mesopotamia, now Iraq, remained unchanged during the past six thousand years of recorded history? In addition, how were the modern inhabitants of Iraq related to their neighbors and, in general, to the peoples of Asia, Africa, and Europe?

Since no anthropometric data from this area were in existence Dr. Buxton and I decided to measure a small series of our Kish workmen. Shortly afterward, we obtained permission from the Officer Commanding the Iraq Army Camp at Hilla to measure some of the soldiers. Thus, Dr. Buxton examined Iraq Army soldiers, while I acted as recorder. These anthropometric data, published by Buxton and Rice (see pp. 81–82), revealed the numerical inadequacy of our samples.

On January 10, 1926, I accompanied Professor Langdon to Jemdet Nasr, which lies in the desert about eighteen miles northeast of Kish. Early in the afternoon we unearthed four complete painted vessels, and several pictographic tablets in linear script (Field, 1926). No human remains were found.

During the season 1927–28 I was attached to the Kish Expedition as physical anthropologist. In March, during excavations at Jemdet Nasr we found several human skeletons (Field, 1932c). At the close of the season I examined 398 Arabs of the Kish area, 231 Iraq Soldiers at Hilla Camp, and 38 Ba'ij Beduins (see pp. 76–89; also Field, 1935a and 1939b).

The results obtained seemed to warrant a continuation of the anthropometric survey of Iraq. Dr. Berthold Laufer, my former chief, approved this project and on April 1, 1934, the Field Museum Anthropological Expedition to the Near East, under my leadership, began work in Baghdad. The Expedition was financed by Mr. Marshall Field. The first four and one-half months of the anthropometric survey were spent in Iraq, where, in addition to our anthropological work, we collected botanical, geological, and zoological specimens. Similar researches were conducted in Iran (Field, 1939b) and among the North Ossetes and Yezidis of the Caucasus, U.S.S.R.

Mr. Richard A. Martin, now Curator of Near Eastern Archaeology at Field Museum, was in charge of collecting zoological specimens (see China; Uvarov; and Schmidt, 1939) and also accompanied me throughout the Expedition in the capacity of photographer. The excellence of the photographs illustrating this publication is entirely due to his technical skill and patience in dealing with these Arabs and Beduins.

Mr. S. Y. Showket, of Basra, acted as interpreter. His knowledge of English, Arabic, Kurdish, Persian, and Chaldean, combined with his finesse in dealing with recalcitrant subjects, made him an invaluable member of the Expedition.

Dr. Walter P. Kennedy, of the Royal College of Medicine in Baghdad, examined the Dulaim and Anaiza blood samples (Field, 1935a, p. 460).

Yusuf Lazar, an Assyrian, was in charge of collecting herbarium specimens and insects (see Uvarov; China).

Technical questions regarding anthropometric measurements and observations were discussed at Harvard with Dr. E. A. Hooton, and in England with Sir Arthur Keith and Dr. L. H. Dudley Buxton.[1]

Prior to our leaving the United States, Mr. Wallace Murray, Chief of the Division of Near Eastern Affairs in the Department of State, had very kindly notified Mr. Paul S. Knabenshue, United States Minister in Baghdad, of our scientific mission. At Mr. Knabenshue's intervention I was granted private audiences with His Majesty the late King Ghazi; the Prime Minister; the Minister of the Interior; the Minister of Education; the Director-General of Health; and the Chief of Police.

[1] Dr. Buxton's premature death from influenza in 1939 came to me as a great shock and personal loss. His students, scattered throughout the world, will always remember his inspiring leadership and stimulus.

As a result of these interviews a special permit was issued enabling members of the Expedition to conduct anthropometric studies throughout Iraq, to collect zoological, botanical, and geological specimens, to take photographs, and to compile tribal maps (see Maps A and B).

During our work in Iraq the Expedition received unusual cooperation from Iraqi officials, as well as from many private individuals. Among the many persons who rendered valuable assistance were: Ali Jaudat Beg, Sir Kinahan Cornwallis, Mr. C. R. Grice, Major W. C. F. Wilson, Sir John Burnett, the late Wing-Commander A. R. M. Richards, Dr. Walter P. Kennedy, Dr. T. H. McLeod, and the *Mutasarrifs* of the Mosul, Kirkuk, Erbil, and Amara *Liwas*.

A letter from the Air Minister in London, Lord Londonderry, to the Air Officer Commanding in Iraq served as an introduction to the members of the British Royal Air Force.

Another valuable letter of introduction was from Mr. John Skliros, Managing Director of the Iraq Petroleum Oil Company in London, to Mr. G. W. Dunkley, General Manager in the Near East, who facilitated our work. During our three weeks in the desert we were guests of the Company.

Appreciation must also be expressed to the late Dr. F. R. S. Shaw, Chief Medical Officer of the Company, and to the late Dr. H. C. Reid, who made possible our work on the Dulaimis at Haditha and to Dr. M. Don Clawson, Chief Dental Surgeon, who rendered assistance in numerous ways.

Through the courtesy of the late Professor James H. Breasted, Director of the Oriental Institute of the University of Chicago, the Expedition was kindly lent a station-wagon by Dr. Henri Frankfort, Director in Iraq of the Oriental Institute Expeditions. This automobile was driven by Mr. H. Mihran. Mr. Gabriel Malak also gave generous assistance.

Dr. B. H. Rassam of the Royal College of Medicine in Baghdad kindly gave me his anthropometric data on 497 individuals measured by him in the Royal Hospital, Baghdad (see Appendix D).

In conclusion, I must record my deep gratitude to His Majesty the late Ghazi ibn Faisal and to his Ministers, who made possible my studies on the physical characters of the modern peoples of Iraq.

At the end of July Mr. Martin, Dr. Kennedy, Yusuf Lazar, and I left Baghdad for Tehran. In Iran we continued our research (Field, 1939). On September 13, we entered the Union of Soviet

Socialist Republics at Baku. The anthropometric data obtained in the Caucasus will appear in a forthcoming Museum publication.

Following our return to Chicago in December, 1934, preparations were begun for the publication of the results obtained by the Expedition.

During the writing of this report I have had the benefit of discussing the general arrangement of the material with Dr. Paul S. Martin, Chief Curator of Anthropology at Field Museum.

Since 2,500 individuals had been studied in Iraq, Iran, and the Caucasus, it was decided to accept the invitation of Dr. Hooton and to have the statistics tabulated on the card system for sorting by the Hollerith machines at the Anthropometric Laboratory in the Peabody Museum at Harvard. During 1935 and part of 1936 the data were prepared for the machines and the introductory sections written. From September, 1936, to June, 1937, I worked on this material at the Peabody Museum. Mr. Donald Scott, Director, facilitated my work in every possible manner.

Throughout this period I had the benefit of numerous conferences with Dr. Hooton, who supervised the preparation of this report and from time to time offered many valuable suggestions, particularly in regard to the methods to be employed in the presentation of these data.

I am also grateful for opportunities to discuss numerous problems with Dr. Carleton S. Coon and with Dr. Carl C. Seltzer, who calculated the statistical tables.

I wish to thank Miss Elizabeth Reniff, my former research assistant, who worked on this report both at Field Museum and at Harvard.

The greater part of the typing was done by Miss Ethel Brady, who arranged the statistical tables, and by Mr. Theodore Scully, who completed the remainder of the manuscript.

Miss Dorothy Pedersen rendered valuable assistance throughout the preparation and proofreading of this publication.

I wish to express gratitude to Miss Eunice Zimmerman, who assisted with the final checking of the report.

I also gratefully acknowledge the aid of Miss Lillian A. Ross, Staff Editor of the Division of Printing, in seeing the manuscript through the press.

My wife has generously assisted in proofreading the greater part of the manuscript.

During the Cambridge meeting of the British Association for the Advancement of Science, in August, 1938, I had the benefit of discussing the preliminary results with Sir Arthur Keith, to whom, because of his encouragement and advice during the past seventeen years, I owe a lasting debt of gratitude.

In Berlin during the same month I had the pleasure of visiting Baron Max Freiherr von Oppenheim, whose first volume on the Beduins has appeared recently (see Bibliography). His chapter on the Anaiza should be read as an introduction to my section on these desert tribesmen.

I wish, also, to record my gratitude to the librarians of the following institutions who facilitated the reference work in every possible manner: Field Museum of Natural History; Oriental Institute, University of Chicago; Peabody Museum, Widener Library, and Institute of Geographical Exploration, Harvard; New York Public Library; Library of Congress; Bodleian Library, Oxford; University Library, Cambridge; London Library; Royal Geographical Society; Royal Asiatic Society; Royal Central Asian Society; Musée de Trocadero, Paris; Instituto di Antropologia della Reale Università, Rome; Palais Azem, Damascus; and Iraq Museum, Baghdad.

Three maps (Frontispiece; Figs. 2, 3) were drawn specially for this publication by Mr. Peter Gerhard, a volunteer assistant. Figure 1 was drawn by Dr. Erwin Raisz, of Harvard University, and Figure 4 by Mr. David Tuch.

The large map (A) showing the distribution of tribes in Iraq has been distributed with the map (B) of Iran since there is an overlap between these two sheets.

Map A, compiled from all available sources, was drawn at Field Museum by Mr. Richard A. Martin.

Wherever possible I have checked the tribal information but in a task of this complexity and magnitude a certain degree of variation must occur, since even the best qualified informants vary in their oral tradition (cf. von Oppenheim).

Furthermore, during the past decade many tribal changes have taken place within the confines of Iraq. To the best of my knowledge, however, there have been no large tribal movements in Iraq comparable to those ordered by Riza Shah Pahlavi in Iran. This does not include the movements of the Assyrians to the Khabur. In Iraq the general trend has been to restrict the wanderings of the nomads in an attempt to make them become settled groups. In this manner conflicts over pasturage or wells can be avoided.

Alphabetical lists of tribal names appearing on these two maps have been prepared by Miss Dorothy Pedersen and by Mr. Peter Gerhard respectively.

The list of the tribes and sub-tribes of the Anaiza (Figs. 5–10) was rewritten by Dr. A. Frayha at the Oriental Institute of the University of Chicago. The transliteration, prepared by Dr. Frayha, was redrawn by Mr. Richard A. Martin.

The place names conform to the spelling adopted by the Permanent Committee on Geographical Names of the Royal Geographical Society of London. As the question of orthography is by no means settled and many names are not yet included in the published lists of the Society, standard practice as adopted by the most recent British map-makers has been used.

All diacritical marks, with but few exceptions, have been omitted throughout the text, but are included in the Glossary (p. 198).

In conclusion, I must express my gratitude to Mr. Abdul-Majid Abbass, and to Mr. Jassim Khalaf, Iraq Government students at the University of Chicago, who checked and made additions to the native names listed in the text and in the Glossary.

HENRY FIELD

THE ANTHROPOLOGY OF IRAQ

PART I, NUMBER 1
THE UPPER EUPHRATES

I. INTRODUCTION

In order to present the results of the anthropometric survey of Iraq it has been decided to arrange the data according to the following plan in the Parts and Numbers of Volume 30 of the Anthropological Series of Field Museum.

THE ANTHROPOLOGY OF IRAQ

PART I
			Males	Females
No. 1.	Upper Euphrates			
	(a)	Dulaim	137	0
	(b)	Anaiza	23	0
	(c)	Individuals in Royal Hospital, Baghdad	439	143
	(d)	Arabs of Kish Area	459	0
	(e)	Iraq Soldiers, Hilla	222	0
	(f)	Ba'ij Beduins, near Kish	35	0
No. 2.	Lower Euphrates–Tigris Region			
	(a)	Marsh Arabs	271	3
	(b)	Subba (Mandeans)	92	33
	(c)	Individuals in An Nasiriya	126	26

PART II
			Males	Females
No. 1.	Northern Jazira			
	(a)	Shammar	299	129
	(b)	Sulubba (Sleyb)	39	10
	(c)	Turkomans	64	31
	(d)	Yezidis	235	77
No. 2.	Kurdistan			
	(a)	Kurds	609	33
	(b)	Assyrians	106	137
	(c)	Jews	111	52
	(d)	Armenians	4	2
	(e)	Gypsies	6	4
	(f)	Chaldeans	1	0
		Total	3278	680

No. 3. Comparative Data
Conclusions

Miss Winifred Smeaton, now Mrs. Homer Thomas, measured 588 females and some of the males. Miss Smeaton was attached to the Expedition from April 1 to July 20, 1934. (See also Appendix E.)

Both parts will be arranged on the same general plan, each section containing chapters on the land and the people, the physical anthropology of the various groups, and a list of the tribes and sub-tribes within the area prescribed.

This report (Part I, No. 1), based on the anthropometric data obtained in May, 1934, is concerned with the physical characters of the peoples of the Upper Euphrates region of Iraq and Syria.

There is no need to compile a chronological survey of references to this area during the past two thousand years,[1] since the reader has ready access to classical sources, to the writings of early travelers,

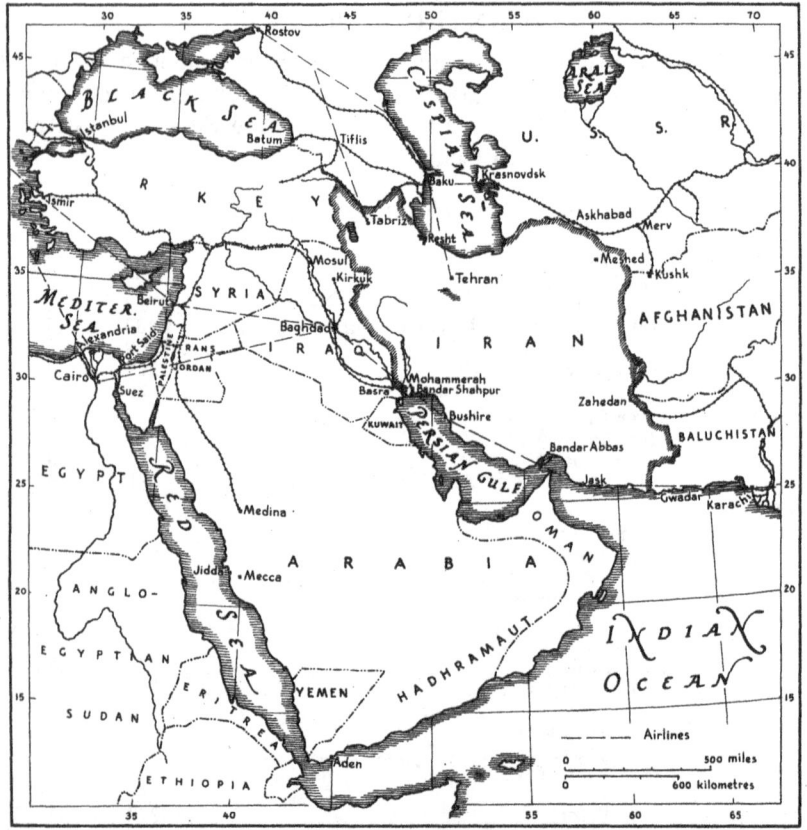

FIG. 1. Geographical position of Iraq.

and to those of Buckingham, Sir Wilfred and Lady Anne Blunt, Mark Sykes, Doughty, Musil, Lawrence, Grant,[2] von Oppenheim, and many others.

[1] For references to the Middle Euphrates during the Assyrian period and down to Ibn Battuta and other Arabic authors see Musil, 1927b, pp. 197 et seq.

[2] Dr. Christina Grant (1937) has compiled almost complete references to the caravans, early travel, and recent exploration of the Syrian Desert.

INTRODUCTION 15

Chapter II deals briefly with the general location of Iraq, and in particular with the boundaries, physical geography, climate, flora, and fauna. There is also an outline of the recent history of the Upper Euphrates area.

Chapter III contains the anthropometric data on the Dulaimis and on the Anaiza tribesmen. The revised tables of the Kish Arabs, Iraq Soldiers, and Ba'ij Beduins, who were measured in 1928, are placed in Chapter IV.

I was fortunate to be granted access to full and unpublished lists of the tribes and sub-tribes in Iraq. The compilers of these data in Chapter V preferred to remain anonymous.

Appendix A contains the figures of registered and unregistered populations to the end of November, 1935. The number of the total population (3,560,456) is based on these data, which were sent from Baghdad by Major C. J. Edmonds.

Appendix B gives the classification of land surface and the population with the mean density per square kilometer of the cultivated region. These figures were compiled in 1930 by Sir Ernest Dowson.

Appendix C, a description of the health conditions among the Arabs of the Kish area, is based on data compiled during 1927–28 when I was attached as physical anthropologist to the Field Museum–Oxford University Joint Expedition to Kish.

Appendix D contains the anthropometric data on 497 individuals obtained during 1932 by Dr. B. H. Rassam in the Royal Hospital, Baghdad.

In Appendix E Miss Smeaton presents the anthropometric data obtained on 32 males and 52 females during 1935 in the Royal Hospital, Baghdad.

Appendix F consists of a list of mammals collected in Iraq either during the 1934 Expedition or as a result of our subsequent appeals for additional specimens for the Museum study collections. The identifications have been made by Mr. Colin C. Sanborn, Curator of Mammals.

A report (Field Mus. Nat. Hist., Zool. Ser., vol. 24, pp. 49–92) on the reptiles and amphibians was published during 1939 by Mr. Karl P. Schmidt, Curator of Amphibians and Reptiles.

The large collections of insects obtained during 1934, and subsequently from Yusuf Lazar, are being determined at the British Museum through the cordial co-operation of Captain N. W. Riley (see Appendix G). Two papers have been published by Field Museum: "Hemiptera from Iraq, Iran, and Arabia," by W. E. China; and "Orthoptera from Iraq and Iran," by B. P. Uvarov.

In Appendix H, Mr. Paul C. Standley, Curator of the Herbarium, has classified the flora collected during the 1934 Expedition and herbarium specimens obtained subsequently from Yusuf Lazar. This list is of particular importance, since in many cases the localities indicate new ranges for genera and species.

In 1937 Field Museum published a report by David Hooper and Henry Field, entitled "Useful Plants and Drugs of Iran and Iraq."

Additional reports on botanical, geological, and zoological specimens are now in preparation.

The reader is referred to a recent publication by Père H. Charles entitled "Tribus Moutonnières du Moyen-Euphrate." (Institut Français de Damas.) This important work deals with the tribes adjoining those referred to in the present report, and for this reason it should be used as a complementary account.

In the same series published in 1934 by the Institut Français de Damas appeared Mr. Albert de Boucheman's monograph entitled "Matériel de la vie bédouine receuilli dans le désert de Syrie (tribu des Arabes Sba'a)." This volume contains an excellent account of the material life of the Sbaa Beduins.

Indexes of the numbers of individuals and plate numbers of the Dulaimis and the Anaiza (p. 207) have been prepared.

The comparative data and the conclusions based on the anthropometric survey of Iraq will be discussed in Part II.

A detailed knowledge of the physical characters of the modern peoples of Iraq and their relationship both to their neighbors and to the ancient dwellers in Mesopotamia not only will throw light on numerous historical problems but also will be of assistance in determining the true racial heritage of the Mediterranean Race.

Furthermore, the European races trace part of their physical and cultural origins to an area extending from the Punjab to the Nile Valley.

Southwestern Asia may well have been one of the nurseries of *Homo sapiens* (Field, 1932b, 1939b).

II. THE LAND[1] AND THE PEOPLE

The Upper Euphrates region may be described as the stretch of the Euphrates River between Raqqa and Al Falluja with an arbitrary boundary in the desert on both the right and left banks of the river (Fig. 3).

In general this area, which covers approximately 22,000 square miles, consists of a steppe-like plateau with rocky outcrops, similar to South African kopjes, some of which rise to a height of 200 or 300 feet above the level of the surrounding country.

Through the center of this inhospitable area flows the Euphrates River, following a general southeasterly course (cf. Ionides, pp. 37–111). Along its banks and those of its tributaries are to be found stretches and patches of cultivated land.

In the course of centuries the river has carved out a trough-like depression through the desert. According to the resistance offered by the geological formation of the land, this valley varies in width from more than ten miles to a narrow precipitous gorge scarcely a mile across.

In the wider sections of the valley, the river meanders, frequently changing its course and forming numerous islands and rapids in the river bed, as well as ledges of rich, alluvial soil near the banks where the land is cultivated.

At Abu Kemal the valley begins to narrow, and the course of the river is due east until it reaches Ana; from here it again flows southeast. The gorge gradually opens out in the neighborhood of Ramadi, where the river flows through a fertile, irrigated, alluvial plain, until the limit of the area is reached at Al Falluja.

As far south as the Tell Aswad reach, the river bed is rocky, with numerous ledges and rapids, but beyond this point the bed of the river and both banks consist of alluvial soil.

The country on the left bank of the river is known to the local inhabitants as the Island (*Al Jazira*),[2] so-called because it lies between the Tigris and the Euphrates, and the country on the right bank is known as *Al Shamiya*, as it is situated on the Damascus (*Sham*) side of the river.

---

[1] For general description see Lyde (pp. 268 et seq.); Carruthers (1918); Blanchard (1925, 1929, especially the bibliography, p. 231); Stamp (1929); and Boesch (1939).

[2] Throughout the remainder of this report *Al Jazira* and *Al Shamiya* are referred to as the Jazira and the Shamiya.

The Euphrates has only two tributaries of any importance, the Belikh and the Khabur, both of which join the parent stream on the left bank, the former in the neighborhood of Deir-ez-Zor and the latter about eight miles upstream from Meyyadin.

Numerous wadis from the desert uplands join the river on both banks. They are dry during the greater part of the year, but after a heavy rain, which may occur miles away in the desert, they are liable to sudden and unexpected floods which render them impassable for an indefinite length of time, from one or two hours up to as much as five days.

The chief canals, few in number, leading from this section of the Euphrates, are the Aziziya, Saqlawiya, Abu Ghuraib, and Ridhwaniya, details of which are as follows:

(1) The Aziziya Canal leaves the right bank of the Euphrates half a mile upstream from Ramadi, and flows in a general south-southeasterly direction into Habbaniya Lake, five miles southeast of Ramadi. Both banks of the canal are extensively cultivated.

(2) The Saqlawiya, one of the largest and most important canals on the Euphrates, is of modern construction. Its intake is six and a half miles upstream from Al Falluja, on the left bank of the river. The canal flows in a general easterly direction, terminating in the Aqarquf, ten miles northwest of Baghdad. The canal head is controlled by sluice gates and has a concrete blockhouse on either bank, where it is crossed by a stone bridge on the main Baghdad-Al Falluja road. This canal, which attracted many sections of the Dulaim tribe from the banks of the Euphrates, waters one of the most fertile tracts of country in the whole area.

(3) The Abu Ghuraib Canal leaves the left bank of the Euphrates four miles downstream from Al Falluja and proceeds in a general easterly direction until due south of Khan Nuqta, when it flows northward. Both banks of the canal are cultivated by Zoba tribesmen.

(4) The Ridhwaniya Canal has its head on the left bank of the Euphrates nine miles downstream from Al Falluja and follows the general direction of the river until it reaches Imam Hamza, where it tails off into a series of distributaries. The Zoba are the chief cultivators on both banks of the canal.

The sudden inundations of the Euphrates are an important factor in the life of the people. There are two flood seasons. Dur-

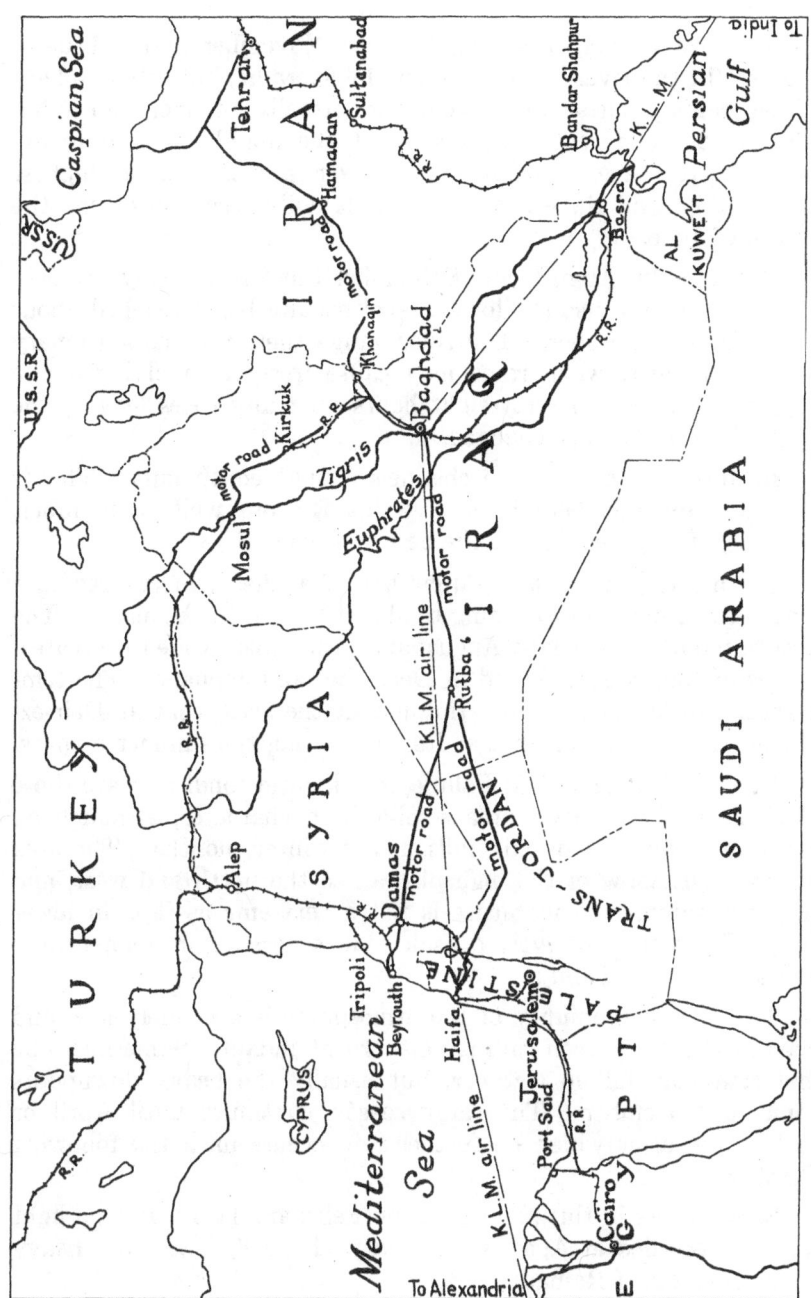

Fig. 2. Communications with Iraq.

ing the first season, occurring between November and February, the rises in the river are caused by the sporadic, but often violent, winter rains. These inundations are usually of short duration. The longer flood season begins about the middle of March and continues to the end of June. The river is usually at its highest during May, and there is a considerable daily recession during the month of June.

During July, August, and September there is a steady decrease of water in the river, the lowest level usually being reached about the middle of October. The river gauge then remains stationary until November, when rains may cause freshets involving a rise of five or six feet in forty-eight hours, in many cases leaving the channels and crossings changed.

In July and August the channels change continually. This is the most difficult period for river navigation, while September, October, and November are the best months.[1]

In this region on the Euphrates, the thermometer readings may range from below freezing to above 120° F. in the shade. The hottest months are usually August and September, while the greatest degree of cold is experienced in December and January. The temperature varies considerably throughout the area, that at Deir-ez-Zor being 10° less than that at Ramadi during the summer months.

Between Raqqa and Al Falluja the climatic conditions are those of a subtropical, inland area semi-arid in character, although an appreciable amount of rain falls in the winter months. The area lies in the shadow of the high plateau to the north and west, and thus the summer temperature is not as extreme as it is in lower Iraq. There is, however, considerable difference in temperature at Raqqa and Al Falluja.

The relative humidity of the atmosphere is extremely low, and even in the wet season rain is not very abundant. Sometimes the first rain may fall in October, but usually the heavy downpours come in November. The rainy season continues until April or early May, after which no further rain occurs until the following October.

Snow is rare in this region, but on February 11–13, 1920, a light fall was recorded at Ana. On January 11, 1926, I was in a heavy hailstorm west of Ramadi.

[1] See Willcocks and Ionides for detailed information on the general hydraulic survey of the Euphrates and Tigris rivers.

The general direction of winds throughout the summer is from the northwest, because atmospheric pressure in the eastern Mediterranean is considerably higher than that in the Persian Gulf during this season. This northwest wind descends from the plateau upon the Jazira like a dry, scorching blast from a furnace, frequently bearing with it a cloud of dust (cf. Coles).

Southern hot winds, from the Persian Gulf, usually alternate with the northwest winds throughout the summer. The influence of these hot winds is particularly noticeable in August and Sep-

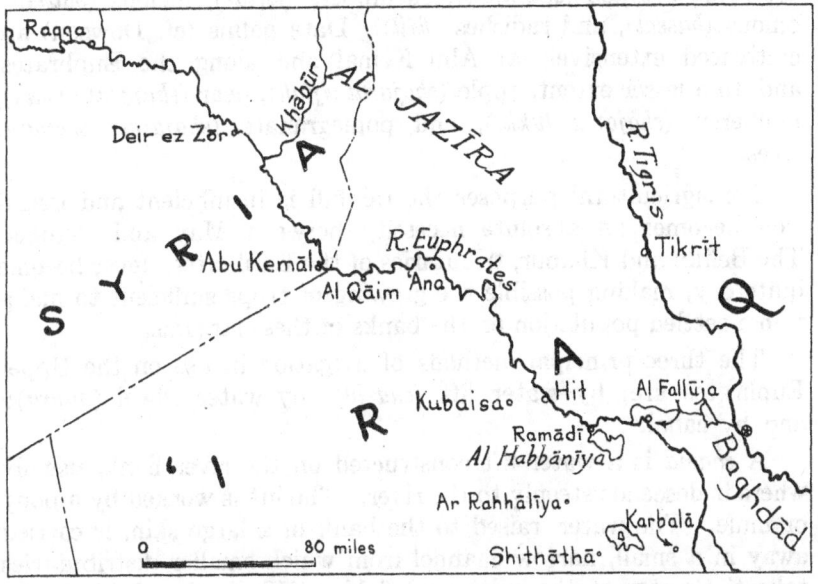

Fig. 3. The Upper Euphrates region. Scale 1:4,000,000.

tember, when they help to ripen the date crop. They are felt as far north as Abu Kemal, the northern limit of the cultivation of *Phoenix dactylifera* (see Dowson). The prevailing wind passes over the plateau of Anatolia and descends on the plains as a dry current of air, rapidly becoming warmer as it descends from the level of the mountains. During the winter months the direction of the wind varies considerably, and breezes often spring up from the south.

Calms rarely occur and the wind generally attains its maximum velocity during the day. In the evening, the wind diminishes to a gentle breeze which gradually gathers speed after dawn on the following morning.

During the summer months, sand storms of considerable intensity frequently occur, and the burning sand, driven along with a cloud of dust, provides a most unpleasant experience (cf. Coles). For hours visibility may be limited to a few hundred feet.

The agricultural crops of this area on the Upper Euphrates, cultivated under the most primitive conditions, comprise chiefly wheat (*huntah*) and barley (*shair*), a certain amount of maize (*ithra*), and a limited quantity of red and white rice (*timmin*), sesame (*simsim*), mash (*mash*), beans (*buqul*), and cotton (*qutn*). There are some brinjals (*badinjan*), cucumbers (*khiar*), melons (*battikh*), onions (*bassal*), and radishes (*fijil*). Date palms (cf. Dowson) are cultivated extensively at Abu Kemal and along the Euphrates, and, to a lesser extent, apple (*shajarat tiffah*), pear (*shajarat armut*), mulberry (*shajarat tukki*), and pomegranate (*shajarat rumman*) trees.

For agricultural purposes the rainfall is insufficient and irrigation becomes an absolute necessity between May and October. The Belikh and Khabur, tributaries of the Euphrates, never become quite dry, making possible the growing of crops sufficient to maintain a settled population on the banks of these streams.

The three principal methods of irrigation in use on the Upper Euphrates are: by water lift (*charid*); by water wheel (*naura*);[1] and by canal.

A *charid* is a water lift constructed on the river bank, usually where it descends steeply to the river. The lift is worked by a pony or mule. The water, raised to the bank in a large skin, is carried away in a small, narrow channel from which smaller distributaries take the water to the cultivated fields. Where the *charid* is the only form of irrigation, water can be carried only from one to one and a half miles inland from the river.

In the construction of a water wheel[2] (*naura*) a series of masonry weirs is built out into the river for a distance of about ten yards, with a masonry trough extending along the top. At the end of this projection into the river is a water wheel (Pl. 48). The force of the current in the stream turns the wheel, on which is fastened a series of small buckets to lift the water. On the turnover of the wheel the water is emptied into an extension from the trough (Pl. 48, upper) and thence conveyed through ordinary channels to

---

[1] Cf. Laufer (1934) for origin and history of the noria or Persian wheel.

[2] Cf. H. Charles, pp. 140–146.

the land to be irrigated. Working day and night, each wheel irrigates about five acres. The cost of maintenance of one wheel is said to be approximately $200 annually. A masonry dam, built out into the river in prolongation of the weirs, raises the water level enough to ensure at low water a sufficient current to turn the water wheels. A series of these weirs and dams built out from both banks toward the center of the river tends to raise the water level and to produce a swift current in the center of the river between the heads of the dams, rendering the passage of boats both difficult and dangerous. In many places the weirs and dams become ruined and submerged, further increasing the dangers of navigation. Norias are not used downstream from Hit.

In spite of their usefulness in cultivation, there are remarkably few canals of any size on the Upper Euphrates (see p. 18). A tribesman will cut a small channel leading from the river to irrigate his crops where this is practicable, but unless the Government displays some interest and activity in the construction of a large canal he will show little initiative in this direction.

The rain produces a desert crop capable of supporting more than a hundred thousand grazing sheep and several thousand camels. During the late autumn, winter, and early spring, after heavy rains, this desert is covered with grass, various desert wild flowers, spinifex, and numerous shrubs which provide excellent grazing for camels.

During this period water can be obtained from depressions in the ground or from the beds of wadis where it collects after rains. At this season, Beduins, principally from the Anaiza and Shammar tribes, wander in well-defined areas grazing their extensive flocks of camels and sheep.

About the end of April or the beginning of May the desert becomes parched, brown, and dry. During the rainless summer months the grazing is thus quickly exhausted and Beduin herdsmen must be continually on the move, compelled to pasture their flocks near the river.

The fauna of Iraq has not yet been studied extensively but numerous papers have been published in the Journal of the Bombay Natural History Society and by specialists of the British Museum (Natural History).

The mammals living in this region include gazelle, hyena, jackal, wild boar, fox, badger, and cheetah. There are many species of birds living beside the Euphrates and the Wadi Thahthar. The reptiles and amphibians probably do not differ from those in other parts of Iraq (cf. Schmidt, 1939).

The insects have not been studied in detail within this area but the reader desirous of additional information on the Hemiptera and Orthoptera should consult the articles by China and Uvarov (see Bibliography).

The mineral resources of the Upper Euphrates are concentrated around the bitumen wells at Hit. Apart from this the area possesses no mineral wealth, with the exception of a negligible quantity of oil[1] from Nafatha, ten miles north of Ramadi. The oil (*mazut*) is used as a remedy for diseases of sheep and camels.

The seven bitumen wells at Hit are said to have been worked for at least 5,000 years, and the supply seems to be almost unlimited. The output in 1920 averaged between 150 and 300 tons per month, most of which was exported up- or downstream in barges (*shakhatir*).

Bitumen is used locally for boat-building, the making of bricks (*tabuq*), caulking of baskets, and as fuel for kilns (*quwar*).

Lime is manufactured at Hit by burning bitumen with limestone from the neighboring quarries, the average output being 300 tons per month, all of which is exported downstream. One of the best quarries lies at Jaladiya, five miles northwest of Hit.

The only controlled salt pans exist at Hit. Three hundred tons of salt were exported during 1920.

The sole manufacturing enterprise of any importance on the Upper Euphrates is also located at Hit, where gufas (Ar. *quffaf*) are constructed. These round boats are made by interlacing tamarisk and mulberry tree branches with basketwork of reeds and straw, and the whole is eventually caulked with a mixture of bitumen and sand. The boats usually draw about twenty-two inches when laden and about six inches when empty. When despatched downstream for sale they are loaded with lime and bitumen, and sold with their cargoes.

A report on economic and commercial conditions in Iraq, by J. P. Summerscale, appeared in 1938.

A brief historical survey shows that this area has seen the rise and fall of some of the most famous empires of the past. As long

[1] This statement was written prior to the activities of the Iraq Petroleum Company, formerly the Turkish Petroleum Company. During 1928 I was attached as a separate archaeological unit to Major A. L. Holt's T. P. C. Survey party operating between Rutba and the Harrat-ar-Rajil. Therefore, all information regarding oil development has been treated as strictly confidential.

ago as 1450 B.C. the Eastern Marches of the Egyptian Empire extended as far as Hit. At a later period the country came in turn under the domination of the Assyrian, Persian, Macedonian, and Roman empires. It was engulfed finally by the tide of the Mohammedan conquest, which swept up from Mecca and Medina in the seventh century as a result of the preaching of Mohammed. Later again, when, under the Omayyad caliphs, the governing center of the Mohammedan world shifted from Mecca and Medina to Damascus, the country again changed masters, and when the Abbassid caliphs in their turn rose to power and Baghdad became their capital, the area in question formed part of their dominions. At length, after other vicissitudes and changes of fortune, the country, in 1534, came under the rule of the Ottoman Turks, who had been ruling over it for nearly 400 years at the time of the outbreak of the World War in 1914.

During the Turkish régime, the country along the Middle Euphrates, although nominally under the control of the Turks, actually became independent of any central authority until comparatively recent years.

The Beduins ranged the country at will, taking toll of the agriculturist and of the caravan. As a result of their depredations, which the central government was not in a position to check, any security or prosperity was rendered impossible, and cultivation of the land existed merely on sufferance.

It was not until the conclusion of the Crimean War (1856), when the Porte found itself with a large army and plenty of money at its disposal, that any serious effort was made to exercise control in the country. Omar Pasha, then governor of Aleppo, at the head of a considerable body of troops, marched down the Euphrates and took possession of Deir-ez-Zor, which was then held by *Fallahin*, who had enjoyed semi-independence under Anaiza protection. It was about this time that the caravan route down the Euphrates from Aleppo to Baghdad was opened to traffic, and traveling by this route, although a somewhat speculative venture, became comparatively safe.

This policy of enforcing the Turkish authority was carried on by Midhat Pasha, who built forts to protect navigation on the Euphrates and the caravan route to Aleppo.

Despite periods of insecurity the Turkish power gradually grew, and the acreage of cultivated land has considerably increased in recent years in the Euphrates Valley. The riverain cultivators usually found it advisable to pay a form of tribute to the larger Beduin tribes in return for protection, or at least for freedom from molestation.

On the outbreak of the World War in 1914 the Euphrates was gradually developed as a line of communication by the Turks, who transported both troops and stores by river from Jerablus to Al Falluja and even to Samawa and An Nasiriya.

As the Turkish domination was replaced by British occupation, Civil Administration was undertaken, and Political Officers were established at Ramadi, Ana, Abu Kemal, and Deir-ez-Zor. The Ramadi division consisted of the old Turkish *Qadhas* of Al Falluja, Ramadi, Hit, Ana, and Abu Kemal, which were administered by Assistant Political Officers. Ana, Hit, and Al Falluja were later placed under the charge of Arab Civil Officials.

The advent of the new Arab government of Iraq has produced a general stabilizing influence on the political situation in the Upper Euphrates region.

The great majority of the inhabitants are Arabs of the Sunni sect.

Christians, Jews, and Shiah Mohammedans are so few in number that they need scarcely be considered as a factor of importance. Owing to the former migratory habits of large sections of the population accurate census figures were difficult to obtain. The following is an approximate estimate of the population derived from various sources in 1920:

Arabs (Sunnis)	331,000
Arabs (Shiahs)	200
Jews	3,600
Christians	1,200
Total	336,000

The Upper Euphrates is the home of four types of Arab, each of which is more or less distinct from the others, possessing its own characteristics.

(1) The Beduins, or purely nomadic wanderers in the desert, are represented in this area by the large and powerful Anaiza confederation (cf. pp. 54–74, 91–93).

(2) The semi-nomads pasture their flocks in the desert, while at the same time they own and cultivate land in the vicinity of the river. The Dulaim (pp. 33–54, 96–101) are a good illustration of this type, approximately 50 per cent of the tribe being semi-nomadic and the remainder settled cultivators.

(3) The settled cultivators reside permanently either on the river bank or in an irrigated area, and engage in purely agricultural pursuits. No tribe on the Upper Euphrates is composed entirely of settled cultivators, and the percentage in each tribe varies. In the Baqqarah, settled cultivators amount to about 75 per cent of the inhabitants.

(4) The town-dweller, engaged in commercial or industrial pursuits, lives on the proceeds of land or houses which he owns, or he may be a government official or a member of the professional classes.

The most important tribal groups living in the Upper Euphrates region are the Anaiza, including the Ruwalla and Amarat sections, and the Dulaim. The Anaiza and the Dulaim are discussed in Chapter III, pages 33–74.

The Amarat, who numbered some 4,500 tents, ranged the eastern portion of the Hamad from west of An Najaf to Deir-ez-Zor. In early spring the Amarat occupied Al Gara depression near Bir Mulussa, eighty miles southwest of Abu Kemal. In summer they migrated to the Euphrates between Ramadi and Deir-ez-Zor, while autumn usually found them encamped on the edge of the desert west of Karbala in the vicinity of Shithatha and Ar Rahhaliya.

In the years following November 11, 1918, the Amarat became friendly with the Dulaim but remained bitter enemies not only of the Shammar Jarba of the Jazira but also of the Southern Shammar of Arabia. They were on bad terms with the Ruwalla, but Fahad Beg and Nuri ibn Shalan came to a friendly agreement in the spring of 1921. The relations of the Amarat with the Sbaa and the Fadan were not cordial.

The chief importance of the Ruwalla was the fact that they commanded the Hit-Damascus road, one of the main trade routes between Syria and Iraq. With their powerful confederates, the Wulud Ali, the Muhallaf, and the Hasanah, who were usually in the closest relations with them, they numbered about 7,000 tents.

The Ruwalla and their allies wandered over the desert from Hama and Homs in the north, where the Hasanah had their summer pasturages. Later they began to settle down as cultivators of the land as far as Qasr-el-Azraq, south of Jebel ed Druze, and down the Wadi Sirhan to Jauf. Their range extended to the east as far as the source of the Wadi Hauran on the Jebel Enaze.[1] In summer they withdrew into the Wadi Sirhan.

---

[1] In the spring of 1928, Mr. W. E. Browne, surveyor for the Iraq Petroleum Company, followed the Wadi Hauran from the wells at Al Mat, north of Rutba,

The Paramount Sheikh of the Ruwalla is Nuri ibn Shalan,[1] one of the most powerful of Beduin chiefs. After his capture of Jauf from the Shammar of Ibn Rashid in 1912 he was the most successful rival of the Southern Shammar. Ibn Rashid, however, succeeded in recapturing Jauf during 1920.

The following information, based on 1920 statistics, is available[2] for this region, passing from northwest to southeast:

Raqqa.—A town in Syria with a population of approximately 2,000 Mohammedans, mainly Arabs and Circassians.

Deir-ez-Zor.—The total population, estimated at 15,000, consisted chiefly of Mohammedan town Arabs. There was a small Christian colony of Syrian Catholics and a few Jews.

Abu Kemal.—The French frontier post, with approximately 750 inhabitants, the majority of whom were Sunnis.

Ana.—Of the 15,000 inhabitants, the majority were Sunnis, with about twenty Jews engaged in trade.

Kubaisa.—The population, numbering about 3,000 Sunnis, was divided into six small tribes or houses.

Hit.—This ancient town stands on the right bank of the Euphrates, 119 miles downstream from Ana. Hit, on the river bank, dominates a mound which is precipitous to the plain but slopes more gradually toward the river. A tall, leaning minaret near the river bank provides a conspicuous landmark which can be seen for many miles. The town, surrounded by a loop-holed wall, gives the impression of being built for defense. There are large gardens of date palms and fruit trees on both banks of the river upstream from the town.

Hit, which is depressing and malodorous, owes these attributes to the bitumen wells and furnaces, the smoke from which causes a hazy atmosphere to hang over the town. The surrounding ground is also redolent of bitumen (*qir*) and sulphur (*kibrit*). Despite the unpleasantness, however, it is said to be decidedly healthful, and

past the Tellul Abaillie, across the Rutba-Amman track to Jebel Enaze. On the southern slopes of this low range of hills we found typologically Paleolithic flint implements on both sides of the small watercourse, which marks the source of the great Wadi Hauran.

[1] When I visited him in Damascus in April, 1928, although he was partly crippled with gout, his commanding presence was felt by all to whom he gave an audience.

[2] Sir William Andrew (p. 73) published the following population figures in 1882: Deir-ez-Zor, 7,000; Ana, 2,000; and Hit, 3,000. For later and more detailed information see Musil, 1927b.

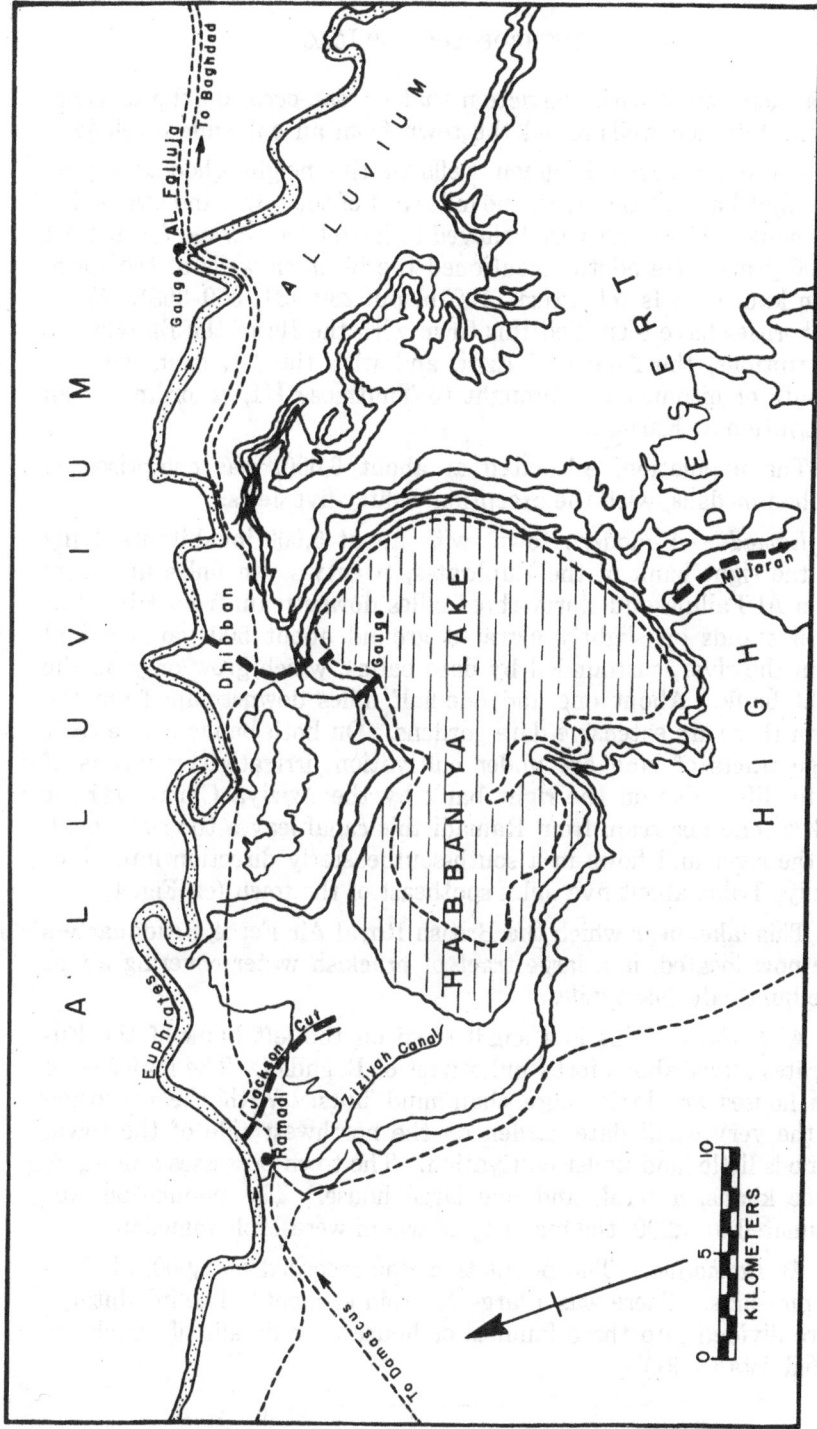

FIG. 4. Environs of Lake Habbaniya (after Ionides).

local sages state with conviction that on one occasion the presence of the bitumen wells saved the town from an epidemic of cholera.

There are seven bitumen wells in the neighborhood, five on the right bank of the river, one mile west of the town, and two on the left bank. These wells are believed to have been worked for at least 5,000 years. Herodotus mentioned the bitumen wells of the town, then known as Is (cf. Musil, 1927b, pp. 230–231, 350–353). Some authorities have identified this town with the Ihi of the Babylonian inscriptions, the Ahava of Ezra, and with the Ist, from which a tribute of bitumen was brought to Thutmose III, according to an inscription at Karnak.

The population, estimated at about 6,000, was comprised of Mohammedans, with the exception of fifty-five Jews.

Ramadi.—A modern town, with about 5,000 inhabitants, lying on the right bank of the Euphrates, twenty-seven miles upstream from Al Falluja and thirty-three miles downstream from Hit. Ramadi stands on slightly elevated ground about 500 yards inland from the river, surrounded by date palms, which grow only on the right bank. About one and one-half miles downstream from the town there are extensive date gardens. On both banks of the river large tracts of land are under cultivation, irrigated by means of water lifts, also on the right bank by the Aziziya Canal. About half a mile upstream from Ramadi this canal leaves the right bank of the river and flows in a south-southeasterly direction into Habbaniya Lake, about five miles southeast of the town (cf. Fig. 4).

This lake, near which the British Royal Air Force Headquarters are now located, is a large tract of brackish water covering about one hundred square miles.

Al Falluja.—This is a small town on the left bank of the Euphrates River about forty miles west of Baghdad. The majority of the houses are little more than mud huts. With the exception of the very small date garden on the northwest side of the town, there is little land under cultivation. The town possesses a mosque, three khans, a serai, and one large house. The population was estimated at 1,200, the majority of whom were Mohammedans.

Ar Rahhaliya.—The population was recorded as 2,000, all Mohammedans. There was a large Negroid element. The inhabitants were divided into three families or houses, for details of which see tribal lists (p. 91).

Little accurate information is available regarding the health conditions among the civil population, where the rule of survival of the fittest holds sway.

As throughout Iraq, eye diseases are extremely common, infection being carried chiefly by flies and dust, and aggravated by the insanitary conditions under which the people live.

Prior to 1925, epidemics of cholera, typhus, and smallpox appeared at intervals and cases of bubonic plague sometimes occurred.

Since the advent of trans-desert travel by automobile and airplane the danger from the spread of epidemics has increased a thousandfold. Medical inspectors were installed at Ramadi but pirate Arab convoys escaped this examination until the Iraq government, realizing the danger, policed all entrances into their territory. The greatest menace came from Pilgrims making the *Haj* to Mecca (cf. Clemow). Present arrangements are more than adequate to safeguard general health interests.

Under the brilliant direction of the Minister of Education and of Dr. M. Jamali, graduate of Columbia University, educational facilities are increasing throughout the country, but the Beduins are little influenced by these changes.

III. THE PHYSICAL ANTHROPOLOGY OF THE DULAIM AND THE ANAIZA

ANTHROPOMETRIC METHODS AND TECHNIQUE

In the previous chapters the land and the peoples of the Upper Euphrates region have been described and a brief summary of historical events has been given.

It seems undesirable to repeat at length the description of the Iraq government permits obtained or the correct procedure involved. My previous work in Iraq, beginning in 1925, facilitated the issuance of all necessary permits and letters of introduction to the *Mutasarrifs* of every *Liwa*.

In general the anthropometric methods and technique follow the procedure adopted by the International Committee at Monaco in 1906. A detailed description of the technique has appeared (pp. 278–288) in my "Contributions to the Anthropology of Iran."

In order to present the statistical data so that they can be compared to my previous figures, obtained in Iraq and Iran, it will be necessary to group the individuals according to the two classificatory systems devised and adopted by Dr. E. A. Hooton in the Laboratory of Anthropology at Harvard and by Sir Arthur Keith.

While the general trends remain the same the greater number of divisions (Keith system) show more clearly any small differences. For the sake of direct comparison, wherever possible, I have grouped the two tables.

Since I am planning to treat each section as but a part of one complete volume there is no necessity to compile comparative tables until the last section. For this reason I am publishing only the vital statistics, morphological characters, statistical analyses, and raw data of the Dulaim and the Anaiza.

On the other hand, since this is the first section of "The Anthropology of Iraq," I have felt it desirable to quote the recalculated tables for my groups of Arabs of the Kish area, Iraq Soldiers, and Ba'ij Beduins. The recalculation was necessary in order that the figures could be sorted and calculated on the Hollerith machines in the Laboratory of Anthropology at Harvard. Further slight differences occurred since some men were eliminated on account of youth or old age, the limits being 18–70 inclusive.

When this plan has been followed, all measurements, indices, and groupings will be directly comparable.

For the sake of comparison the series of 100 Arabs measured at Kish by Dr. L. H. Dudley Buxton during the first week of January, 1926, has been added. These tables were recalculated at Harvard from the raw data.

LIST OF ANTHROPOMETRIC ABBREVIATIONS

B = head breadth
B' = minimum frontal diameter
B'/B = fronto-parietal index
B'/J = zygo-frontal index
B/L = cephalic index
Big. B. = bigonial breadth
Biz. B. = bizygomatic breadth
C.I. = cephalic index
E.B. = ear breadth
EB/EL = ear index
E.I. = ear index
E.L. = ear length
F.P.I. = fronto-parietal index
G.B. = greatest breadth
G.H. = total facial height
G'H = upper facial height
GH/J = facial index
G'H/J = upper facial index

Go-Go = bigonial breadth
Go-Go/J = zygo-gonial index
G.O.L. = glabello-occipital length
J = bizygomatic breadth
L = glabello-occipital length
L.L. = lower limb length
M.F.D. = minimum frontal diameter
N.B. = nasal breadth
N.H. = nasal height
NB/NH = nasal index
N.I. = nasal index
R.S.H. = relative sitting height
S.H. = sitting height
T.F.H. = total facial height
T.F.I. = total facial index
U.F.H. = upper facial height
U.F.I. = upper facial index
Zyg.fr.I. = zygo-frontal index
Zyg.go.I. = zygo-gonial index

THE DULAIM[1]

The Dulaim, the largest semi-nomadic tribe in this area, state that they came to Iraq under the leadership of one Thamir, from the Dulaimiyat Springs in central Arabia. They are Sunnis of the Shafiite sect. Numbering approximately 26,000 men, they possessed cultivated lands on both banks of the Euphrates from Imam Hamza to Al Qaim.

About 50 per cent of the tribe were settled agriculturists, the remainder being nomads who raised sheep and camels, moving both into the eastern Shamiya and into the Jazira for their winter grazing. The nomadic sections usually left their summer habitat on the Euphrates about September and returned in April. No definite area or routes could be laid down for the migration of the nomad element as their movements were governed by the quantity of grazing available in the various areas.

[1] This introductory section is based on data obtained prior to 1921. During 1934, wherever possible, I checked this information. See also "A Handbook of Arabia" (vol. 1, pp. 53–54, London, 1920); Ashkenazi (1938); Ayrout (1938); Charles (1939); and von Oppenheim (vol. 1, pp. 186–189, 1939).

The Dulaim shared the pastures of the Amarat, with whom they were on friendly terms, in the eastern Shamiya. In the Jazira the nomad portion of the tribe sometimes moved as far north as Tikrit on the right bank of the Tigris.

The agricultural portions of the Dulaim cultivate a strip of land on both banks of the Euphrates, and along the Aziziya, Abu Ghuraib, Saqlawiya, and subsidiary canals.

The crops produced by the Dulaim are chiefly wheat, barley, rice, mash, maize, and millet (*dukhn*). Dates and other fruit such as apples, figs (*tin*), and pomegranates are grown in gardens surrounding the towns. The Dulaim export grain both up and down the Euphrates to the large market towns on the river, and also to Kubaisa and Ar Rahhaliya for sale to the desert tribes and for trans-desert market towns.

Toward the end of 1918 the Dulaim were closely allied with the Amarat section of the Anaiza, and at enmity with the Shammar Jarba and the settled Shiah tribes of the Lower Euphrates.

When the insurrection of 1920 finally had been subdued, and Sheikh Dhari ibn Dhahir of the Zoba tribe had fled, many sections of that tribe agreed to acknowledge Ali Sulaiman of the Dulaim as their Paramount Chief and became part of the Dulaim.

A list of Zoba sections, which either affiliated themselves with the Dulaim or set up as independent tribes, follows:

DULAIM		INDEPENDENT	
Luhaib	Saadan	Chitadah	Shaar
Shuwartan	Shiti	Faddaghah	Dulaim
Bani Zaid	Subaihat	Haiwat	Qartan
Qara-Ghul	Sumailat	Hitawiyin	
Khurushiyin			

The main part of the Qara-Ghul tribe, which was located on the left bank of the Euphrates about six miles downstream from Imam Hamza, had been independent since about 1840. The Qara-Ghul of the Zoba was a small colony from this tribe.

DULAIM TRIBESMEN MEASURED AT HADITHA

At Haditha on May 21 and 22, 1934, I examined 137 Dulaim tribesmen. The arrangements were made by the late Dr. H. C. Reid, Medical Officer of the Iraq Petroleum Company, whose guests we were.

Age.—The average age for 136 Dulaimis was 32.40 (range 20–64). Sixty-six per cent of the individuals were under thirty-five years of age. On the basis of age grouping the sample obtained should be a representative series of these tribesmen. No. 1076 was omitted.

AGE DISTRIBUTION

Age	No.	Per cent	Age	No.	Per cent
18–19	0	....	45–49	12	8.82
20–24	28	20.59	50–54	3	2.21
25–29	40	29.41	55–59	0	....
30–34	22	16.18	60–64	4	2.94
35–39	19	13.97	65–69	0	....
40–44	8	5.88	70–x	0	....
			Total	136	100.00

MORPHOLOGICAL CHARACTERS OF DULAIMIS

Skin.—The color was darker than that of the average Arab of the Kish area. Individually it ranged from that of a typical southern European to dark brown. The constant exposure to the weather, combined with the general neglect of washing except for ritual ablutions in which sand often replaced water, tended to give the older individuals a weather-beaten appearance. In general, the Dulaimis possessed a skin color little different from that of the Arabs from the "Fertile Crescent" to Morocco.

Nos. 1062 and 1124 (Pl. 36) had some Negro blood. No. 1109 had very dark hands and the color of his body was considerably darker than that of the average individual.

Hair.—The hair color varied from dark brown to black, which I now think should have been classified as very dark brown. No trace of blondism was present. In form the hair had low waves, seven individuals (5.30 per cent) possessing deep wavy hair. The three men recorded as having curly-frizzly hair indicate the presence of Negro blood, a feature which appears in the photographic analyses. Ninety-five men (72.52 per cent) had hair of medium texture. An almost equal proportion of the remainder occurred at both extremes of the scale. The coarser element might also be associated with a Negroid element. Sixty-six hair samples were obtained.

Hair on the head was abundant. No. 1124, who was completely bald, had no hair on his entire body. He stated that he had always been hairless, as were his three brothers, but that his parents possessed the normal amount of hair (Pl. 36).

On the other hand abnormal hairiness of the body was not recorded, and the general impression retained was that the amount of body hair was average for any group of Arabs in Iraq.

HAIR

Color	No.	Per cent	Form	No.	Per cent
Black	93	70.99	Straight	0	
Very dark brown	2	1.53	Very low waves	0	
Dark brown	19	14.50	Low waves	122	92.42
Brown	0		Deep waves	7	5.30
Reddish brown	0		Curly-frizzly	3	2.27
Light brown	0		Woolly	0	
Red	0				
Black and gray	14	10.69	Total	132	99.99
Dark brown and gray	0		Texture	No.	Per cent
Light brown and gray	0		Coarse	19	14.50
Gray	3	2.29	Coarse-medium	0	
White	0		Medium	95	72.52
			Medium-fine	3	2.29
Total	131	100.00	Fine	14	10.69
			Total	131	100.00

Eyes.—The majority of the eyes were dark brown but one-third of the individuals had mixed eyes, indicating submerged blondism. Two men had blue eyes. The majority of irises were homogeneous, although rather more than one-third were zoned. The few rayed irises could only have been recorded on the light eyes. The sclera were clear, with the exception of twelve men (8.82 per cent), four of whom were recorded as bloodshot.

EYES

Color	No.	Per cent	Iris	No.	Per cent
Black	0		Homogeneous	78	57.35
Dark brown	78	56.93	Rayed	6	4.41
Blue-brown	14	10.22	Zoned	52	38.24
Blue-brown	6	4.38			
Green-brown	31	22.63	Total	136	100.00
Green-brown	2	1.46	Sclera	No.	Per cent
Gray-brown	4	2.92	Clear	124	91.18
Blue	2	1.46	Yellow	0	
Gray	0		Speckled	8	5.88
Light brown	0		Bloodshot	4	2.94
Blue-gray	0		Speckled and bloodshot	0	
Blue-green	0		Speckled and yellow	0	
			Yellow and bloodshot	0	
Total	137	100.00			
			Total	136	100.00

The eyes, or more properly the eye slits, were horizontal as in Europeans.

No. 1046 had bright blue eyes. He stated that in the village of Khraair more than half the population have blue eyes. He explained his own case by saying that when his mother was pregnant she saw a man with blue eyes which influenced the eye color of her unborn child. Many Dulaimis agreed that there were numerous persons with blue eyes among this tribe. The blue element was present in Nos. 1016, 1047, 1090, 1108, 1112, 1119, and 1120. Nos. 1021, 1036, and 1037 had light green-brown eyes. No. 1065 was almost blind in the right eye. No. 1074 had poor vision in his left eye. No. 1076 had poor vision in both eyes. He had applied *kubeli* mixed with sugar in both eyes. This gave them a red color. No. 1105 was slightly cross-eyed, the right eye being out of alignment.

Nose.—The majority (70.80 per cent) of the noses were straight in profile, with only 13.87 per cent convex. Half of the Dulaimis had medium nasal wings, with 30.37 per cent in the narrowest categories. The remainder (16.29 per cent) of the alae were medium-flaring or flaring, once again indicating the presence of a Negroid element. Two men had thicker than average nasal tips and one man was recorded in the double plus classification.

NOSE

Profile	No.	Per cent	Wings	No.	Per cent
Wavy	10	7.30	Compressed	32	23.70
Straight	97	70.80	Compressed-medium	9	6.67
Concave	3	2.19	Medium	72	53.33
Convex	19	13.87	Medium-flaring	17	12.59
Concavo-convex	8	5.84	Flaring	5	3.70
			Flaring plus	0	.....
Total	137	100.00	Total	135	99.99

Mouth.—The lips varied from thin (No. 1081) to thick (No. 1080). Some individuals (Nos. 1027, 1058, 1087, and 1092) showed marked lower lip eversion. Nos. 1022 and 1030 had thin upper lips.

Teeth.—The occlusion was normal slight over for the entire group, with the exception of four men (2.96 per cent) each of whom had a marked-over bite.

Since half of the group was under thirty years of age the good condition of the teeth is not unusual. The average age of the group was 32.40 (range 20–64). There were relatively few teeth lost and 85.18 per cent of the Dulaimis possessed either good or excellent teeth.

TEETH

Bite	No.	Per cent	Loss	No.	Per cent	Condition	No.	Per cent
Under	0	.....	None	0	.....	Very bad	2	1.86
Edge-to-edge	0	.....	1–4	11	91.67	Bad	7	6.48
Slight over	131	97.04	5–8	0	.....	Fair	7	6.48
Marked over	4	2.96	9–16	1	8.33	Good	66	61.11
			17–	0	.....	Excellent	26	24.07
Total	135	100.00	Total	12	100.00	Total	108	100.00

The following men had very good teeth: Nos. 1012, 1013, 1029–1031, 1033–1035, 1037, 1040, 1043, 1045, 1061, 1064, 1066,

1067, 1078, 1079, 1086, 1089, 1097, 1098, 1113, 1122, 1123, and 1142. Nos. 1018, 1055, 1058, 1063, 1110, and 1121 had fair teeth, while Nos. 1053, 1060, 1075, 1083, and 1101 were poor. The teeth were very bad in Nos. 1059, 1065, and 1119. No. 1054 had marked-over occlusion. No. 1085 had teeth markedly sloping inward. No. 1114 had lower front teeth showing much wear. No. 1095 had gold fillings in his front teeth and No. 1119 had three teeth covered with gold. No. 1062 had a broken right upper incisor as a result of a gun accident.

Musculature.—In general this was either good or excellent, although there were a few obvious cases of malnutrition. The outdoor activities of these tribesmen who, to some extent, are pastoral nomads as well as agriculturists, tend to produce a healthy and virile group.

Musculature	No.	Per cent
Poor	0	
Fair	0	
Average	0	
Good	121	92.37
Excellent	10	7.63
Total	131	100.00

Nos. 1056, 1097, 1108, and 1110 had well-developed muscles, but Nos. 1017, 1048, and 1125 were in poor physical condition.

Health.—The majority (91.91 per cent) were in good health. Nine Dulaimis (6.62 per cent) were recorded as being in fair health.

Health	No.	Per cent
Poor	0	
Fair	9	6.62
Average	0	
Good	125	91.91
Excellent	2	1.47
Total	136	100.00

Disease.—Twenty-three men had smallpox scars. In 1924, No. 1044 had smallpox, causing a cataract in the left eye, but ten years later he was not totally blind. No. 1021 also had chicken pox scars. No. 1042 had a skin disease on the head. No. 1047 had a scar on the right cheek, the result of a dog bite. No. 1067 had a large lump over the left temple, which he said was a birthmark. No. 1124, the hairless man, has been described on page 35.

Blood Groupings.—Twenty blood samples were sent to Dr. Walter P. Kennedy in Baghdad. These are included in his report (1935, pp. 475–480).

Branding Scars.—Among 137 individuals forty-six Dulaimis (66.42 per cent) bore *kawi* or *chawi* scars.

Tattooing.—Fifty-eight (51.33 per cent) out of 113 individuals bore simple tattooed designs. These will be examined in detail in a forthcoming publication dealing with body-marking in Southwestern Asia (cf. Field, 1935a, pp. 455–456, and Charles, pp. 109–111).

Henna.—Nos. 1026, 1037, and 1123 had applied henna (Ar. *henna*), *Lawsonia* sp., to the palms of the hands "to harden them." No. 1032 had "decorated" his nails with henna.

Kohl.—No. 1021 had applied kohl (*kuhl*), finely powdered antimony, below his eyes "to cool them from the desert heat and the burning dust." Nos. 1047, 1081, and 1132 had used kohl below their eyes "because of pain due to the brightness of the sun."

SUMMARY

The average Dulaimi had low wavy hair, medium in texture, and extremely dark brown merging into black in color. The eyes were various shades of brown but two individuals had definitely blue eyes. The sclera were clear and the iris mainly homogeneous. The nose was straight in profile with medium or compressed wings, although there was a group with medium-flaring wings. The occlusion was normal. The musculature and health were good.

STATISTICAL ANALYSES OF DULAIMIS

There now remains the task of grouping the total series of Dulaimis[1] according to the Harvard and Keith classificatory systems for stature, sitting height (trunk length), minimum frontal diameter, head breadth, cephalic index, nasal height, nasal breadth, and nasal index.

Stature.—The Dulaimis were medium to tall according to both systems. There is remarkably little difference in the groupings. The average stature for 136 individuals was 167.67 (range 152–181), which is slightly higher than the average for Southwestern Asia.

STATURE

Harvard system	No.	Per cent	Keith system	No.	Per cent
Short (x–160.5)	11	8.09	Short (x–159.9)	8	5.88
Medium (160.6–169.4)	78	57.35	Medium (160.0–169.9)	80	58.82
Tall (169.5–x)	47	34.56	Tall (170.0–179.9)	47	34.56
			Very tall (180.0–x)	1	0.74
Total	136	100.00	Total	136	100.00

[1] No. 1029 was omitted.

Sitting Height (Trunk Length).—The Keith system shows that the majority (58.82 per cent) have medium to long trunk lengths. The six men (4.41 per cent) with very long (90.0+) trunk lengths and the one with a very short (x–74.9) trunk indicate the maximum of variation. The relative sitting height index of 50.08 (range 44–59) together with the stature groupings reveals that the trunk length and leg length are approximately equal but an increase in trunk length is followed by an advance in stature.

SITTING HEIGHT (Trunk Length)

Group	No.	Per cent
Very short (x–74.9)	1	0.74
Short (75.0–79.9)	14	10.29
Medium (80.0–84.9)	80	58.82
Long (85.0–89.9)	35	25.73
Very long (90.0–x)	6	4.41
Total	136	99.99

Minimum Frontal Diameter.—The forehead was narrow or very narrow in 73.53 per cent of the cases. The majority (64.71 per cent) fall into the narrow category, the next greatest number (25 per cent) being wide. Two distinct elements appear to be present.

MINIMUM FRONTAL DIAMETER

Group	No.	Per cent
Very narrow (x–99)	12	8.82
Narrow (100–109)	88	64.71
Wide (110–119)	34	25.00
Very wide (120–x)	2	1.47
Total	136	100.00

Head Breadth.—The mean for this measurement was 141.34 (range 132–155) with 191.04 for the head length. The Keith system reveals no Dulaimi in the very narrow category and only six Dulaimis in the very wide division. The majority (57.35 per cent) possessed wide heads but 38.24 per cent were narrow. Two distinct elements appear to be present here. These may well be the straight-nosed and convex-nosed dolichocephals.

HEAD BREADTH

Group	No.	Per cent
Very narrow (120–129)	0	.....
Narrow (130–139)	52	38.24
Wide (140–149)	78	57.35
Very wide (150–x)	6	4.41
Total	136	100.00

Cephalic Index.—According to the Harvard system the majority (79.41 per cent) were dolichocephalic, with only one brachycephal in the entire series of 136 Dulaimis.

The Keith classificatory system reveals a rather different grouping. The majority (56.62 per cent) were dolichocephalic but there were six brachycephals and no ultrabrachycephals. The most interesting new group was formed by the thirteen (9.56 per cent) ultradolichocephals (x–70.0).

The mean cephalic index was 74.04 (range 65–84.9). Therefore the Dulaimis were dolicho-mesocephals with a strong tendency toward ultradolichocephaly.

CEPHALIC INDEX

Harvard system	No.	Per cent	Keith system	No.	Per cent
Dolichocephalic (x–76.5)	108	79.41	Ultradolichocephalic (x–70.0)	13	9.56
Mesocephalic (76.6–82.5)	27	19.85	Dolichocephalic (70.1–75.0)	77	56.62
Brachycephalic (82.6–x)	1	0.74	Mesocephalic (75.1–79.9)	40	29.41
Total	136	100.00	Brachycephalic (80.0–84.9)	6	4.41
			Ultrabrachycephalic (85.0–x)	0	.....
			Total	136	100.00

Facial Measurements.—The upper facial height was medium long (48.53 per cent) or medium short (31.62 per cent). Twenty-five Dulaimis (18.38 per cent) had long (76–x) upper faces.

The total length of the face was either medium long (55.15 per cent) or medium short (34.56 per cent). It is remarkable that only eleven men (8.09 per cent) fell into the long face (130–x) category.

The majority (56.62 per cent) of the Dulaimis were leptoprosopic with 8.82 per cent in the euryprosopic classification.

Thus the faces were long, primarily the result of long upper faces.

FACIAL MEASUREMENTS

Upper facial height	No.	Per cent	Total facial height	No.	Per cent
Short (x–63)	2	1.47	Short (x–109)	3	2.21
Medium short (64–69)	43	31.62	Medium short (110–119)	47	34.56
Medium long (70–75)	66	48.53	Medium long (120–129)	75	55.15
Long (76–x)	25	18.38	Long (130–x)	11	8.09
Total	136	100.00	Total	136	100.01

TOTAL FACIAL INDEX

Group	No.	Per cent
Euryprosopic (x–84.5)	12	8.82
Mesoprosopic (84.6–89.4)	47	34.56
Leptoprosopic (89.5–x)	77	56.62
Total	136	100.00

Nasal Measurement and Indices.—The nose is one of the most significant racial criteria in Southwestern Asia. This fact was demonstrated clearly in my studies of the modern peoples of Iran (Field, 1939).

The Dulaimis possessed medium or short noses, there being seven men (5.15 per cent) in the long nose (60–x) category.

The nose was medium narrow (51.47 per cent) or medium wide (38.24 per cent). Ten men had very narrow noses and four possessed wide noses, indicating Negro blood.

The majority (64.71 per cent) of the Dulaimis were leptorrhine. Forty-five men (33.09 per cent) were mesorrhine but only three fell into the platyrrhine category. This latter again suggests the presence of Negro blood.

NASAL MEASUREMENTS

Nasal height	No.	Per cent	Nasal width	No.	Per cent
Short (x–49)	30	22.06	Very narrow (x–29)	10	7.35
Medium (50–59)	99	72.79	Medium narrow (30–35)	70	51.47
Long (60–x)	7	5.15	Medium wide (36–41)	52	38.24
Total	136	100.00	Wide (42–x)	4	2.94
			Total	136	100.00

NASAL INDEX

Group	No.	Per cent
Leptorrhine (x–67.4)	88	64.71
Mesorrhine (67.5–83.4)	45	33.09
Platyrrhine (83.5–x)	3	2.21
Total	136	100.01

In order to furnish additional statistical data for comparison with those in my Iran Report the following tables have been calculated:

SITTING HEIGHT (Trunk Length)

Standing height	900–x		899–850		849–800		799–750		749–x		Totals	
	No.	%	No.	%	No.	%	No.	%	No.	%	No.	%
1800–x	1	0.74	0	....	0	....	0	....	0	....	1	0.74
1799–1700	2	1.47	25	18.38	17	12.50	3	2.21	0	....	47	34.56
1699–1600	3	2.21	11	8.09	58	42.65	9	6.62	0	....	81	59.57
x–1599	1	0.74	0	....	4	2.94	2	1.47	0	....	7	5.14
											136	100.01

Minimum Frontal Diameter

Head breadth	x–99 No.	%	100–109 No.	%	110–119 No.	%	120–x No.	%	Totals No.	%
120–129	0	....	0	....	0	....	0	....	0	.....
130–139	0	....	16	11.76	34	25.00	2	1.47	52	38.23
140–149	0	....	7	5.15	68	50.00	3	2.21	78	57.36
150–x	0	....	0	....	4	2.94	2	1.47	6	4.41
									136	100.00

Bizygomatic Breadth

Total face length	x–124 No.	%	125–134 No.	%	135–x No.	%	Totals No.	%
x–114	0	....	7	5.15	6	4.41	13	9.56
115–124	4	2.94	37	27.21	43	31.62	84	61.77
125–x	0	....	15	11.03	24	17.65	39	28.68
							136	100.01

Upper Facial Length

Total face length	x–63 No.	%	64–69 No.	%	70–75 No.	%	76–81 No.	%	82–x No.	%	Totals No.	%
x–109	1	0.74	2	1.47	0	....	0	....	0	....	3	2.21
110–119	1	0.74	30	22.06	14	10.29	2	1.47	0	....	47	34.56
120–129	0	....	11	8.09	51	37.50	12	8.82	1	0.74	75	55.15
130–x	0	....	0	....	1	0.74	8	5.88	2	1.47	11	8.09
											136	100.01

Nasal Width

Nasal length	x–29 No.	%	30–35 No.	%	36–41 No.	%	42–x No.	%	Totals No.	%
x–49	2	1.47	13	9.56	14	10.29	1	0.74	30	22.06
50–59	8	5.88	54	39.71	35	25.74	2	1.47	99	72.80
60–x	0	....	5	3.68	2	1.47	0	....	7	5.15
									136	100.01

Measurements and Indices of Dulaimis

Measurements	No.	Range	Mean	S.D.	C.V.
Age	136	20–64	32.40±0.56	9.60±0.39	29.63±1.21
Stature	136	152–181	167.67±0.30	5.25±0.21	3.13±0.13
Sitting height	136	75–92	84.07±0.20	3.42±0.14	4.07±0.17
Head length	136	167–208	191.04±0.39	6.69±0.27	3.50±0.14
Head breadth	136	132–155	141.34±0.28	4.83±0.20	3.42±0.14
Minimum frontal diameter	136	101–128	113.02±0.22	3.88±0.16	3.43±0.14
Bizygomatic diameter	136	120–149	134.95±0.30	5.25±0.21	3.89±0.16
Bigonial diameter	136	94–125	106.66±0.30	5.16±0.21	4.84±0.20
Total facial height	136	105–139	121.50±0.34	5.90±0.24	4.86±0.20
Upper facial height	136	55–89	71.55±0.27	4.65±0.19	6.50±0.27
Nasal height	136	44–67	52.86±0.25	4.24±0.17	8.02±0.33
Nasal breadth	136	25–48	34.70±0.22	3.87±0.16	11.15±0.46
Ear length	136	44–71	57.94±0.27	4.68±0.19	8.08±0.33
Ear breadth	136	26–43	34.41±0.17	2.94±0.12	8.54±0.35
Indices					
Relative sitting height	136	44–59	50.08±0.12	2.08±0.09	4.15±0.17
Cephalic	136	65–85	74.04±0.20	3.51±0.14	4.74±0.19
Fronto-parietal	136	72–89	80.20±0.17	2.88±0.12	3.59±0.15
Zygo-frontal	136	76–91	84.18±0.16	2.80±0.11	3.33±0.14
Zygo-gonial	136	66–92	79.39±0.20	3.45±0.14	4.35±0.18
Total facial	136	75–109	90.35±0.30	5.20±0.21	5.76±0.24
Upper facial	136	43–66	53.15±0.20	3.54±0.14	6.66±0.27
Nasal	136	44–99	65.66±0.51	8.80±0.36	13.40±0.55
Ear	136	45–76	60.02±0.34	5.92±0.24	9.86±0.40

PHOTOGRAPHIC ANALYSES

When the Dulaimis had been sorted according to racial and physical types the following results were obtained:

>Classic Mediterranean: No. 1013 (Plates 2, 3)
>Fine Mediterranean: No. 1052 (Plate 4)
>Coarse Mediterranean: No. 1080 (Plate 4)
>Iraqo-Mediterranean: Nos. 1039, 1037 (Plate 5)
>Dolichocephals: Nos. 1011, 1053, 1054, 1044 (Plates 6, 7)
>Brachycephals: Nos. 1048, 1010 (Plate 8)
>Short-faced: No. 1049 (Plate 9)
>Long-faced: No. 1018 (Plate 9)
>Short and narrow-faced: No. 1050 (Plate 10)
>Short and broad-faced: No. 1065 (Plate 10)
>Mixed-eyed: Nos. 1021, 1023 (Plate 11)
>Blue-eyed: No. 1046 (Plate 12)
>Green-brown-eyed: No. 1059 (Plate 12)
>Straight-nosed: No. 1034 (Plate 13)
>Very slightly convex-nosed: No. 1019 (Plate 13)
>Slightly convex-nosed: Nos. 1093, 1045 (Plate 14)
>Convex-nosed: Nos. 1041, 1017, 1055 (Plates 15–17)
>Very low wavy hair: No. 1084 (Plate 18)
>Low wavy hair: No. 1092 (Plate 18)
>Deep wavy hair: No. 1066 (Plate 19)
>Very deep wavy hair: No. 1028 (Plate 19)
>Hairless Dulaimi (Negroid): No. 1124 (Plate 36)

Examination of the photographs reveals that while the Dulaimis are considerably mixed in racial characters they still belong to the Mediterranean Race. They show less variation than the Arabs of the Kish area or the Iraq Soldiers but more variation than either the Ba'ij or the Anaiza Beduins.

The Dulaimis appear to belong to the straight-nosed, leptoprosopic and dolichocephalic division of the Mediterranean Race which may be termed the Iraqo-Mediterranean group in contradistinction to the convex-nosed, leptoprosopic, and dolichocephalic Iranian Plateau Race (cf. Field, 1939).

These speculations will be examined in detail in the final part of this volume when all my anthropometric data can be utilized for discussion.

SUMMARY

The average Dulaimi is medium in stature, and medium to long in trunk length, and possesses a narrow forehead, a wide to narrow head breadth, a dolicho-mesocephalic index, a long upper face, a medium total facial height and a leptoprosopic index, a nose medium in length, medium narrow or medium wide and a leptorrhine to mesorrhine index.

The Dulaimis are believed to be of mixed blood and the general impression obtained during the study of them suggests that they belong neither to the pure Beduin type of the North Arabian and Syrian Deserts, nor to the sedentary Arab groups of central and southern Iraq. The average Dulaimi is thus, from physical aspect, not pure in type, but this group is particularly interesting because it appears to combine the physical features of the Beduin and the Arab.

Measurements of Dulaimis

No.	Age	Stature	SH	L	B	B'	J	go-go	GH	G'H	NH	NB
1007	30	1692	820	194	136	113	134	111	124	72	53	34
1008	32	1710	830	188	141	110	138	108	128	71	56	33
1009	25	1745	776	194	141	113	133	105	122	67	48	29
1010	25	1644	805	178	145	114	138	104	128	73	53	30
1011	20	1598	843	196	135	111	128	98	115	67	48	34
1012	20	1668	835	194	148	118	141	111	120	68	44	38
1013	30	1710	775	188	148	116	136	110	128	75	57	32
1014	35	1653	840	184	146	108	131	104	126	72	53	34
1015	25	1710	835	194	136	107	124	105	118	70	49	33
1016	45	1673	885	196	148	115	140	108	138	86	58	39
1017	50	1540	760	193	138	116	135	107	121	72	51	35
1018	30	1720	843	192	138	113	130	105	131	78	62	32
1019	27	1647	842	189	140	112	133	108	118	68	52	38
1020	35	1760	925	194	138	109	132	105	128	70	52	33
1021	25	1765	853	194	139	113	137	105	126	73	58	35
1022	35	1597	847	190	139	110	132	111	116	74	55	34
1023	30	1604	823	190	142	108	129	101	119	76	58	37
1024	32	1640	847	193	139	116	132	108	123	71	50	34
1025	20	1640	795	188	134	106	126	101	108	58	44	34
1026	30	1654	836	189	144	111	135	108	127	72	51	35
1027	22	1701	834	196	140	115	136	115	138	79	60	37
1028	40	1701	837	194	135	116	134	105	114	70	47	39
1029	28	....	...	...	...	...	...	...	...	..	..	..
1030	25	1640	817	197	142	108	122	108	123	73	54	30
1031	23	1636	837	205	151	115	140	115	121	68	49	37
1032	20	1593	845	187	137	106	128	104	119	74	52	40
1033	25	1655	853	193	144	113	134	108	118	67	49	38
1034	22	1681	844	188	144	115	138	112	126	76	58	33
1035	25	1725	888	193	147	117	142	108	123	70	48	35
1036	25	1682	874	185	144	113	138	109	122	73	53	38
1037	27	1630	834	197	142	113	125	99	116	68	55	41
1038	35	1676	835	184	147	112	132	110	115	68	52	40
1039	27	1627	805	195	137	106	124	114	124	72	55	37
1040	20	1728	844	187	151	113	132	108	114	68	48	31
1041	35	1650	780	184	132	108	143	97	116	72	55	30
1042	45	1687	816	190	136	112	138	105	130	78	61	31
1043	20	1780	909	196	144	117	140	107	123	70	50	37
1044	45	1672	828	197	135	113	134	110	125	76	58	42
1045	20	1600	793	197	142	118	134	108	119	74	56	40
1046	35	1676	878	196	139	113	134	107	126	75	54	36
1047	20	1730	853	183	147	115	135	107	115	61	46	37
1048	30	1720	831	167	141	107	136	107	128	74	58	28
1049	30	1677	892	195	143	118	134	105	114	65	47	34
1050	45	1662	846	191	134	108	127	95	117	69	50	37
1051	60	1700	870	195	147	115	137	107	127	82	54	40
1052	20	1605	845	181	147	112	128	105	118	66	51	34

Indices of Dulaimis

No.	EL	EB	RSH	B/L	B'/B	GH/J	G'H/J	NB/NH	EB/EL	go-go/J	B'/J
1007	51	34	48.5	70.1	83.1	92.5	53.7	64.2	66.7	82.8	84.3
1008	58	32	48.5	75.0	78.0	92.8	51.4	58.9	55.2	78.3	79.7
1009	56	36	44.5	72.7	80.1	91.7	50.4	60.4	64.3	79.0	85.0
1010	55	36	48.9	81.5	78.7	92.8	52.9	56.6	65.5	75.4	82.6
1011	55	30	52.8	68.9	82.2	89.8	52.3	70.8	54.6	76.6	86.7
1012	58	39	50.0	76.3	80.0	85.1	48.2	86.4	67.2	78.7	83.7
1013	58	28	45.3	78.7	78.4	94.1	55.1	56.1	48.3	80.9	85.3
1014	60	33	50.8	79.4	74.0	96.2	55.0	64.2	55.0	79.4	82.4
1015	64	34	48.8	70.1	78.7	95.2	56.5	55.9	53.1	84.7	86.3
1016	67	36	52.9	75.5	77.7	98.6	61.4	67.2	53.7	77.1	82.1
1017	64	30	49.4	71.5	84.1	89.6	53.3	68.6	46.9	79.3	85.9
1018	56	33	49.0	71.9	81.9	100.8	60.0	51.6	58.9	80.8	86.9
1019	58	33	51.1	74.1	80.0	88.7	51.1	73.1	56.9	81.2	84.2
1020	67	32	52.6	71.1	79.0	97.0	53.0	63.5	47.8	80.0	82.6
1021	60	32	48.3	71.7	81.3	92.0	53.3	60.3	53.3	76.6	82.5
1022	62	32	53.0	73.2	79.1	87.9	56.1	61.8	51.6	84.1	83.3
1023	52	31	51.3	74.7	76.1	92.3	58.9	63.8	59.6	78.3	83.7
1024	52	33	51.6	72.0	83.5	93.2	53.8	68.0	63.5	81.8	87.9
1025	57	34	48.5	71.3	79.1	85.7	46.0	77.3	59.7	80.2	84.1
1026	58	34	50.5	76.2	77.1	94.1	53.3	68.6	58.6	80.0	82.2
1027	59	41	49.0	71.4	82.1	101.5	58.1	61.7	69.5	84.6	84.6
1028	63	40	49.4	69.6	85.9	85.1	52.2	83.0	63.5	78.4	86.6
1029	..	..	....	....	....	....	....	....	....	....	....
1030	60	32	49.8	72.1	76.1	100.8	59.8	55.6	53.3	88.5	88.5
1031	56	35	51.2	73.7	76.2	86.4	48.6	75.5	62.5	82.1	82.1
1032	58	34	53.0	73.3	77.4	93.0	57.8	76.9	58.6	81.3	82.8
1033	64	36	51.5	74.6	78.5	88.1	50.0	77.6	56.3	80.6	84.3
1034	57	35	50.2	76.6	79.9	91.3	55.1	56.9	61.4	81.2	83.3
1035	64	35	51.5	76.2	79.6	86.6	49.3	72.9	54.7	76.1	82.4
1036	59	35	52.0	77.8	78.5	88.4	52.9	71.7	59.3	79.0	81.9
1037	61	28	51.2	72.1	79.6	92.8	54.4	74.6	45.9	79.2	90.4
1038	63	34	49.8	79.9	76.2	87.1	51.5	76.9	54.0	83.3	84.9
1039	48	33	49.5	70.3	77.4	100.0	58.1	67.3	68.8	91.9	85.5
1040	58	36	48.8	80.8	74.8	86.4	51.5	64.6	62.1	81.8	85.6
1041	66	38	47.3	71.7	81.8	81.1	50.3	54.6	57.6	67.8	75.5
1042	61	40	48.4	71.6	82.4	94.2	56.5	50.8	65.6	76.1	81.2
1043	52	34	51.1	73.5	81.3	87.9	50.0	74.0	65.4	76.4	83.6
1044	58	38	49.5	68.5	83.7	93.3	56.7	72.4	65.5	82.1	84.3
1045	60	31	49.6	72.1	83.1	88.8	55.2	71.4	51.7	80.6	88.1
1046	58	35	52.4	70.9	81.3	94.0	56.0	66.7	60.3	79.9	84.3
1047	60	37	49.3	80.3	78.2	85.2	45.2	80.4	61.7	79.3	85.2
1048	60	33	48.3	84.4	75.9	94.1	54.4	48.3	55.0	78.7	78.7
1049	53	34	53.2	73.3	82.5	85.1	48.5	72.3	64.2	78.4	88.1
1050	60	32	50.9	70.2	80.6	92.1	54.3	74.0	53.3	74.8	85.0
1051	57	31	51.2	75.4	78.2	92.7	59.9	74.1	54.4	78.1	83.9
1052	57	36	52.6	81.2	76.2	92.2	51.6	66.7	63.2	82.0	87.5

Measurements of Dulaimis—continued

No.	Age	Stature	SH	L	B	B'	J	go-go	GH	G'H	NH	NB
1053	30	1720	863	200	137	116	136	108	123	70	51	38
1054	40	1732	787	197	137	120	137	115	125	70	53	36
1055	42	1725	857	190	145	116	142	110	132	75	58	37
1056	30	1804	917	207	148	121	145	125	123	72	55	36
1057	35	1574	925	190	143	112	141	109	123	71	53	35
1058	40	1710	832	195	144	116	138	110	127	78	59	34
1059	60	1710	840	194	145	110	137	103	133	78	63	40
1060	60	1665	770	194	147	110	135	98	123	71	54	29
1061	26	1703	804	180	144	114	134	113	122	68	52	40
1062	28	1668	837	193	141	113	131	103	124	73	53	31
1063	30	1672	757	191	148	116	132	102	123	71	51	30
1064	50	1670	832	190	142	111	133	117	125	75	51	36
1065	40	1738	880	185	145	115	140	108	109	66	54	41
1066	23	1593	820	182	145	121	138	104	116	70	50	28
1067	21	1693	910	194	155	114	139	108	122	77	58	35
1068	45	1720	895	206	139	114	136	104	132	79	55	39
1069	25	1615	805	187	138	110	128	107	116	66	47	36
1070	25	1725	832	195	136	115	132	103	116	69	56	34
1071	26	1666	856	190	147	115	137	110	132	80	62	33
1072	24	1648	850	183	136	110	131	116	119	66	46	33
1073	35	1766	878	181	138	109	130	103	138	86	67	30
1074	20	1675	844	192	142	113	136	107	121	73	59	33
1075	45	1736	878	194	140	110	136	108	119	70	48	41
1076	..	1673	820	176	138	111	136	104	118	65	48	29
1077	45	1643	842	200	138	112	134	104	120	70	51	35
1078	45	1703	872	193	133	108	130	101	117	64	47	36
1079	25	1760	876	197	142	118	138	107	119	69	50	35
1080	35	1670	815	185	146	116	138	107	107	67	44	39
1081	20	1635	845	192	138	116	133	107	116	71	51	36
1082	25	1640	843	190	141	107	129	106	110	66	57	37
1083	60	1716	825	194	142	115	139	113	126	78	57	46
1084	25	1705	803	184	145	110	140	110	123	69	50	30
1085	20	1624	838	193	147	118	135	102	114	68	51	30
1086	21	1670	828	198	139	115	126	100	123	72	52	31
1087	25	1685	825	188	142	110	130	102	121	68	49	35
1088	25	1625	817	177	133	109	130	100	114	67	55	32
1089	25	1773	800	195	143	118	140	105	120	68	47	39
1090	20	1783	876	196	138	120	140	110	118	70	54	32
1091	25	1730	860	200	141	115	132	107	126	67	48	38
1092	25	1607	858	190	137	107	125	97	117	68	54	28
1093	25	1632	830	178	138	107	134	104	122	72	53	27
1094	25	1628	847	187	138	110	128	103	119	68	53	32
1095	30	1635	840	192	144	114	134	102	126	73	58	31
1096	30	1685	858	202	138	115	133	111	126	79	58	36
1097	29	1642	825	188	150	120	140	103	121	71	51	33
1098	35	1680	838	197	145	116	135	109	123	78	58	34

Indices of Dulaimis—continued

No.	EL	EB	RSH	B/L	B'/B	GH/J	G'H/J	NB/NH	EB/EL	go-go/J	B'/J
1053	53	34	50.2	68.5	84.7	90.4	51.5	74.5	64.2	79.4	85.3
1054	67	35	45.4	69.5	87.6	91.2	51.1	67.9	52.2	82.9	87.6
1055	54	36	49.7	76.3	80.0	93.0	52.8	63.8	66.7	77.5	81.7
1056	63	33	50.8	71.5	81.8	84.8	49.6	65.5	52.4	86.2	83.5
1057	56	34	....	75.3	78.3	87.2	50.4	66.0	60.7	77.3	79.4
1058	60	37	48.7	73.9	80.6	92.0	56.5	57.6	61.7	79.7	84.1
1059	61	35	49.1	74.7	75.9	97.1	56.9	63.5	57.4	75.2	80.3
1060	60	34	46.2	75.8	74.8	91.1	52.6	53.7	73.3	72.6	81.5
1061	60	32	47.2	80.0	79.2	91.0	50.8	76.9	53.3	84.3	85.1
1062	50	30	50.2	73.1	80.1	94.7	55.7	58.5	60.0	78.6	86.3
1063	57	37	45.3	77.5	78.4	93.2	53.8	58.8	64.9	77.3	87.9
1064	52	35	49.8	74.7	78.2	94.0	56.4	70.6	67.3	88.6	84.1
1065	58	36	50.6	78.4	79.3	77.9	47.1	75.9	62.1	77.1	82.1
1066	60	33	51.5	79.7	83.5	84.1	50.7	56.0	55.0	75.4	87.7
1067	55	31	53.8	79.9	73.6	87.8	55.4	60.3	56.4	77.7	82.0
1068	63	38	52.0	67.5	82.0	97.1	58.1	70.9	60.3	77.5	83.8
1069	51	36	49.8	73.8	79.7	90.6	51.6	76.6	70.6	83.6	85.9
1070	54	36	48.2	69.7	84.6	87.9	52.3	60.7	66.7	78.0	87.1
1071	58	37	51.4	77.4	78.2	96.4	58.4	53.2	63.8	80.3	83.9
1072	57	35	51.6	74.3	80.9	90.8	50.4	71.7	61.4	88.6	84.0
1073	52	33	49.7	76.2	79.0	106.2	66.2	44.8	63.5	79.2	83.9
1074	61	40	50.4	74.0	79.6	89.0	53.7	55.9	65.6	78.7	83.1
1075	70	39	50.6	72.2	78.6	87.5	51.5	85.4	55.7	79.4	80.9
1076	62	39	49.0	78.4	80.4	86.8	47.8	60.4	62.9	76.5	81.6
1077	55	36	51.2	69.0	81.2	89.6	52.2	68.6	65.5	77.6	83.6
1078	55	33	51.2	68.9	81.2	90.0	49.2	76.6	60.0	77.7	83.1
1079	50	35	49.8	72.1	83.1	86.2	50.0	70.0	70.0	77.5	85.5
1080	53	35	48.8	78.9	79.5	77.5	48.6	72.2	66.0	77.5	84.1
1081	51	32	51.7	71.9	84.1	87.2	53.4	70.6	62.8	80.5	87.2
1082	54	34	51.4	74.2	75.9	85.3	51.2	64.9	63.0	82.2	83.0
1083	59	34	48.1	73.2	81.0	90.7	56.1	80.7	57.6	81.3	82.7
1084	53	27	47.1	78.8	75.9	87.9	49.3	60.0	50.9	78.6	78.6
1085	62	31	51.6	76.6	80.3	84.4	50.4	58.8	50.0	75.6	87.4
1086	62	34	49.6	70.2	82.7	97.6	57.1	59.6	54.8	79.4	91.3
1087	54	33	49.0	75.5	77.5	93.1	52.3	71.4	61.1	78.5	84.6
1088	62	35	50.3	75.1	82.0	87.7	51.5	58.2	56.5	76.9	83.9
1089	53	30	45.1	73.3	82.5	85.7	48.6	83.0	56.6	75.0	84.3
1090	58	32	49.1	70.4	87.0	84.3	50.0	59.3	55.2	78.6	85.7
1091	62	38	49.7	70.5	81.6	95.5	50.8	65.5	61.3	81.1	87.1
1092	50	34	53.4	72.1	78.1	93.6	54.4	51.9	68.0	77.6	85.6
1093	59	32	50.9	77.5	77.5	91.0	53.7	50.9	54.2	77.6	79.9
1094	56	30	52.0	73.8	79.7	93.0	53.1	60.4	53.6	80.5	85.9
1095	57	37	51.4	75.0	79.2	94.0	54.5	53.5	64.9	76.1	85.1
1096	59	36	50.9	68.3	83.3	94.7	59.4	62.1	61.0	83.5	86.5
1097	56	33	50.2	79.8	80.0	86.4	50.7	64.7	58.9	73.6	85.7
1098	57	37	49.9	73.6	80.0	91.1	57.8	58.6	64.9	81.5	85.9

Measurements of Dulaimis—*concluded*

No.	Age	Stature	SH	L	B	B'	J	go-go	GH	G'H	NH	NB
1099	35	1728	888	192	137	113	133	106	124	73	50	32
1100	40	1780	872	191	138	114	137	116	125	74	58	28
1101	30	1775	882	198	132	111	131	103	127	73	55	32
1102	30	1722	882	193	144	114	141	111	111	65	47	35
1103	35	1691	904	196	140	114	135	105	124	70	49	33
1104	25	1671	830	192	145	112	134	104	115	69	51	31
1105	45	1685	840	178	141	116	136	104	122	69	50	30
1106	35	1631	820	193	140	115	136	112	121	73	48	35
1107	30	1682	840	196	140	115	141	107	116	67	51	38
1108	25	1670	820	195	140	116	143	113	116	67	48	35
1109	20	1680	840	195	144	109	131	102	128	76	55	36
1110	35	1643	832	190	145	114	138	110	114	65	47	37
1111	20	1662	812	187	135	109	128	98	118	67	50	26
1112	45	1623	838	189	136	115	136	98	123	68	49	36
1113	25	1566	812	186	141	111	137	107	116	64	49	30
1114	40	1620	820	190	141	111	136	107	123	71	53	35
1115	45	1680	814	180	140	111	134	116	134	80	59	33
1116	35	1655	817	201	148	116	142	112	121	78	56	33
1117	25	1760	860	203	151	123	147	115	124	78	59	34
1118	30	1622	830	194	143	111	131	107	123	72	54	36
1119	50	1713	880	193	136	112	143	116	125	74	50	37
1120	25	1673	872	184	133	111	131	102	118	70	56	30
1121	25	1784	858	184	140	108	132	96	125	71	53	34
1122	25	1688	838	188	143	113	137	108	122	72	53	28
1123	28	1750	836	201	149	127	146	120	121	74	56	36
1124	22	1642	796	192	132	109	130	98	127	72	48	46
1125	40	1655	842	181	139	115	136	105	123	72	57	33
1126	35	1685	776	198	140	113	141	111	126	75	61	34
1127	24	1703	820	192	137	112	132	111	121	72	51	34
1128	25	1620	815	200	145	118	140	111	122	74	48	37
1129	30	1650	800	192	142	113	137	96	116	75	58	35
1130	30	1685	900	194	146	112	140	110	110	69	56	37
1131	45	1648	814	198	148	114	147	111	126	77	57	38
1132	35	1736	885	195	140	112	136	108	118	68	52	37
1133	20	1673	827	186	137	118	136	105	126	67	52	33
1134	25	1615	822	184	133	102	128	104	112	68	53	35
1135	20	1536	780	185	132	108	123	94	117	71	50	31
1136	23	1706	893	187	138	116	137	107	122	76	51	40
1137	24	1650	798	187	142	118	131	104	123	72	51	37
1138	30	1622	844	193	138	110	138	109	133	78	59	32
1139	35	1722	848	201	140	113	132	104	118	70	53	32
1140	30	1703	882	188	132	111	138	108	124	71	54	35
1141	25	1660	844	179	142	112	132	103	121	71	54	37
1142	35	1722	898	193	150	118	139	105	120	71	52	32
1143	25	1623	790	184	145	113	135	110	118	76	56	32

INDICES OF DULAIMIS—*concluded*

No.	EL	EB	RSH	B/L	B'/B	GH/J	G'H/J	NB/NH	EB/EL	go-go/J	B'/J
1099	60	34	51.3	71.4	82.5	93.2	54.9	64.0	56.7	79.7	85.0
1100	52	36	49.0	72.3	82.6	91.2	54.0	48.3	69.2	84.7	83.2
1101	57	34	49.7	66.7	84.1	97.0	55.7	58.2	59.7	78.6	84.7
1102	58	35	51.2	74.6	79.2	78.7	46.1	74.5	60.3	78.7	80.9
1103	56	35	53.5	71.4	81.4	91.9	51.9	67.4	62.5	77.8	84.4
1104	59	33	49.7	75.5	77.2	85.8	51.5	60.8	55.9	77.6	83.6
1105	59	38	49.8	79.2	82.3	89.7	50.7	60.0	64.4	76.5	85.3
1106	54	33	50.3	72.5	82.1	89.0	53.7	72.9	61.1	82.4	84.6
1107	56	34	49.9	71.4	82.1	82.3	47.5	74.5	60.7	75.9	81.6
1108	55	31	49.1	71.8	82.9	81.1	46.9	60.3	56.4	79.0	81.1
1109	55	36	50.0	73.9	75.7	97.7	58.0	65.5	65.5	77.9	83.2
1110	55	34	50.6	76.3	78.6	82.6	47.1	78.7	61.8	79.7	82.6
1111	52	28	48.9	72.2	80.7	92.2	52.3	52.0	53.9	76.6	85.2
1112	70	33	51.6	72.0	84.6	90.4	50.0	73.5	47.1	72.1	84.6
1113	46	35	51.9	75.8	78.7	84.7	46.7	61.2	76.1	78.1	81.0
1114	51	32	50.6	74.2	78.7	90.4	52.2	66.0	62.8	78.7	81.6
1115	60	37	48.5	77.8	79.3	100.0	59.7	55.9	61.7	86.6	82.8
1116	59	36	49.4	73.6	78.4	85.2	54.9	58.9	61.0	78.9	81.7
1117	67	37	48.9	74.4	81.5	84.4	53.0	57.6	55.2	78.2	83.7
1118	51	31	51.2	73.7	77.6	93.9	55.0	66.7	60.8	81.7	84.7
1119	69	37	51.4	70.5	82.4	87.4	51.7	74.0	53.6	81.1	78.3
1120	64	38	52.1	72.3	83.5	90.1	53.4	53.6	59.4	77.9	84.7
1121	56	32	48.1	76.1	77.1	94.7	53.8	64.2	57.1	72.7	81.8
1122	53	33	49.6	76.1	79.0	89.1	52.6	52.8	62.3	78.8	82.5
1123	57	34	47.8	74.1	85.2	82.9	50.7	64.3	59.7	82.2	87.0
1124	61	35	48.5	68.8	82.6	97.7	55.4	95.8	57.4	75.4	83.9
1125	64	38	50.9	76.8	82.7	90.4	52.9	57.9	59.4	77.2	84.6
1126	62	31	46.1	70.7	80.1	89.4	53.2	55.7	50.0	78.7	80.1
1127	56	36	48.2	71.4	81.8	91.7	54.6	66.7	64.3	84.1	84.9
1128	56	33	50.3	72.5	81.4	87.1	52.9	77.1	58.9	79.3	84.3
1129	54	35	48.5	74.0	79.6	84.7	54.7	60.3	64.8	70.1	82.5
1130	60	32	53.4	75.3	76.7	78.6	49.3	66.1	53.3	78.6	80.0
1131	64	39	49.4	74.8	77.0	85.7	52.4	66.7	60.9	75.5	77.6
1132	56	34	51.0	71.8	80.0	86.8	50.0	71.2	60.7	79.4	82.4
1133	62	40	49.4	73.7	86.1	92.7	49.3	63.5	64.5	77.2	86.8
1134	56	31	50.9	72.3	76.7	87.5	53.1	66.0	55.4	81.3	79.7
1135	55	36	50.8	71.4	81.8	95.1	57.7	62.0	65.5	76.4	87.8
1136	57	35	52.3	73.8	84.1	89.1	55.5	78.4	61.4	78.1	84.6
1137	63	35	48.4	75.9	83.1	93.9	55.0	72.6	55.6	79.4	90.1
1138	50	34	52.0	71.5	79.7	96.4	56.5	54.2	68.0	79.0	79.7
1139	56	32	49.2	69.7	80.7	89.4	53.0	60.4	57.1	78.8	85.6
1140	60	38	51.8	70.2	84.1	89.9	51.4	64.8	63.3	78.3	80.4
1141	60	38	50.8	79.3	78.9	91.7	53.8	68.5	63.3	78.0	84.9
1142	56	40	52.2	77.7	78.7	86.3	51.1	61.5	71.4	75.5	84.9
1143	54	38	48.7	78.8	77.9	87.4	56.3	57.1	70.4	81.5	83.7

Anthropology of Iraq

Morphological Characters of Dulaimis

No.	Hair Form	Hair Texture	Hair Color	Eyes Color	Eyes Sclera	Eyes Iris	Nose Profile	Nose Wings
1007	l w	medium	blk, gray	dk br	clear	hom	wavy	medium
1008	l w	fine	black	bl-br	clear	zon	c-c	cp-m
1009	l w	medium	black	dk br	clear	hom	str	comp
1010	l w	fine	black	dk br	clear	hom	str	comp
1011	l w	coarse	black	dk br	clear	hom	conc	medium
1012*	...	.....	.....	dk br	clear	hom	str	m-fl
1013	l w	medium	black	dk br	clear	zon	str	medium
1014	l w	fine	black	dk br	clear	hom	c-c	medium
1015	l w	coarse	v dk br	gray-br	clear	zon	str	cp-m
1016	c-f	medium	blk, gray	*bl*-br †	clear	zon	str	medium
1017	l w	m-fine	blk, gray	gr-br	speck	ray	conv	medium
1018	l w	medium	dk br	bl-br	clear	hom	wavy	cp-m
1019	l w	medium	dk br	gr-br	clear	zon	conv	m-flar
1020	l w	medium	black	gr-br	clear	hom	str	medium
1021	l w	medium	black	gr-br	blood	zon	str	medium
1022	l w	coarse	blk, gray	gr-br	clear	hom	conv	medium
1023	l w	medium	black	gr-br	clear	ray	str	medium
1024	d w	coarse	black	gr-br	clear	zon	str	medium
1025	l w	medium	dk br	dk br	clear	hom	conc	medium
1026	l w	coarse	black	dk br	clear	zon	str	comp
1027	l w	medium	black	gr-br	clear	zon	str	medium
1028	d w	coarse	black	gr-br	clear	zon	conv	m-fl
1029	l w	coarse	black	gr-br	clear	zon	wavy	m-fl
1030	l w	medium	black	bl-br	clear	hom	str	comp
1031	l w	coarse	black	gr-br	clear	zon	str	medium
1032	l w	medium	black	dk br	clear	hom	str	m-fl
1033	l w	medium	black	bl-br	clear	zon	str	medium
1034	d w	fine	black	*bl*-br	clear	zon	str	cp-m
1035	l w	medium	black	dk br	clear	hom	conc	m-fl
1036	l w	coarse	black	*gr*-br	clear	zon	conv	comp
1037	l w	fine	black	*gr*-br	clear	zon	conv	flar
1038	l w	medium	black	dk br	clear	hom	str	medium
1039	l w	medium	dk br	gr-br	clear	zon	str	medium
1040	l w	medium	black	dk br	clear	hom	wavy	comp
1041	l w	medium	blk, gray	bl-br	clear	zon	conv	comp
1042	l w	medium	black	gr-br	clear	zon	conv	comp
1043	l w	medium	black	dk br	clear	hom	str	medium
1044	l w	medium	blk, gray	gr-br	speck	zon	wavy	medium
1045	l w	fine	black	dk br	clear	hom	conv	medium
1046	l w	medium	black	blue	.....	...	str	medium
1047	l w	medium	black	*bl*-br	clear	zon	conv	m-fl
1048	l w	medium	black	gray-br	clear	zon	wavy	comp
1049	l w	medium	black	dk br	clear	hom	str	medium
1050	l w	medium	blk, gray	gr-br	clear	zon	str	medium
1051	l w	medium	blk, gray	gray-br	clear	zon	str	medium
1052	l w	medium	black	dk br	clear	hom	conv	medium
1053	l w	medium	black	gr-br	clear	zon	conv	medium
1054	l w	medium	blk, gray	dk br	clear	hom	wavy	medium
1055	l w	medium	blk, gray	gr-br	clear	zon	str	comp
1056	l w	medium	black	dk br	clear	hom	str	medium
1057	l w	medium	black	dk br	clear	hom	conv	comp
1058	l w	medium	black	gr-br	clear	zon	str	medium
1059	l w	medium	gray	gr-br	clear	zon	str	medium
1060	l w	coarse	blk, gray	dk br	clear	hom	str	cp-m
1061	l w	medium	black	gray-br	clear	zon	str	m-fl
1062	l w	medium	black	gr-br	clear	zon	str	m-fl

*Shaved
† Almost blue

PHYSICAL ANTHROPOLOGY: DULAIM AND ANAIZA 53

MORPHOLOGICAL CHARACTERS OF DULAIMIS—*continued*

No.	\<HAIR\>			\<EYES\>			\<NOSE\>	
	Form	Texture	Color	Color	Sclera	Iris	Profile	Wings
1063	d w	medium	black	dk br	clear	hom	str	cp-m
1064	d w	medium	dk br	gr-br	clear	zon	wavy	medium
1065	l w	medium	black	gr-br	blood	zon	conv	m-fl
1066	d w	medium	dk br	gr-br	clear	zon	str	medium
1067	l w	medium	black	dk br	clear	hom	str	comp
1068	l w	medium	black	dk br	clear	hom	str	medium
1069	l w	medium	dk br	dk br	speck	ray	c-c	m-fl
1070	l w	medium	dk br	dk br	clear	hom	str	medium
1071	l w	medium	black	dk br	clear	hom	str	medium
1072	l w	medium	black	dk br	clear	zon	str	comp
1073	l w	medium	dk br	dk br	clear	hom	str	comp
1074	l w	coarse	black	bl-br	blood	ray	str	medium
1075	l w	fine	black	dk br	clear	hom	str	comp
1076	l w	medium	black	dk br	clear	hom	str	medium
1077	l w	medium	dk br	gr-br	clear	ray	str	medium
1078	l w	medium	black	dk br	clear	hom	c-c	medium
1079	l w	medium	black	dk br	clear	hom	str	medium
1080	l w	medium	black	dk br	clear	hom	str	m-fl
1081	l w	medium	black	dk br	clear	hom	str	medium
1082	l w	medium	black	dk br	clear	hom	str	medium
1083	l w	coarse	blk, gray	dk br	speck	hom	conv	medium
1084	l w	medium	black	dk br	clear	hom	str	comp
1085	d w	medium	black	gr-br	clear	zon	str	medium
1086	l w	medium	black	dk br	clear	hom	conv	comp
1087	c-f	medium	black	dk br	clear	hom	wavy	medium
1088	l w	medium	dk br	dk br	clear	hom	str	comp
1089	c-f	medium	black	dk br	clear	hom	str	medium
1090	l w	coarse	black	*bl*-br*	speck	zon	str	medium
1091	l w	medium	black	dk br	clear	hom	str	m-fl
1092	l w	medium	black	dk br	clear	hom	wavy	comp
1093	l w	medium	dk br	dk br	clear	hom	conv	comp
1094	l w	coarse	black	dk br	clear	hom	str	medium
1095	l w	medium	black	dk br	clear	hom	str	m-fl
1096	l w	medium	black	dk br	clear	hom	c-c	m-fl
1097	l w	medium	black	dk br	blood	hom	str	medium
1098	l w	fine	dk br	bl-br	clear	hom	str	comp
1099	l w	fine	dk br	gr-br	clear	zon	str	medium
1100	l w	medium	black	dk br	clear	zon	str	comp
1101	l w	fine	black	dk br	clear	hom	str	comp
1102	l w	medium	v dk br	dk br	clear	hom	str	cp-m
1103	l w	m-fine	black	dk br	clear	hom	str	medium
1104	l w	medium	black	dk br	clear	hom	str	medium
1105	l w	medium	blk, gray	gr-br	speck	zon	str	comp
1106	...	......	.....	gr-br	speck	zon	str	medium
1107	l w	medium	black	dk br	clear	hom	str	medium
1108	l w	medium	dk br	blue	clear	zon	str	medium
1109	...	....	black	dk br	clear	hom	str	medium
1110	l w	fine	black	bl-br	clear	zon	str	medium
1111	l w	medium	black	dk br	clear	hom	str	medium
1112	l w	medium	dk br	*bl*-br	clear	zon	str	medium
1113	l w	medium	black	bl-br	clear	zon	str	medium
1114	l w	fine	dk br	bl-br	clear	zon	str	medium
1115	l w	medium	blk, gray	dk br	clear	hom	str	comp
1116	l w	medium	black	bl-br	clear	hom	str	medium
1117	l w	medium	black	gr-br	speck	zon	str	medium
1118	l w	medium	black	gr-br	clear	zon	str	medium
1119	l w	coarse	gray	*bl*-br*	clear	ray	str	medium

*Almost blue

MORPHOLOGICAL CHARACTERS OF DULAIMIS—*concluded*

	HAIR			EYES			NOSE	
No.	Form	Texture	Color	Color	Sclera	Iris	Profile	Wings
1120	...	....	black	bl-br	clear	zon	conv	comp
1121	l w	coarse	black	dk br	clear	hom	str	medium
1122	l w	medium	black	dk br	clear	hom	str	medium
1123	l w	fine	black	dk br	clear	hom	str	medium
1124*	...	....	.....	bl-br	clear	hom	str	flar
1125	l w	medium	dk br	bl-br	clear	zon	conv	....
1126	l w	medium	.....	dk br	clear	hom	str	comp
1127	l w	medium	black	dk br	clear	hom	str	m-fl
1128	l w	medium	black	dk br	clear	hom	str	flar
1129	l w	medium	black	dk br	clear	zon	str	cp-m
1130	l w	m-fine	black	dk br	clear	hom	str	medium
1131	l w	medium	gray	dk br	clear	hom	str	m-fl
1132	l w	coarse	black	dk br	clear	hom	str	comp
1133	l w	medium	black	dk br	clear	zon	str	comp
1134	l w	fine	black	dk br	clear	hom	str	flar
1135	l w	medium	black	dk br	clear	hom	str	medium
1136	l w	medium	dk br	gr-br	clear	zon	c-c	flar
1137	l w	fine	black	dk br	clear	hom	str	medium
1138	l w	coarse	black	dk br	clear	hom	str	comp
1139	l w	medium	black	dk br	clear	hom	str	comp
1140	l w	coarse	black	dk br	clear	hom	str	cp-m
1141	l w	medium	black	dk br	clear	hom	str	comp
1142	l w	medium	black	dk br	clear	hom	c-c	medium
1143	l w	medium	black	dk br	clear	hom	c-c	......

* Hairless

THE ANAIZA[1]

The Anaiza tribesman states that he is a descendant of Wail, who belonged to a younger branch of the Asad group, and further claims that Anaz, son of Wail, was the founder of the tribe.

The original home of the Anaiza is believed to have been just north of Medina on the watershed between the Red Sea and the basin of the Wadi al Rumma (cf. Doughty, vol. 2, p. 392). In the latter half of the eighteenth century the Anaiza started to move northward. The Fadan and the Hasanah pushed the Shammar before them across the Euphrates and established themselves on the northern steppes. The Amarat, Wulud Ali, and Sbaa appear to have been the next to move, and later came the Ruwalla.

The great group of the Anaiza, numerically probably the largest group in the nomad Arab tribes, occupied the triangle of the North Arabian or Syrian Desert, often called the Hamad, which has its base

[1] This introductory section is based on data compiled prior to 1921. As leader of the Field Museum North Arabian Desert Expedition, 1927, 1928, and 1934, I checked this information whenever possible. During 1928 in Damascus I had the privilege of discussing these matters with Nuri ibn Shalan, Sheikh of the Ruwalla and Paramount Sheikh of the Anaiza. For selected references to the Anaiza see Carruthers (1918), "A Handbook of Arabia" (1920), Doughty (1926), Musil (1927a, 1927b, 1928), de Boucheman (1934), Lawrence (1926), Raswan (1930, 1935, 1936), Grant (1937), Guarmani (1938), and von Oppenheim (1939).

on Lat. 30° N., with Jauf about at its center, and its apex at Alep. On the left bank of the Euphrates the pastures north of Deir-ez-Zor and along the Khabur River were also visited by the Anaiza. A smaller group of kindred tribes lived near Taima between the Hejaz Railway and the southwest borders of the Nefud. The tribe was not united under one head, but divided into several large sections which maintained a generally friendly attitude, which did not exclude, however, raids and feuds between the sections.

The most famous stocks of horses and the greatest number of camels were found among the northern Anaiza. Their camel herds, estimated at 600,000 head, supplied the markets of Egypt, Syria, and Iraq. Beduins of the purest blood and tradition, the Anaiza remained entirely beyond the control of the Turkish Government. Except for a few palm gardens on the Euphrates and a village near Damascus, their sheikhs never acquired settled land nor did they attempt to cultivate the Hamad or stony desert. Their geographical position gave them command of the main trade route between Syria and Iraq, and at the same time compelled them to keep on good terms with those who controlled their commercial markets; namely, the larger towns on both edges of the Syrian Desert.

The Anaiza are hereditary foes of the Shammar, and northern Arabia during the last 150 years has been dominated by the feuds of these two tribal confederations.

During the past fifteen years conditions have changed entirely as a result of trans-desert automobile and air routes, followed by the construction of the Iraq Petroleum Company's bifurcated pipe-lines. Large-scale raids of Beduin tribes upon each other are now virtually impossible. Armored cars, airplane bombs, and, to quote the Beduins, "the-gun-that-never-stops," are serious deterrents not only to raiding of any kind but also to any digression from British, French, or Iraqi prescribed areas of migration.

During construction of the pipe-lines many thousands of Beduin tribesmen were employed in numerous capacities. Personal observation and the reports of labor officers show that the tribesmen were capable, conscientious, and often skilful workmen. They obeyed orders cheerfully and followed instructions unhesitatingly. In May, 1934, I was most astonished to find Anaiza tribesmen, with shaven heads, washed and disinfected bodies, engaged in pipe-line construction near H-3 station, where we were the guests of the Iraq Petroleum Company.

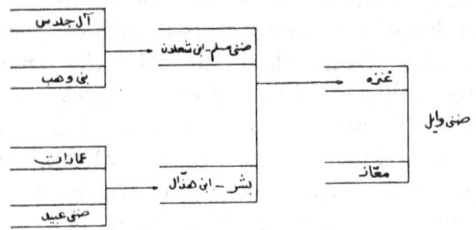
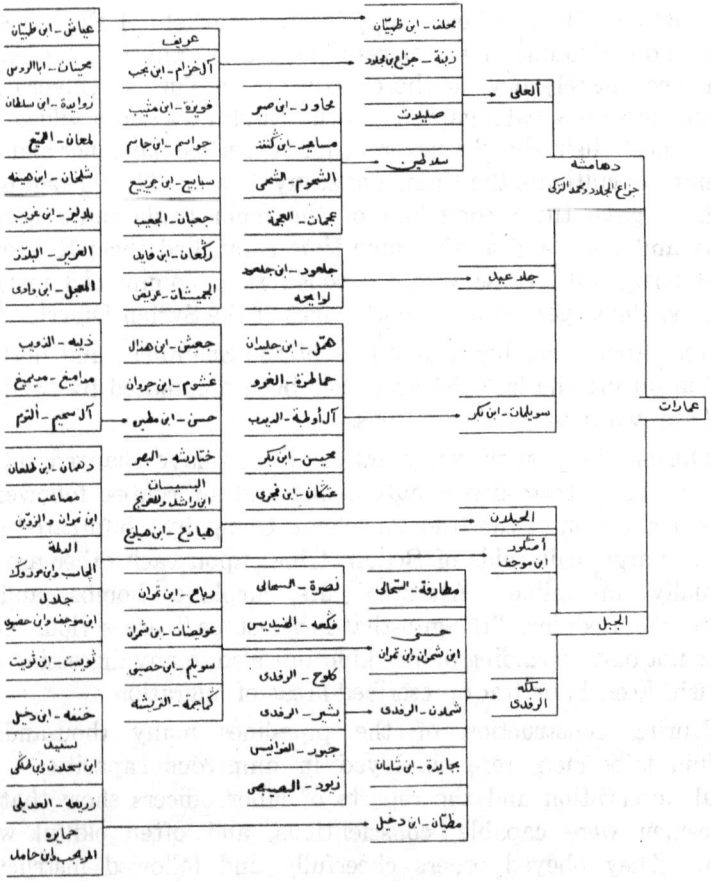

FIG. 5. Tribes and sub-tribes of the Anaiza Beduins.

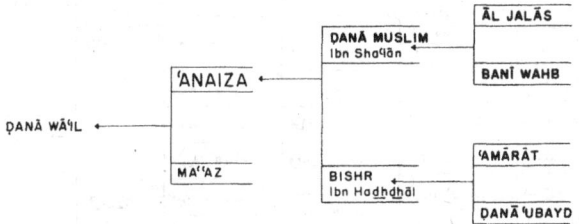

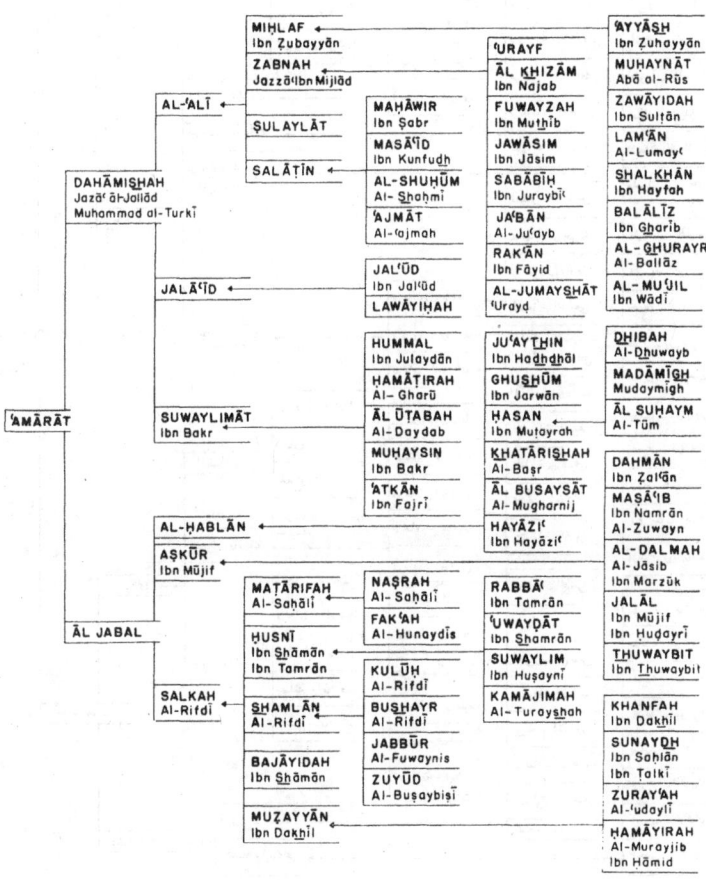

Fig. 6. Tribes and sub-tribes of the Anaiza Beduins.

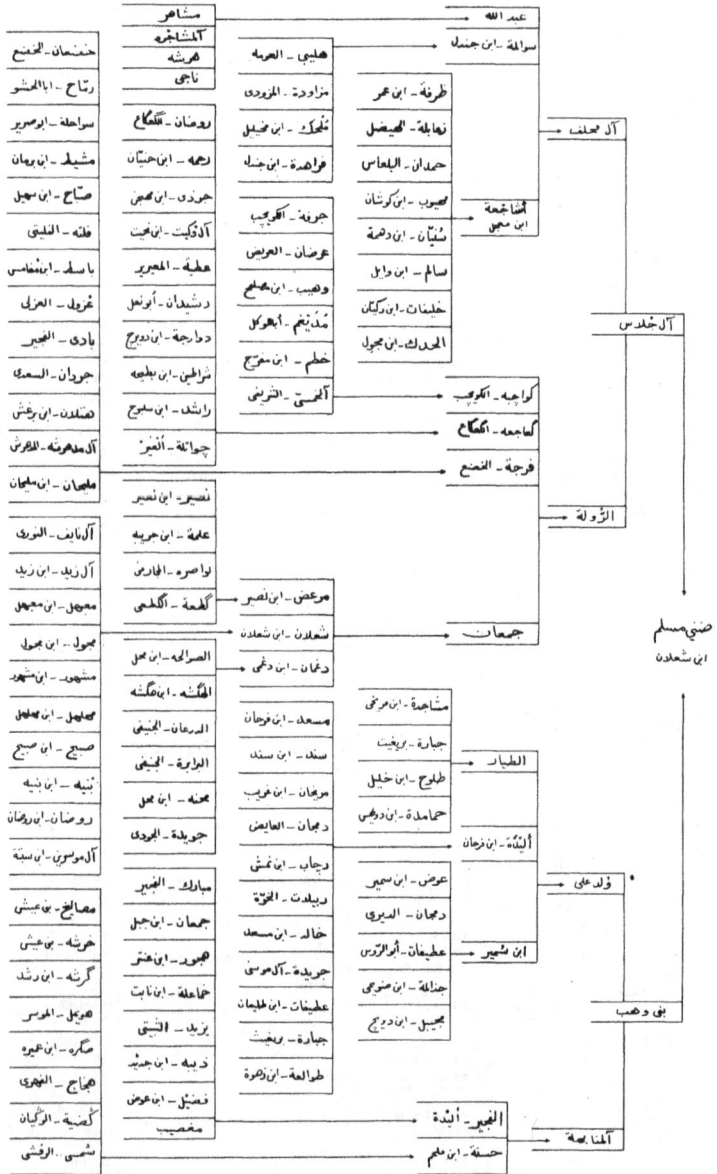

FIG. 7. Tribes and sub-tribes of the Anaiza Beduins.

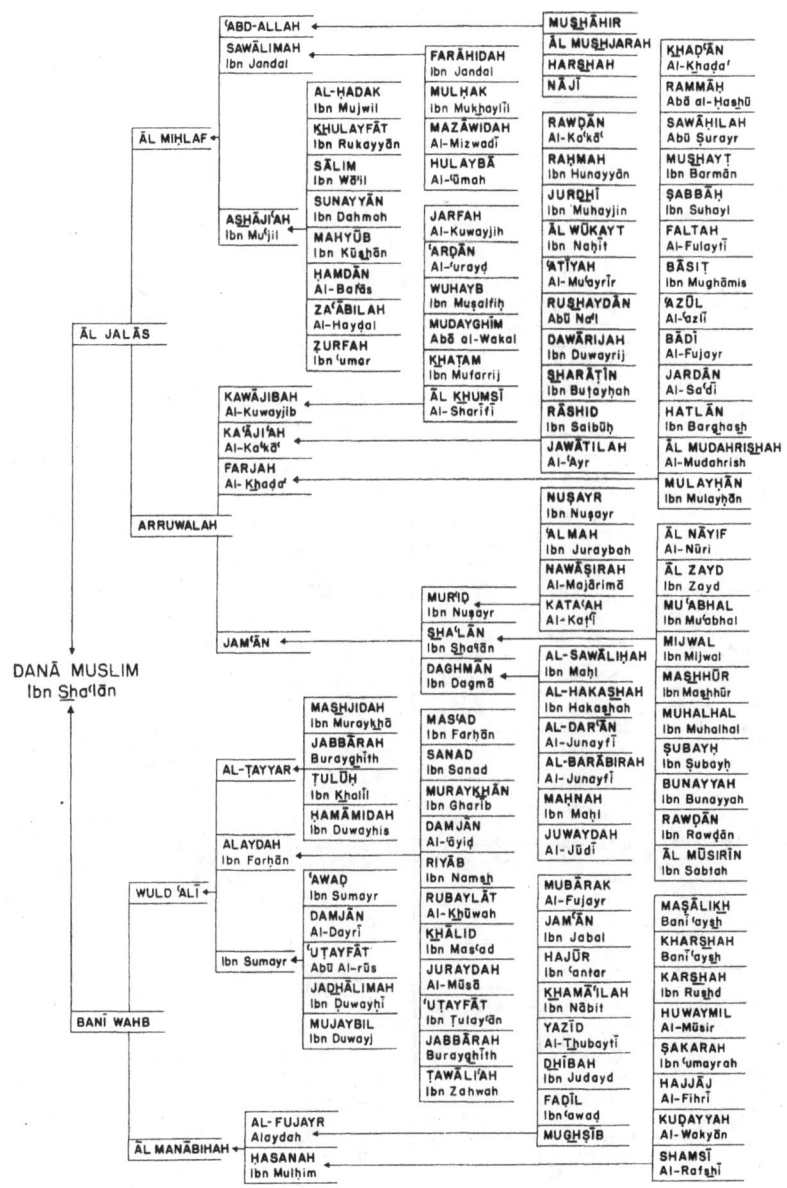

FIG. 8. Tribes and sub-tribes of the Anaiza Beduins.

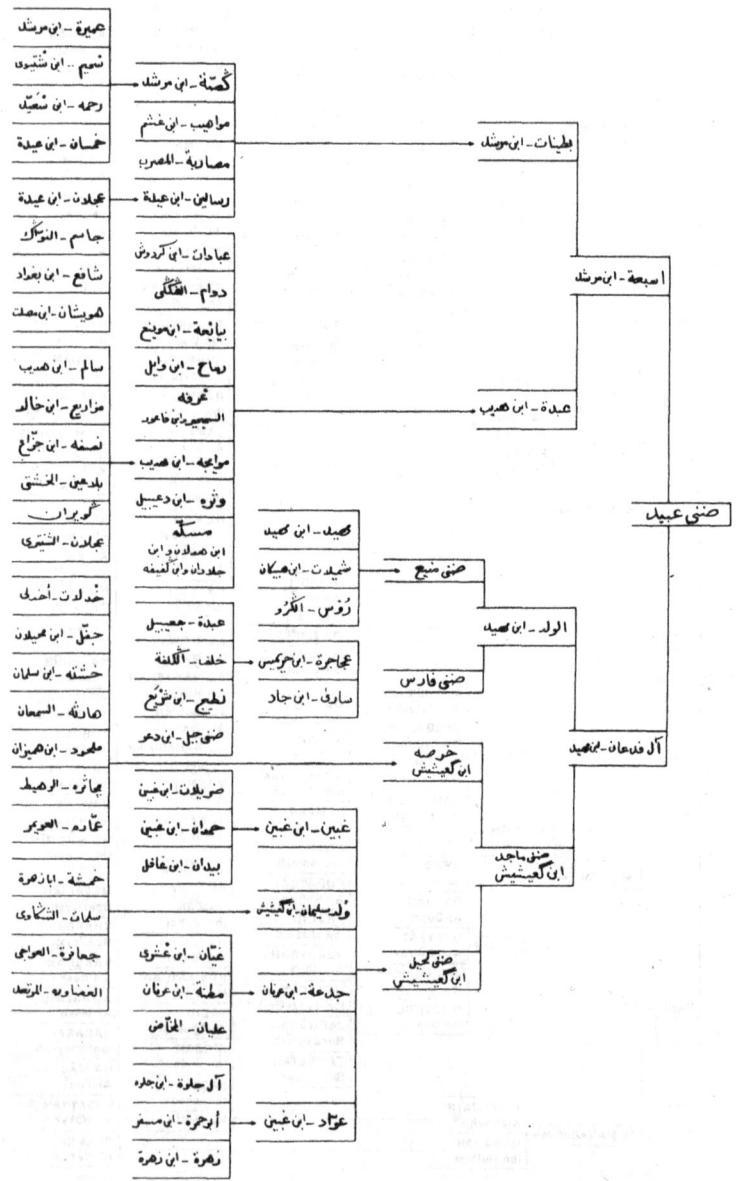

FIG. 9. Tribes and sub-tribes of the Anaiza Beduins.

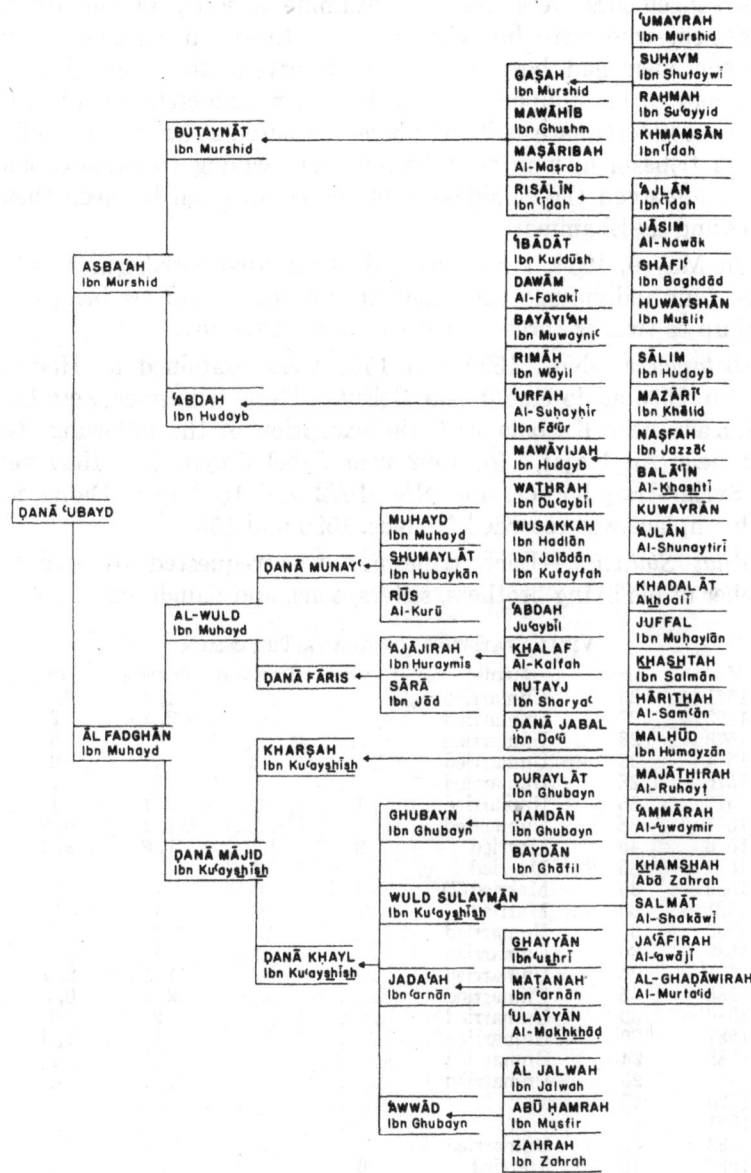

FIG. 10. Tribes and sub-tribes of the Anaiza Beduins.

Arrangements were made to examine a series of the Anaiza tribesmen, who were brought into the station dispensary. After nineteen men had been measured, observed, and photographed, work had to be stopped as a result of a misunderstanding.[1] This was most unfortunate as I could have measured at least one hundred Anaiza tribesmen, who were friendly and willing to submit, since they understood the significance of the comparison between themselves and the Shammar.

On May 9, 1934, I measured nineteen Anaiza tribesmen. The other four individuals examined at different localities brings the total up to twenty-three—a most inadequate series.

Birthplaces.—Nos. 1590 and 1591 were examined at Haditha and No. 1593 at de Kuani near Beirut. These tribesmen were born at Razaza near Karbala with the exception of the following: No. 1571 near An Najaf, No. 1592 near Jebel Sinjar, No. 1589 near the Syrian-Iraq border, and Nos. 1572 and 1593 near Damascus. No birthplace was recorded for Nos. 1590 and 1591.

Vital Statistics.—Each tribesman was requested to give the number of his living brothers, sisters, sons, and daughters.

VITAL STATISTICS* OF ANAIZA TRIBESMEN

No.	Age	Married	Sons	Daughters	Brothers	Sisters
1571	28	Unmarried	....	....	1, *1*	2, *3*
1572	27	Unmarried	....	....	2, *1*	2
1573	28	Unmarried	....	....	2	0
1574	32	Unmarried	....	....	1	0
1575	28	Unmarried	....	....	3	1
1576	35	Unmarried	1	0	0, *1*	1
1577	38	Unmarried	....	....	1, *1*	0, *2*
1578	45	Married	2	1, *1*	1, *2*	2, *1*
1579	36	Married	1	0	0, *2*	1
1580	30	Married(2)	0	0	5	3
1581	30	Married	2, *1*	0	1	0
1582	35	Unmarried	....	....	1	1
1583	30	Unmarried	....	....	1	2
1584	25	Unmarried	....	....	1, *2*	2, *1*
1585	35	Unmarried	....	....	2, *2*	0, *3*
1586	30	Unmarried	....	....	2	0
1587	30	Unmarried	....	....	5	1, *3*
1588	24	Unmarried	....	....	1	1
1589	25	Unmarried	....	....	2, *3*	4
1590	45	....	....	....	....	....
1591	50	....	....	....	....	....
1592	35	Unmarried	....	....	2, *4*	1, *1*
1593	40	Married	0	0	....	....

*The italicized numbers refer to the deceased relatives.

[1] I sincerely hope never to encounter again a man as abysmally ignorant, superciliously arrogant, and deliberately obstructive as the individual who stopped this important piece of research through inciting the tribesmen to object to examination by fabricating such falsehoods and lies as that we were using powerful magic and casting spells over them.

DEMOGRAPHY

Brothers	No.	Per cent	Sisters	No.	Per cent
None	0	....	None	4	20.00
1	6	30.00	1	5	25.00
2	5	25.00	2	4	20.00
3–4	5	25.00	3–4	6	30.00
5–6	4	20.00	5–6	1	5.00
7 or more	0	....	7 or more	0	....
Total	20	100.00	Total	20	100.00

Sons	No.	Per cent	Daughters	No.	Per cent
None	2	33.33	None	5	83.33
1	2	33.33	1	0	....
2	1	16.67	2	1	16.67
3–4	1	16.67	3–4	0	....
5–6	0	....	5–6	0	....
7 or more	0	....	7 or more	0	....
Total	6	100.00	Total	6	100.00

The size of the families, as indicated by these unreliable figures, tends to be large, especially when there is every reason to suppose a high rate of infant mortality. Few tribesmen admitted having children, probably because of the innate fear of evil spells.

Age.—The average age for the group was 34.15, with a range of 20 to 54. Eighteen men (78.27 per cent) were between the ages of 25 and 39.

AGE DISTRIBUTION

Age	No.	Per cent	Age	No.	Per cent
18–19	0	....	45–49	2	8.70
20–24	1	4.35	50–54	1	4.35
25–29	6	26.09	55–59	0	....
30–34	6	26.09	60–64	0	....
35–39	6	26.09	65–69	0	....
40–44	1	4.35	70–x	0	....
			Total	23	100.02

MORPHOLOGICAL CHARACTERS OF ANAIZA BEDUINS

Skin.—As a result of exposure to wind and to other vagaries of climate, the skin was slightly darker than that of the average Arab of Iraq. The secondary shadings of different parts of the body were in no way peculiar, but the exposed parts were slightly darker than those habitually clothed. On the head, which is always covered, the skin was considerably lighter in color in many cases but never as white as in Europeans. No. 1572 (Plate 38), who had a dark skin, appeared to possess some Negro blood.

Hair.—The hair was black or very dark brown. In form the hair had low waves and in texture was either coarse or medium. Nine men had shaven heads.

HAIR

Color	No.	Per cent	Form	No.	Per cent
Black	19	86.36	Straight	0	.....
Very dark brown	1	4.55	Very low waves	0	.....
Dark brown	0	.....	Low waves	11	91.67
Brown	0	.....	Deep waves	1	8.33
Reddish brown	0	.....	Curly-frizzly	0	.....
Light brown	0	.....	Woolly	0	.....
Red	0	.....			
Black and gray	0	.....	Total	12	100.00
Dark brown and gray	0	.....			
Light brown and gray	0	.....	Texture	No.	Per cent
Gray	2	9.09	Coarse	9	69.23
White	0	.....	Coarse-medium	0	.....
			Medium	4	30.77
Total	22	100.00	Medium-fine	0	.....
			Fine	0	.....
			Total	13	100.00

Abnormal hairiness of the body was not observed and the general impression retained was that the Anaiza had about the same amount of body hair as the Arabs of central Iraq.

Eyes.—In general the eyes were brown in color, varying from gray-brown to dark brown. The presence of individuals with mixed eyes indicates a submerged blondism. The sclera were clear, with the exception of three men with bloodshot eyes. The iris was homogeneous or zoned with three individuals in the rayed classification.

EYES

Color	No.	Per cent	Iris	No.	Per cent
Black	0	.....	Homogeneous	9	42.86
Dark brown	7	31.82	Rayed	3	14.29
Blue-brown	7	31.82	Zoned	9	42.86
Blue-brown	0	.....			
Green-brown	5	22.73	Total	21	100.01
Green-brown	0	.....	Sclera	No.	Per cent
Gray-brown	3	13.64	Clear	19	86.36
Blue	0	.....	Yellow	0	.....
Gray	0	.....	Speckled	0	.....
Light brown	0	.....	Bloodshot	3	13.64
Blue-gray	0	.....	Speckled and bloodshot	0	.....
Blue-green	0	.....	Speckled and yellow	0	.....
Total	22	100.01	Yellow and bloodshot	0	.....
			Total	22	100.00

The eyes, or more properly the eye slits, were horizontal as in Europeans. In general, the eyes were clear and the vision was keen, features characteristic of the nomads of this region.

Nose.—The nasal profile was convex or straight in about equal proportions. The alae were medium to compressed with but four Anaiza tribesmen slightly above the average. One man had a wider nasal tip than the average and one individual appeared in the double plus category, indicating the presence of Negro blood.

NOSE

Profile	No.	Per cent	Wings	No.	Per cent
Wavy	0	....	Compressed	3	13.04
Concave	0	....	Compressed-medium	2	8.70
Straight	8	36.36	Medium	14	60.87
Convex	9	40.91	Medium-flaring	4	17.39
Concavo-convex	5	22.73	Flaring	0	....
			Flaring plus	0	....
Total	22	100.00	Total	23	100.00

Mouth.—The majority of the lips were thicker than those of the average European, and there was considerable lower lip eversion in a number of individuals, especially Nos. 1573, 1575, and 1583. The relatively thin lips of No. 1589 appeared to be exceptional.

Teeth.—The occlusion was recorded as marked-over bite but this seems hardly probable and I think this should have been slight-over bite, a far more normal occlusion.

TEETH

Bite	No.	Per cent	Condition	No.	Per cent
Under	0	....	Very bad	0	....
Edge-to-edge	0	....	Bad	1	6.25
Slight over	0	....	Fair	2	12.50
Marked over	22	100.00	Good	9	56.25
			Excellent	4	25.00
Total	22	100.00	Total	16	100.00

The dental condition was either good or excellent with but three exceptions. Nos. 1574, 1577, 1585, 1586, 1589, and 1592 were excellent; Nos. 1575, 1576, 1579, 1582, 1583, 1587, 1588, and 1590 were good; and Nos. 1573 and 1578 were fair. No. 1580 had irregular front teeth.

Musculature and Health.—The Anaiza Beduins had well-developed musculature and those examined were in good health.

Musculature	No.	Per cent	Health	No.	Per cent
Poor	0	....	Poor	0	....
Fair	1	4.55	Fair	1	4.55
Average	0	....	Average	0	....
Good	19	86.36	Good	20	90.91
Excellent	2	9.09	Excellent	1	4.55
Total	22	100.00	Total	22	100.01

Disease.—Nos. 1574 and 1585 had smallpox scars. No. 1584 had ringworm on his face. No. 1583 had scars on his head as a result of a fall from a camel. No. 1591 was blind in the left eye and his vision was poor in the right eye.

Tattooing.—Nos. 1585, 1589, 1592, and 1593 had simple tattooed designs and twelve were recorded as bearing none (cf. Charles, pp. 109–111).

Branding.—Each individual, with the exception of Nos. 1576, 1579, 1582, 1585, 1589, 1591, and 1593, bore circular branded marks on his arms or wrists. Each brand is referred to as a *chawi* or *kawi*. No. 1572 said that branding was used "to prevent smallpox." No. 1580 had a large *chawi* scar on the inside of his left wrist "to cure a racking cough." No. 1581 had five large, circular marks on his right wrist "to make it strong for stone throwing."

Kohl.—No. 1580 had applied kohl beneath his eyes "to strengthen them."

Unrecorded.—No morphological observations were recorded on No. 1593.

Summary.—The average Anaiza tribesman had low wavy hair, coarse or medium in texture, and extremely dark brown merging into black in color. The eyes were various shades of brown, but fifteen men (68.19 per cent) had mixed eyes. The sclera were clear, but the iris was either homogeneous or rayed. The nose was convex or straight in almost equal proportions, with medium wings. The lips were thicker than those of the average European. The teeth, musculature, and health were good.

STATISTICAL ANALYSES OF ANAIZA BEDUINS

There now remains the task of grouping the twenty-three Anaiza tribesmen according to the Harvard and Keith classificatory systems for stature, sitting height (trunk length), minimum frontal diameter, head breadth, cephalic index, nasal height, nasal breadth, and nasal index.

Stature.—The Anaiza were medium to short according to both systems. The results of the two groupings happen to be identical. The average stature for twenty-two men was 162.96 (range 146.0–178.0), which is well below the average for Southwestern Asia (about 166.0). No. 1593 was omitted.

STATURE

Harvard system	No.	Per cent	Keith system	No.	Per cent
Short (x–160.5)	6	27.27	Short (x–159.9)	6	27.27
Medium (160.6–169.4)	12	54.55	Medium (160.0–169.9)	12	54.55
Tall (169.5–x)	4	18.18	Tall (170.0–179.9)	4	18.18
			Very tall (180.0–x)	0	.....
Total	22	100.00	Total	22	100.00

Sitting Height (Trunk Length).—The Keith system shows that the majority (81.81 per cent) had trunk lengths greater than 84.9. They were almost equally divided between the long (85.0–89.9) and the very long (90.0–x) categories. No. 1593 was omitted.

SITTING HEIGHT (Trunk Length)

Group	No.	Per cent
Very short (x–74.9)	2	9.09
Short (75.0–79.9)	0	.....
Medium (80.0–84.9)	2	9.09
Long (85.0–89.9)	10	45.45
Very long (90.0–x)	8	36.36
Total	22	99.99

In the preceding table, which follows the Keith system, we see that whereas the stature was medium to short the trunk length was either long or very long. This reveals an unbalanced proportion between the length of the trunk and that of the legs. The Anaiza had very short legs combined with long trunks. The average relative sitting height was 53.68.

Minimum Frontal Diameter.—The head was wide (110–119) or narrow (100–109), there being no individuals in the categories above and below these ranges. The mean was 110.30 (range 101–120).

MINIMUM FRONTAL DIAMETER

Group	No.	Per cent
Very narrow (x–99)	0	.....
Narrow (100–109)	9	40.90
Wide (110–119)	13	59.09
Very wide (120–x)	0	.....
Total	22	99.99

Head Breadth.—The head varied from narrow to wide with a mean of 137.50 (range 123–149). There were more Anaiza tribesmen in the narrow-headed categories than at the other end of the scale.

HEAD BREADTH

Group	No.	Per cent
Very narrow (120–129)	1	4.34
Narrow (130–139)	12	52.18
Wide (140–149)	9	39.13
Very wide (150–x)	1	4.34
Total	23	99.99

Cephalic Index.—According to the Harvard system the majority (82.60 per cent) were dolichocephalic, with only one brachycephal in the series.

The Keith classificatory system reveals that the Anaiza were dolichocephalic with a strong tendency toward ultradolichocephaly.

The mean head length was 191.22, which, combined with the relatively narrow breadth (137.50) gave a cephalic index of 72.72, a figure which I believe to be close to that of the Proto-Mediterranean mean.

CEPHALIC INDEX

Keith system	No.	Per cent	Harvard system	No.	Per cent
Ultradolichocephalic (x–70.0)	6	26.09	Dolichocephalic (x–76.5)	19	82.60
Dolichocephalic (70.1–75.0)	13	56.52	Mesocephalic (76.6–82.5)	3	13.04
Mesocephalic (75.1–79.9)	3	13.04	Brachycephalic (82.6–x)	1	4.35
Brachycephalic (80.0–84.9)	1	4.35	Total	23	99.99
Ultrabrachycephalic (85.0–x)	0	.....			
Total	23	100.00			

The Anaiza tribesmen were long-headed with a trend toward accentuation of this head proportion.

Facial Measurements.—The upper part of the face tended to be long (70+) but 43.47 per cent were below this arbitrary figure. The largest groupings were either medium short or medium long. The mean was 70.25 (range 60–84).

The total facial length was either medium short or medium long. No. 1586 had a very long face (132.0). The mean was 120.50 (range 110–132).

A grouping of the total facial indices places 77.27 per cent in the leptoprosopic category with only one tribesman recorded as euryprosopic.

FACIAL MEASUREMENTS

Upper facial height	No.	Per cent	Total facial height	No.	Per cent
Short (x–63)	1	4.34	Short (x–109)	0	.....
Medium short (64–69)	9	39.13	Medium short (110–119)	11	47.83
Medium long (70–75)	8	34.78	Medium long (120–129)	11	47.83
Long (76–x)	5	21.74	Long (130–x)	1	4.35
Total	23	99.99	Total	23	100.01

Total Facial Index

Group	No.	Per cent
Euryprosopic (x–84.5)	1	4.55
Mesoprosopic (84.6–89.4)	4	18.18
Leptoprosopic (89.5–x)	17	77.27
Total	22	100.00

In general the face was long, actually and relatively, the result of an elongated upper facial height combined with a medium wide face.

Nasal Measurements and Indices.—The Anaiza tribesmen possessed noses medium in height, medium narrow or medium wide in breadth, and a leptorrhine or mesorrhine index. The mean height was 53.66 (range 44–63), the breadth 34.61 (range 24–45), and the nasal index 66.18 (range 44–95). One man was Negroid.

Nasal Measurements

Nasal height	No.	Per cent	Nasal breadth	No.	Per cent
Short (x–49)	4	17.39	Very narrow (x–29)	2	8.70
Medium (50–59)	17	73.91	Medium narrow (30–35)	12	52.18
Long (60–x)	2	8.70	Medium wide (36–41)	8	34.78
Total	23	100.00	Wide (42–x)	1	4.34
			Total	23	100.00

Nasal Index

Group	No.	Per cent
Leptorrhine (x–67.4)	14	60.87
Mesorrhine (67.5–83.4)	8	34.78
Platyrrhine (83.5–x)	1	4.34
Total	23	99.99

To furnish additional statistical data for comparison with those in my Report on Iran the following tables have been calculated:

Sitting Height (Trunk Length)

Standing height	900–x No.	%	899–850 No.	%	849–800 No.	%	799–750 No.	%	749–x No.	%	Totals No.	%
1800–x	0		0		0		0		0		0	
1799–1700	4	18.18	0		0		0		0		4	18.18
1699–1600	4	18.18	7	31.82	0		0		1	4.55	12	54.55
x–1599	0		3	13.64	2	9.09	0		1	4.55	6	27.28
											22	100.01

No. 1593 omitted.

Minimum Frontal Diameter

Head breadth	x–99 No.	%	100–109 No.	%	110–119 No.	%	120–x No.	%	Totals No.	%
120–129	0		1	4.55	0		0		1	4.55
130–139	0		4	18.18	7	31.82	0		11	50.00
140–149	0		4	18.18	5	22.73	0		9	40.91
150–x	0		0		1	4.55	0		1	4.55
									22	100.01

No. 1587 omitted.

ANTHROPOLOGY OF IRAQ

BIZYGOMATIC BREADTH

Total facial length	x-124 No.	%	125-134 No.	%	135-x No.	%	Totals No.	%
x-114	0	....	4	18.18	0	....	4	18.18
115-124	1	4.55	8	36.36	2	9.09	11	50.00
125-x	0	....	4	18.18	3	13.64	7	31.82
							22	100.00

No. 1587 omitted.

UPPER FACIAL LENGTH

Total facial length	x-63 No.	%	64-69 No.	%	70-75 No.	%	76-81 No.	%	82-x No.	%	Totals No.	%
x-109	0	....	0	....	0	....	0	....	0	....	0	....
110-119	1	4.35	9	39.13	1	4.35	0	....	0	....	11	47.83
120-129	0	....	0	....	9	39.13	2	8.70	0	....	11	47.83
130-x	0	....	0	....	0	....	1	4.35	0	....	1	4.35
											23	100.01

NASAL WIDTH

Nasal length	x-29 No.	%	30-35 No.	%	36-41 No.	%	42-x No.	%	Totals No.	%
x-49	0	....	2	8.70	2	8.70	0	....	4	17.40
50-59	1	4.35	10	43.47	5	21.73	1	4.35	17	73.90
60-x	1	4.35	0	....	1	4.35	0	....	2	8.70
									23	100.00

MEASUREMENTS AND INDICES OF ANAIZA BEDUINS

Measurements	No.	Range	Mean	S.D.	C.V.
Age	23	20-54	34.15±1.03	7.35±0.73	21.52±2.14
Stature	22	146-178	162.96±0.95	6.60±0.67	4.05±0.41
Sitting height	22	72-98	87.85±0.83	5.76±0.59	6.56±0.67
Head length	23	176-202	191.22±0.93	6.63±0.66	3.47±0.35
Head breadth	22	123-149	137.50±0.69	4.77±0.49	3.47±0.35
Minimum frontal diameter	22	101-120	110.30±0.59	4.08±0.41	3.70±0.38
Bizygomatic diameter	22	120-139	130.20±0.63	4.40±0.42	3.38±0.34
Bigonial diameter	22	90-117	100.38±0.74	5.28±0.53	5.26±0.52
Total facial height	23	110-134	120.50±0.82	5.80±0.58	4.81±0.48
Upper facial height	23	60-84	70.25±0.70	5.00±0.50	7.12±0.71
Nasal height	23	44-63	53.66±0.65	4.64±0.46	8.65±0.86
Nasal breadth	23	25-45	34.61±0.56	3.99±0.40	11.53±1.15
Ear length	23	48-67	56.82±0.64	4.52±0.45	7.95±0.79
Ear breadth	23	29-40	33.78±0.40	2.82±0.28	8.35±0.83
Indices					
Relative sitting height	22	44-57	53.68±0.43	3.00±0.31	5.59±0.57
Cephalic	23	65-85	71.91±0.57	4.05±0.40	5.60±0.56
Fronto-parietal	22	72-86	79.81±0.44	3.03±0.31	3.80±0.39
Zygo-frontal	22	80-91	84.42±0.36	2.48±0.25	2.94±0.30
Zygo-gonial	22	69-83	77.08±0.42	2.94±0.30	3.81±0.39
Total facial	22	80-104	92.70±0.63	4.35±0.44	4.69±0.48
Upper facial	22	46-60	53.96±0.47	3.30±0.34	6.12±0.62
Nasal	23	44-95	66.18±1.41	10.00±0.99	15.11±1.50
Ear	23	47-76	59.54±0.67	4.76±0.47	7.99±0.79

PHOTOGRAPHIC ANALYSES OF ANAIZA BEDUINS

The photographs of the Anaiza tribesmen have been arranged in order of ascending age from 24 to 45.

In general, the Anaiza were far more homogeneous in the physical characters of the head and face than the Dulaim. The basic element of which No. 1571 (Plates 40 and 41) is an excellent example, probably approaches the Proto-Mediterranean type.

Since we are dealing with but twenty-three tribesmen this series can not be described as adequate in any sense. We must therefore proceed with extra caution in attempting to analyze and segregate the racial elements within this small group.

Among the Anaiza the following variations occur:

   Basic Mediterranean: No. 1571 (Plates 40, 41)
   Iraqo-Mediterranean: No. 1589 (Plate 37)
   Very long-headed (G.O.L. 201): No. 1573 (Plate 39)
   Ultradolichocephal (C.I. 67.0): No. 1571 (Plates 40, 41)
   Brachycephal (C.I. 83.3): No. 1592 (Plate 46)
   Short-faced: No. 1582 (Plate 46)
   Long-faced: No. 1586 (Plate 42)
   Green-brown-eyed: No. 1585 (Plate 45)
   Gray-brown-eyed: No. 1589 (Plate 37)
   Blue-brown-eyed: No. 1587 (Plate 42)
   Straight-nosed: No. 1575 (Plate 39)
   Very slightly convex-nosed: No. 1589 (Plate 37)
   Slightly convex-nosed: No. 1578 (Plate 47)
   Convex-nosed: No. 1579 (Plate 47)
   Markedly convex-nosed: Nos. 1573, 1576 (Plates 39, 45)
   Negroid: No. 1572 (Plate 38)

Examination of the photographs reveals that the Anaiza tribesmen belong to a relatively homogeneous Mediterranean type. They show considerably less variation in racial characters than the Dulaimis.

SUMMARY

The average Anaiza tribesman is medium to short in stature, long to very long in trunk length, and possesses a wide or narrow forehead, a wide or narrow head, dolichocephalic or ultradolichocephalic index, medium short or medium long upper and total facial heights with a leptoprosopic index, a nose medium in length, medium narrow or medium wide, and a leptorrhine or mesorrhine index.

The Anaiza tribesmen appear to belong to the straight-nosed, leptoprosopic, leptorrhine, and dolichocephalic division of the Mediterranean Race. Furthermore, they are racially distinct since nomadic life in the desert restricts intermarriage. The infiltration of Negro blood through the age-old custom of a Negro bodyguard for the Sheikh is the solitary factor which has permeated every large Beduin encampment. In my forthcoming report on the Shammar Beduins of northwestern Iraq, the racial significance of this Negroid element will be discussed in the part entitled "The Northern Jazira."

ANTHROPOLOGY OF IRAQ

MEASUREMENTS OF ANAIZA BEDUINS

No.	Age	Stature	SH	L	B	B'	J	go-go	GH	G'H	NH	NB
1571	28	1635	908	197	132	108	124	94	117	67	50	33
1572	27	1666	905	198	138	114	131	104	113	60	45	38
1573	28	1620	926	201	140	110	134	106	128	74	58	34
1574	32	1756	932	196	137	107	128	96	123	75	54	35
1575	28	1543	850	197	140	107	125	103	118	67	48	36
1576	35	1711	979	194	138	115	138	104	126	70	56	43
1577	38	1623	867	198	141	108	128	98	111	64	45	34
1578	45	1640	873	191	134	105	126	101	126	74	56	34
1579	36	1615	878	194	136	110	130	97	122	72	56	32
1580	30	1709	922	192	137	113	128	99	123	71	54	41
1581	30	1570	842	193	140	113	128	104	118	68	50	31
1582	35	1580	880	189	137	108	129	94	112	68	51	35
1583	30	1658	895	198	147	117	138	103	124	78	57	36
1584	25	1483	823	180	141	108	125	95	115	64	50	36
1585	35	1677	891	177	140	110	128	101	124	77	62	27
1586	30	1628	857	193	142	114	134	105	132	80	56	34
1587	30	1610	860	188	135	...	...	98	117	72	50	32
1588	24	1570	857	182	141	108	127	92	119	68	52	30
1589	25	1720	970	196	138	112	137	114	128	72	52	36
1590	45	1593	715	180	125	102	125	95	114	67	56	28
1591	50	1602	747	184	135	113	138	97	115	68	48	32
1592	35	1643	908	186	155	114	137	107	125	74	57	39
1593	40	....	...	191	130	110	132	100	128	74	60	40

MORPHOLOGICAL CHARACTERS OF ANAIZA BEDUINS

No.	HAIR			EYES			NOSE	
	Form	Texture	Color	Color	Sclera	Iris	Profile	Wings
1571	l w	coarse	black	gr-br	blood	ray	c-c	medium
1572*	...	......	black	gr-br	clear	zon	c-c	m-fl
1573*	...	......	black	dk br	clear	zon	conv	comp
1574	l w	coarse	black	bl-br	clear	zon	conv	medium
1575*	...	......	black	dk br	clear	hom	str	m-fl
1576	...	coarse	v dk br	bl-br	blood	....	conv	medium
1577*	...	......	black	gr-br	clear	zon	c-c	medium
1578	l w	coarse	gray	bl-br	clear	zon	conv	comp
1579*	...	......	black	bl-br	clear	zon	conv	cp-m
1580	l w	coarse	black	bl-br	clear	hom	str	medium
1581	l w	coarse	black	bl-br	clear	hom	str	medium
1582*	...	......	black	dk br	clear	hom	conv	medium
1583	l w	medium	black	dk br	clear	ray	conv	medium
1584*	...	......	black	gray-br	clear	ray	str	medium
1585	l w	coarse	black	gr-br	clear	zon	conv	medium
1586*	...	......	black	gray-br	clear	hom	c-c	medium
1587	l w	medium	black	bl-br	clear	hom	c-c	m-fl
1588*	...	......	black	dk br	clear	hom	str	comp
1589	d w	coarse	black	gray-br	clear	hom	conv	medium
1590	l w	medium	black	dk br	clear	hom	str	cp-m
1591	l w	medium	gray	dk br	blood	zon	str	m-fl
1592	l w	coarse	black	gr-br	clear	zon	str	medium
1593	...	....	.....	.....	.....	....	.....	....

*Shaved.

Indices of Anaiza Beduins

No.	EL	EB	RSH	B/L	B'/B	GH/J	G'H/J	NB/NH	EB/EL	go-go/J	B'/J
1571	58	35	55.6	67.0	81.8	94.4	54.0	66.0	60.3	75.8	87.1
1572	58	37	54.3	69.7	82.6	86.3	45.8	84.4	63.8	79.4	87.0
1573	58	30	57.2	69.7	78.6	95.5	55.2	58.6	51.7	79.1	82.1
1574	63	34	53.1	69.9	78.1	96.1	58.6	64.8	54.0	75.0	83.6
1575	50	30	55.1	71.1	76.4	94.4	53.6	75.0	60.0	82.4	85.6
1576	57	35	57.2	71.1	83.3	91.3	50.7	76.8	61.4	75.4	83.3
1577	56	32	53.4	71.2	76.6	86.7	50.0	75.6	57.1	76.6	84.4
1578	53	34	53.2	70.2	78.4	100.0	58.7	60.7	64.2	80.2	83.3
1579	64	37	54.4	70.1	80.9	93.9	55.4	57.1	57.8	74.6	84.6
1580	64	38	54.5	71.4	82.5	96.1	55.5	75.9	59.4	77.3	88.3
1581	53	32	53.6	72.5	80.7	92.2	53.1	62.0	60.4	81.3	88.3
1582	52	30	55.7	72.5	78.8	86.8	52.7	68.6	57.7	72.9	83.7
1583	53	35	54.0	74.2	79.6	89.9	56.5	63.2	66.0	74.6	84.8
1584	55	34	55.5	78.3	76.6	92.0	51.2	72.0	61.8	76.0	86.4
1585	60	34	53.2	79.1	78.6	96.9	60.2	43.6	56.7	78.9	85.9
1586	56	31	52.6	73.6	80.3	98.5	59.7	60.7	55.4	78.4	85.1
1587	51	31	53.4	71.8	....	....	....	64.0	60.8	....	....
1588	56	35	54.6	77.5	76.6	93.7	53.5	57.7	62.5	72.4	85.0
1589	58	35	56.4	70.4	81.2	93.4	52.6	69.2	60.3	83.2	81.8
1590	60	33	44.9	69.4	81.6	91.2	53.6	50.0	55.0	76.0	81.6
1591	51	39	46.6	73.4	83.7	83.3	49.3	66.7	76.5	70.3	81.9
1592	55	30	55.2	83.3	73.6	91.2	54.0	68.4	54.6	78.1	83.2
1593	63	36	....	68.1	84.6	97.0	56.1	66.6	57.1	75.8	83.3

Ram-Faced Types Among the Dulaim and the Anaiza

According to Keith (pp. 52–53), "among eastern peoples distributed in the southwestern part of Asia from the Pamir to Asia Minor, there occurs a type of face which seizes upon the attention of the student of human races. People with this type of countenance are sometimes described as 'ram-faced'; the upper face carrying the nose is long, while the mandibular part of the face is short."

This criterion is important so we must tabulate my Iraq groups.

Facial Measurements and Indices

Group	U.F.H.	T.F.H.	U.F.I.	Biz.B.	T.F.I.
Dulaim	71.55	121.50	53.15	134.95	90.35
Anaiza	70.25	120.50	53.96	130.20	92.70
Ba'ij Beduins	73.63	117.2	57.37	128.5	91.4
Kish Arabs	72.97	119.8	56.62	129.5	92.73
Iraq Soldiers	73.88	120.92	55.23	133.85	90.5

When the Dulaim and the Anaiza are grouped according to the Keith system, the following tables result.

Dulaim

Total facial height	Upper facial height			
	x–63	64–69	70–75	76–x
x–109	1	2	0	0
110–119	1	30	14	2
120–129	0	11	51	13
130–x	0	0	1	10

ANAIZA

Total facial height	x–63	Upper facial height 64–69	70–75	76–x
x–109	0	0	0	0
110–119	1	9	1	0
120–129	0	0	9	2
130–x	0	0	0	1

Direct comparisons can be made (Field, 1935, pp. 51 et seq.) between various groups of Arabs of the Kish area, Iraq Army Soldiers and the Ba'ij Beduins on the one hand and the Dulaim and the Anaiza tribesmen on the other. The relative frequency of occurrence of this "ram-faced" type can thus be determined.

IV. ADDITIONAL ANTHROPOMETRIC DATA FROM IRAQ

The examination of the metric and morphological data on the Dulaim and the Anaiza has been completed in the preceding chapter.

Since this report on the Upper Euphrates area forms the first part of the volume entitled "The Anthropology of Iraq," it will not be out of place to add the recalculated statistics on my Iraq figures and observations.

It is necessary to explain that when my anthropometric data were placed on punch cards for the Hollerith sorting machines certain omissions and rearrangements had to be made in order that the results might conform to the methods standardized by Dr. Hooton in the Laboratory of Anthropology at Harvard. For example, only individuals between the ages of eighteen and seventy were included. In addition, the grouping, according to cephalic, facial, and nasal indices, and stature conforms to the Harvard classificatory system.

In the following pages I have added these new means for the measurements and indices together with the regrouped morphological characters. In this manner the anthropometric data on Dulaim, Anaiza, Kish Arabs, Iraq Soldiers and Ba'ij Beduins are directly comparable.

There is no need to analyze the material on the last three groups, since they form the basis for my monograph, "Arabs of Central Iraq, Their History, Ethnology, and Physical Characters." Furthermore, these data have been discussed by Sir Arthur Keith and W. M. Krogman (1932, pp. 301-333), Keith (1935, pp. 11-76), Coon (pp. 411-413), and Field (1939a).

Measurements and Indices of Kish Arabs, Iraq Soldiers, and Ba'ij Beduins

Measurements	Kish Arabs		Iraq Soldiers		Ba'ij Beduins	
	No.	Mean	No.	Mean	No.	Mean
Age	359	33.75	221	23.75	35	36.45
Stature	340	168.30	222	172.56	35	168.18
Sitting height	342	82.51	222	85.09	35	83.38
Leg length	340*	85.79	222	87.47	35	84.80
Head length	358	188.76	222	186.24	35	191.31
Head breadth	359	141.91	221	143.71	35	139.93
Minimum frontal diameter	358	111.50	221	114.10	35	110.86
Bizygomatic breadth	357	129.90	222	133.95	35	128.15
Bigonial breadth	357	103.10	221	107.10	35	101.34
Total facial height	355	119.95	221	121.10	35	116.70
Upper facial height	355	73.00	221	74.15	35	73.30
Nasal height	358	58.50	221	57.02	35	59.90
Nasal breadth	359	35.42	222	34.76	35	34.82
Ear length	359	62.26	221	59.82	35	62.42
Ear breadth	359	35.31	222	36.06	35	36.51
Indices						
Relative sitting height	340	49.08	222	49.30	35	49.76
Cephalic	358	75.33	221	76.62	35	73.29
Fronto-parietal	358	78.67	221	79.33	35	79.60
Zygo-frontal	355	85.98	221	84.94	35	86.30
Zygo-gonial	355	79.27	220	79.69	35	79.51
Total facial	354	92.65	220	90.45	35	91.30
Upper facial	354	56.51	222	55.43	35	57.29
Nasal	358	61.14	221	61.62	35	58.06
Ear	359	57.06	221	60.94	35	59.06

*Derived from means.

Measurements and Indices of Kish Arabs

(Observed at Jemdet Nasr and Kish, March–June, 1928)

Measurements	No.	Range	Mean	S.D.	C.V.
Age	359	18–70	33.75±0.46	12.95±0.33	38.37±0.97
Stature	340	149–193	168.30±0.22	6.15±0.16	3.65±0.09
Sitting height	342	66–95	82.51±0.17	4.53±0.12	5.49±0.14
Head length	358	167–208	188.76±0.25	7.14±0.18	3.78±0.10
Head breadth	359	120–158	141.91±0.21	5.79±0.15	4.08±0.10
Minimum frontal diameter	358	93–124	111.50±0.19	5.32±0.13	4.77±0.12
Bizygomatic diameter	357	105–149	129.90±0.27	7.45±0.19	5.74±0.14
Bigonial diameter	357	72–130	103.10±0.27	7.68±0.19	7.45±0.19
Total facial height	355	100–144	119.95±0.26	7.25±0.18	6.04±0.15
Upper facial height	355	60–94	73.00±0.20	5.55±0.14	7.60±0.19
Nasal height	358	44–79	58.50±0.17	4.88±0.12	8.34±0.21
Nasal breadth	359	25–54	35.42±0.12	3.42±0.09	9.66±0.24
Ear length	359	44–79	62.26±0.18	4.92±0.12	7.90±0.20
Ear breadth	359	26–46	35.31±0.13	3.60±0.09	10.20±0.26
Indices					
Relative sitting height	340	42–55	49.08±0.08	2.12±0.05	4.32±0.11
Cephalic	358	62–88	75.33±0.14	3.93±0.10	5.22±0.13
Fronto-parietal	358	66–95	78.67±0.15	4.29±0.11	5.45±0.14
Zygo-frontal	355	76–99	85.98±0.16	4.60±0.12	5.35±0.14
Zygo-gonial	355	63–98	79.27±0.18	4.92±0.12	6.21±0.16
Total facial	354	70–124	92.65±0.27	7.45±0.19	8.04±0.20
Upper facial	354	46–75	56.51±0.18	4.89±0.12	8.65±0.22
Nasal	358	36–91	61.14±0.26	7.24±0.18	11.84±0.30
Ear	359	41–80	57.06±0.21	6.00±0.15	10.52±0.26

HARVARD CLASSIFICATIONS OF KISH ARABS

Stature

	Short (x–160.5)	Medium (160.6–169.4)	Tall (169.5–x)	Total
Number	39	148	153	340
Per cent	11.47	43.53	45.00	100.00

Cephalic Index

	Dolichocephalic (x–76.5)	Mesocephalic (76.6–82.5)	Brachycephalic (82.6–x)	Total
Number	224	125	9	358
Per cent	62.57	34.92	2.51	100.00

Facial Index

	Euryprosopic (x–84.5)	Mesoprosopic (84.6–89.4)	Leptoprosopic (89.5–x)	Total
Number	43	77	234	354
Per cent	12.15	21.75	66.10	100.00

Nasal Index

	Leptorrhine (x–76.4)	Mesorrhine (76.5–83.4)	Platyrrhine (83.5–x)	Total
Number	292	64	2	358
Per cent	81.56	17.88	0.56	100.00

Vital Statistics of Kish Arabs

Brothers	No.	Per cent	Sisters	No.	Per cent
None	79	22.13	None	98	27.37
1	103	28.85	1	113	31.56
2	79	22.13	2	77	21.51
3–4	74	20.73	3–4	53	14.80
5–6	19	5.32	5–6	12	3.35
7 or more	3	0.84	7 or more	5	1.40
Total	357	100.00	Total	358	99.99

Sons	No.	Per cent	Daughters	No.	Per cent
None	55	27.09	None	65	32.02
1	56	27.59	1	46	22.66
2	42	20.69	2	47	23.15
3–4	41	20.20	3–4	33	16.26
5–6	8	3.94	5–6	6	2.96
7 or more	1	0.49	7 or more	6	2.96
Total	203	100.00	Total	203	100.01

Morphological Characters of Kish Arabs

Skin Color

	No.	Per cent
Very light	0	....
Light	0	....
Dark	1	20.00
Very dark	4	80.00
Total	5	100.00

HAIR

Color	No.	Per cent	Mustache	No.	Per cent
Black	40	13.38	Black	0	....
Very dark brown	10	3.34	Very dark brown	0	....
Dark brown	197	65.89	Dark brown	1	8.33
Brown	0	....	Brown	0	....
Reddish brown	8	2.68	Reddish brown	2	16.67
Light brown	2	0.67	Light brown	2	16.67
Red	0	....	Red	0	....
Black and gray	5	1.67	Black and gray	1	8.33
Dark brown and gray	23	7.69	Dark brown and gray	5	41.67
Light brown and gray	1	0.33	Light brown and gray	0	....
Gray	12	4.01	Gray	0	....
White	1	0.33	White	1	8.33
Total	299	99.99	Total	12	100.00

Form	No.	Per cent	Texture	No.	Per cent
Straight	12	4.76	Coarse	35	12.03
Very low waves	5	1.98	Coarse-medium	1	0.34
Low waves	208	82.54	Medium	178	61.17
Deep waves	12	4.76	Medium-fine	9	3.09
Curly-frizzly	14	5.56	Fine	68	23.37
Woolly	1	0.40	Total	291	100.00
Total	252	100.00			

Head hair (quantity)	No.	Per cent	Face hair	No.	Per cent
− −	9	3.70	Mustache	40	57.97
−	27	11.11	Beard	5	7.25
Average	1	0.41	Mustache and beard	24	34.78
+	103	42.39	Total	69	100.00
+ +	96	39.51			
+ + +	7	2.88			
Total	243	100.00			

Beard (quantity)	No.	Per cent	Body hair	No.	Per cent
− −	8	3.36	− −	6	3.26
−	80	33.61	−	44	23.91
Average	0	....	Average	1	0.54
+	94	39.50	+	104	56.52
+ +	52	21.85	+ +	26	14.13
+ + +	4	1.68	+ + +	3	1.63
Total	238	100.00	Total	184	99.99

FACIAL FEATURES

Brow-ridges	No.	Per cent	Glabella	No.	Per cent
Continuous	1	1.49	− −	0	....
Median	66	98.51	−	0	....
Total	67	100.00	Average	0	....
			+	7	100.00
			+ +	0	....
			+ + +	0	....
			Total	7	100.00

Malars (projection)	No.	Per cent	Prognathism	No.	Per cent
Average	0	....	Alveolar	5	71.43
+	20	50.00	Facial	2	28.57
+ +	19	47.50	Total	7	100.00
+ + +	1	2.50			
Total	40	100.00			

Additional Anthropometric Data

Lip eversion	No.	Per cent
− −	0	....
−	6	21.43
Average	0	....
+	13	46.43
+ +	9	32.14
+ + +	0	....
Total	28	100.00

Eyebrows

Concurrency	No.	Per cent	Thickness	No.	Per cent
−	6	18.18	− −	1	1.89
Average	0	....	−	13	24.53
+	25	75.76	Average	0	....
+ +	2	6.06	+	22	41.51
			+ +	17	32.08
			+ + +	0	....
Total	33	100.00	Total	53	100.01

Lateral extension	No.	Per cent
−	3	10.00
Average	0	....
+	26	86.67
+ +	1	3.33
Total	30	100.00

Eyes

Color	No.	Per cent	Iris	No.	Per cent
Black	2	0.60	Homogeneous	93	27.60
Dark brown	258	77.25	Rayed	10	2.97
Blue-brown	22	6.59	Zoned	234	69.44
Blue-brown	1	0.30			
Green-brown	40	11.98	Total	337	100.01
Green-brown	0	....			
Gray-brown	10	2.99	Sclera	No.	Per cent
Blue	0	....	Clear	218	64.31
Gray	0	....	Yellow	2	0.59
Light brown	1	0.30	Speckled	25	7.37
Blue-gray	0	....	Bloodshot	83	24.48
Blue-green	0	....	Speckled and bloodshot	10	2.95
			Speckled and yellow	1	0.29
Total	334	100.01	Yellow and bloodshot	0	....
			Total	339	99.99

Nose

Bridge	No.	Per cent	Septum	No.	Per cent
Height +	4	20.00	Straight	64	96.97
Breadth +	16	80.00	Convex	2	3.03
Total	20	100.00	Total	66	100.00

Septum inclination	No.	Per cent
Up	3	27.27
Down	8	72.73
Total	11	100.00

Profile	No.	Per cent	Tip thickness	No.	Per cent
Wavy	0	.....	− −	21	21.21
Concave	39	11.27	Average	1	1.01
Straight	198	57.23	+	45	45.45
Convex	66	19.08	+ +	32	32.32
Concavo-convex	43	12.43			
Total	346	100.01	Total	99	99.99

Tip elevation	No.	Per cent	Wings	No.	Per cent
Elevated	23	18.70	Compressed	34	12.69
Horizontal	19	15.45	Compressed-medium	11	4.10
Depressed	81	65.85	Medium	159	59.33
			Medium-flaring	30	11.19
Total	123	100.00	Flaring	28	10.45
			Flaring plus	6	2.24
			Total	268	100.00

TEETH

Bite	No.	Per cent	Loss	No.	Per cent
Under	3	0.96	None	228	69.09
Edge-to-edge	8	2.56	1–4	85	25.76
Slight over	208	66.67	5–8	7	2.12
Marked over	93	29.81	9–16	8	2.42
			17–	2	0.61
Total	312	100.00	All	0	.....
			Total	330	100.00

Condition	No.	Per cent	Eruption	No.	Per cent
Very bad	2	11.76	Complete	335	97.67
Bad	2	11.76	Incomplete	8	2.33
Fair	0	.....			
Good	10	58.82	Total	343	100.00
Excellent	3	17.65			
Total	17	99.99			

Wear	No.	Per cent	Caries	No.	Per cent
None	39	26.00	None	76	51.01
Slight	12	8.00	−	6	4.03
Average	19	12.67	+	27	18.12
+	35	23.33	+ +	23	15.44
+ +	28	18.67	+ + +	17	11.41
+ + +	17	11.33			
Total	150	100.00	Total	149	100.01

BODY DEVELOPMENT

Musculature	No.	Per cent	Chest	No.	Per cent
Poor	19	5.44	− −	3	0.89
Fair	40	11.46	−	15	4.46
Average	6	1.72	Average	43	12.80
Good	189	54.15	+	231	68.75
Excellent	95	27.22	+ +	44	13.10
Total	349	99.99	Total	336	100.00

Scapulae (vertebral borders)	No.	Per cent
Concave	0	.....
Straight	55	88.71
Convex	7	11.29
Total	62	100.00

Additional Anthropometric Data

Ears

Helix	No.	Per cent	Lobe	No.	Per cent
−	3	5.56	Attached	49	44.95
Average	1	1.85	Free	60	55.05
+	38	70.37			
+ +	11	20.37	Total	109	100.00
+ + +	1	1.85			
Total	54	100.00			

Darwin's Point	No.	Per cent
− −	1	1.67
−	5	8.33
Average	0	....
+	36	60.00
+ +	16	26.67
+ + +	2	3.33
Total	60	100.00

Health

	No.	Per cent	Disease	No.	Per cent
Poor	14	4.02	Smallpox	19	22.09
Fair	12	3.45	Fever	48	55.81
Average	1	0.29	Headache	4	4.65
Good	208	59.77	Stomach pain	3	3.49
Excellent	113	32.47	Scalp	1	1.16
			Cataract	9	10.47
Total	348	100.00	Trachoma	0	....
			Baghdad Boil	2	2.33
			Chicken pox	0	....
			Total	86	100.00

Eyes

Blindness	No.	Per cent
Right eye	4	40.00
Left eye	3	30.00
Both eyes	3	30.00
Total	10	100.00

Tattooing

Quantity	No.	Per cent
None	151	43.39
Some	197	56.61
Extensive	0	....
Total	348	100.00

Henna

	No.	Per cent
Hair	8	88.89
Body	0	....
Hands	1	11.11
Feet	0	....
Total	9	100.00

In December, 1925, and during the first part of the following month Dr. L. H. Dudley Buxton and I were attached as volunteer assistants to the Field Museum–Oxford University Joint Expedition at Kish.

While Dr. Buxton measured 100 Arab workmen employed at the excavations I acted as recorder. He also examined sixty-four Iraq Army Soldiers at Hilla camp (cf. Buxton and Rice, 1931; and Field, 1935a, p. 101).

With the permission of Dr. Buxton the figures for Kish workmen were recalculated at Harvard and the following tables resulted.

MEASUREMENTS AND INDICES OF KISH WORKMEN
(*After Buxton*)

Measurements	No.	Range	Mean	S.D.	C.V.
Stature	95	152–193	168.39±0.47	6.78±0.33	4.03±0.20
Head length	100	173–205	190.14±0.43	6.39±0.30	3.36±0.16
Head breadth	100	126–155	142.75±0.35	5.16±0.25	3.61±0.17
Minimum frontal diameter	100	97–116	107.86±0.26	3.88±0.19	3.60±0.17
Bizygomatic diameter	100	120–144	135.10±0.34	5.00±0.24	3.70±0.18
Bigonial diameter	100	90–117	105.06±0.37	5.44±0.26	5.18±0.25
Total facial height	100	90–134	114.30±0.50	7.45±0.36	6.52±0.31
Upper facial height	100	55–84	67.30±0.32	4.80±0.23	7.13±0.34
Nasal height	100	36–63	47.58±0.32	4.72±0.23	9.86±0.47
Nasal breadth	100	25–45	33.74±0.22	3.21±0.15	9.51±0.45
Indices					
Cephalic	100	65–85	75.30±0.22	3.30±0.16	4.38±0.21
Fronto-parietal	100	69–86	75.70±0.21	3.09±0.15	4.08±0.19
Zygo-frontal	100	72–91	79.74±0.21	3.16±0.15	3.96±0.19
Zygo-gonial	100	66–92	77.56±0.26	3.81±0.18	4.91±0.23
Total facial	100	70–109	84.40±0.34	5.05±0.24	5.98±0.29
Upper facial	100	43–60	49.55±0.22	3.30±0.16	6.66±0.32
Nasal	100	48–95	71.74±0.62	9.16±0.44	12.77±0.61

MORPHOLOGICAL CHARACTERS OF KISH WORKMEN
(*After Buxton*)

HAIR

Color	No.	Per cent	Form	No.	Per cent
Black	86	92.47	Straight	1	5.26
Very dark brown	1	1.08	Very low waves	16	84.21
Dark brown	0	....	Low waves	2	10.53
Brown	1	1.08	Deep waves	0	....
Reddish brown	0	....	Curly-frizzly	0	....
Light brown	1	1.08	Woolly	0	....
Red	0	....			
Black and gray	0	....	Total	19	100.00
Dark brown and gray	0	....			
Light brown and gray	0	....			
Gray	0	....			
White	4	4.30			
Total	93	100.01			

EYES

Color	No.	Per cent
Black	0	....
Dark brown	77	80.21
Blue-brown	0	....
Blue-brown	0	....
Green-brown	16	16.67
Green-brown	0	....
Gray-brown	0	....
Blue	0	....
Gray	0	....
Light brown	3	3.13
Blue-gray	0	....
Blue-green	0	....
Total	96	100.01

ADDITIONAL ANTHROPOMETRIC DATA

MEASUREMENTS, INDICES, AND OBSERVATIONS OF IRAQ SOLDIERS

With the generous permission of the Officer Commanding Hilla Army Camp, 222 soldiers were measured, from June 14 to 17, 1928. Mr. S. Y. Showket obtained the front and profile photographs of each individual.

MEASUREMENTS AND INDICES OF IRAQ SOLDIERS
(Observed at Hilla Camp, June 14-17, 1928)

Measurements	No.	Range	Mean	S.D.	C.V.
Age	221	18–49	23.75±0.19	4.20±0.13	17.68±0.57
Stature	222	158–190	172.56±0.24	5.25±0.17	3.04±0.10
Sitting height	222	72–98	85.09±0.19	4.26±0.14	5.01±0.16
Head length	222	167–208	186.24±0.32	7.08±0.23	3.80±0.12
Head breadth	221	126–161	143.71±0.25	5.46±0.18	3.80±0.12
Minimum frontal diameter	221	101–128	114.10±0.23	4.96±0.16	4.35±0.14
Bizygomatic diameter	222	105–149	133.95±0.25	5.55±0.18	4.14±0.13
Bigonial diameter	221	90–133	107.10±0.28	6.28±0.20	5.86±0.19
Total facial height	221	100–144	121.10±0.31	6.80±0.22	5.62±0.18
Upper facial height	221	60–89	74.15±0.22	4.80±0.15	6.47±0.21
Nasal height	221	44–75	57.02±0.23	4.96±0.16	8.70±0.28
Nasal breadth	222	28–57	34.76±0.16	3.60±0.12	10.36±0.33
Ear length	221	48–75	59.82±0.19	4.20±0.13	7.02±0.23
Ear breadth	222	29–46	36.06±0.15	3.39±0.11	9.40±0.30
Indices					
Relative sitting height	222	44–55	49.30±0.10	2.28±0.07	4.62±0.15
Cephalic	221	65–91	76.62±0.18	3.99±0.13	5.21±0.17
Fronto-parietal	221	69–92	79.33±0.18	3.90±0.13	4.92±0.16
Zygo-frontal	221	76–99	84.94±0.16	3.44±0.11	4.05±0.13
Zygo-gonial	220	66–95	79.69±0.20	4.35±0.14	5.46±0.18
Total facial	220	75–109	90.45±0.26	5.70±0.18	6.30±0.20
Upper facial	222	46–72	55.43±0.18	4.08±0.13	7.36±0.24
Nasal	221	44–83	61.62±0.32	7.00±0.22	11.36±0.36
Ear	221	45–80	60.94±0.27	5.92±0.19	9.71±0.31

HARVARD CLASSIFICATIONS OF IRAQ SOLDIERS

STATURE

	Short (x–160.5)	Medium (160.6–169.4)	Tall (169.5–x)	Total
Number	2	66	154	222
Per cent	**0.90**	**29.73**	**69.37**	**100.00**

CEPHALIC INDEX

	Dolichocephalic (x–76.5)	Mesocephalic (76.6–82.5)	Brachycephalic (82.6–x)	Total
Number	110	97	14	221
Per cent	**49.77**	**43.89**	**6.33**	**99.99**

FACIAL INDEX

	Euryprosopic (x–84.5)	Mesoprosopic (84.6–89.4)	Leptoprosopic (89.5–x)	Total
Number	25	79	116	220
Per cent	**11.36**	**35.91**	**52.73**	**100.00**

NASAL INDEX

	Leptorrhine (x–67.4)	Mesorrhine (67.5–83.4)	Platyrrhine (83.5–x)	Total
Number	183	38	0	221
Per cent	82.81	17.19	.....	100.00

VITAL STATISTICS OF IRAQ SOLDIERS

Brothers	No.	Per cent	Sisters	No.	Per cent
None	16	7.21	None	40	18.02
1	44	19.82	1	74	33.33
2	79	35.59	2	56	25.23
3–4	61	27.48	3–4	43	19.37
5–6	12	5.41	5–6	6	2.70
7 or more	10	4.50	7 or more	3	1.35
Total	222	100.01	Total	222	100.00

Sons	No.	Per cent	Daughters	No.	Per cent
None	29	49.15	None	44	74.58
1	21	35.59	1	12	20.34
2	8	13.56	2	3	5.08
3–4	1	1.69	3–4	0	.....
5–6	0	.....	5–6	0	.....
7 or more	0	.....	7 or more	0	.....
Total	59	99.99	Total	59	100.00

MORPHOLOGICAL CHARACTERS OF IRAQ SOLDIERS

HAIR

Color	No.	Per cent	Form	No.	Per cent
Black	4	5.41	Straight	0	.....
Very dark brown	1	1.35	Very low waves	0	.....
Dark brown	68	91.89	Low waves	5	83.33
Brown	0	.....	Deep waves	0	.....
Reddish brown	0	.....	Curly-frizzly	1	16.67
Light brown	0	.....	Woolly	0	.....
Red	0	.....			
Black and gray	0	.....	Total	6	100.00
Dark brown and gray	1	1.35	Texture	No.	Per cent
Light brown and gray	0	.....	Coarse	0	.....
Gray	0	.....	Coarse-medium	0	.....
White	0	.....	Medium	4	80.00
Total	74	100.00	Medium-fine	0	.....
			Fine	1	20.00
			Total	5	100.00

NOSE

Profile	No.	Per cent	Wings	No.	Per cent
Wavy	0	.....	Compressed	7	4.96
Concave	14	9.52	Compressed-medium	20	14.18
Straight	83	56.46	Medium	67	47.52
Convex	46	31.29	Medium-flaring	29	20.57
Concavo-convex	4	2.72	Flaring	16	11.35
			Flaring plus	2	1.42
Total	147	99.99	Total	141	100.00

Additional Anthropometric Data

Eyes

Color	No.	Per cent	Iris	No.	Per cent
Black	2	1.34	Homogeneous	129	87.16
Dark brown	126	84.56	Rayed	0	.....
Blue-brown	7	4.70	Zoned	19	12.84
Blue-brown	1	0.67			
Green-brown	12	8.05	Total	148	100.00
Green-brown	0	.....			
Gray-brown	1	0.67	Sclera	No.	Per cent
Blue	0	.....	Clear	146	97.99
Gray	0	.....	Yellow	0	.....
Light brown	0	.....	Speckled	0	.....
Blue-gray	0	.....	Bloodshot	3	2.01
Blue-green	0	.....	Speckled and bloodshot	0	.....
			Speckled and yellow	0	.....
Total	149	99.99	Yellow and bloodshot	0	.....
			Total	149	100.00

Teeth

Bite	No.	Per cent	Eruption	No.	Per cent
Under	1	0.73	Complete	131	93.57
Edge-to-edge	2	1.46	Incomplete	9	6.43
Slight over	123	89.78			
Marked over	11	8.03	Total	140	100.00
Total	137	100.00			

Loss	No.	Per cent	Caries	No.	Per cent
None	112	80.58	None	1	14.29
1–4	27	19.42	−	0	.....
5–8	0	.....	+	5	71.43
9–16	0	.....	+ +	1	14.29
17+	0	.....	+ + +	0	.....
All	0	.....			
Total	139	100.00	Total	7	100.01

Body Development

Musculature	No.	Per cent	Chest	No.	Per cent
Poor	0	.....	− −	0	.....
Fair	0	.....	−	0	.....
Average	6	4.23	Average	6	4.23
Good	132	92.96	+	132	92.96
Excellent	4	2.82	+ +	4	2.82
Total	142	100.01	Total	142	100.01

Health

	No.	Per cent	Disease	No.	Per cent
Poor	0	.....	Smallpox	21	91.30
Fair	0	.....	Fever	0	.....
Average	0	.....	Headache	0	.....
Good	140	100.00	Stomach pain	0	.....
Excellent	0	.....	Scalp	2	8.70
			Cataract	0	.....
Total	140	100.00	Trachoma	0	.....
			Baghdad Boil	0	.....
			Chicken pox	0	.....
			Total	23	100.00

Tattooing

Quantity	No.	Per cent
None	103	46.40
Some	119	53.60
Extensive	0	.....
Total	222	100.00

Measurements and Indices of Ba'ij Beduins
(Observed between Kish and Jemdet Nasr, July 10, 1928)

Measurements	No.	Range	Mean	S.D.	C.V.
Age	35	18–69	36.45±1.31	11.45±0.92	31.41±0.25
Stature	35	155–178	168.18±0.60	5.22±0.42	3.10±0.25
Sitting height	35	75–92	83.38±0.39	3.45±0.28	4.14±0.33
Head length	35	179–202	191.31±0.61	5.37±0.43	2.81±0.23
Head breadth	35	123–152	139.93±0.74	6.51±0.52	4.65±0.37
Minimum frontal diameter	35	101–124	110.86±0.59	5.20±0.42	4.69±0.38
Bizygomatic diameter	35	115–144	128.15±0.72	6.35±0.51	4.96±0.40
Bigonial diameter	35	90–113	101.34±0.66	5.76±0.46	5.68±0.46
Total facial height	35	100–129	116.70±0.61	5.35±0.43	4.58±0.37
Upper facial height	35	65–84	73.30±0.48	4.20±0.34	5.73±0.46
Nasal height	35	52–71	59.90±0.49	4.28±0.35	7.15±0.58
Nasal breadth	35	28–48	34.82±0.40	3.51±0.28	10.08±0.81
Ear length	35	56–71	62.42±0.40	3.48±0.28	5.58±0.45
Ear breadth	35	29–43	36.51±0.35	3.09±0.25	8.46±0.68
Indices					
Relative sitting height	35	44–55	49.76±0.26	2.24±0.18	4.50±0.36
Cephalic	35	65–85	73.29±0.45	3.96±0.32	5.40±0.44
Fronto-parietal	35	72–89	79.60±0.45	3.99±0.32	5.01±0.40
Zygo-frontal	35	76–99	86.30±0.55	4.84±0.39	5.61±0.45
Zygo-gonial	35	69–95	79.51±0.61	5.37±0.43	6.75±0.54
Facial	35	80–104	91.30±0.55	4.80±0.39	5.26±0.42
Upper facial	35	49–66	57.29±0.47	4.14±0.33	7.23±0.58
Nasal	35	44–75	58.06±0.70	6.12±0.49	10.54±0.85
Ear	35	49–68	59.06±0.45	3.96±0.32	6.71±0.54

Harvard Classifications of Ba'ij Beduins

Stature

	Short (x–160.5)	Medium (160.6–169.4)	Tall (169.5–x)	Total
Number	3	18	14	35
Per cent	8.57	51.43	40.00	100.00

Cephalic Index

	Dolichocephalic (x–76.5)	Mesocephalic (76.6–82.5)	Brachycephalic (82.6–x)	Total
Number	29	5	1	35
Per cent	82.86	14.29	2.86	100.01

Facial Index

	Euryprosopic (x–84.5)	Mesoprosopic (84.6–89.4)	Leptoprosopic (89.5–x)	Total
Number	3	9	23	35
Per cent	8.57	25.71	65.71	99.99

Nasal Index

	Leptorrhine (x–67.4)	Mesorrhine (67.5–83.4)	Platyrrhine (83.5–x)	Total
Number	32	3	0	35
Per cent	91.43	8.57	.....	100.00

Additional Anthropometric Data

Vital Statistics of Ba'ij Beduins

Brothers	No.	Per cent	Sisters	No.	Per cent
None	23	65.71	None	25	71.43
1	4	11.43	1	6	17.14
2	4	11.43	2	1	2.86
3–4	4	11.43	3–4	2	5.71
5–6	0	....	5–6	1	2.86
7 or more	0	....	7 or more	0	....
Total	35	100.00	Total	35	100.00

Sons	No.	Per cent	Daughters	No.	Per cent
None	13	56.52	None	9	39.13
1	3	13.04	1	6	26.09
2	3	13.04	2	5	21.74
3–4	3	13.04	3–4	1	4.35
5–6	0	....	5–6	2	8.70
7 or more	1	4.35	7 or more	0	....
Total	23	99.99	Total	23	100.01

Morphological Characters of Ba'ij Beduins

Hair

Color	No.	Per cent	Form	No.	Per cent
Black	13	43.33	Straight	0	....
Very dark brown	0	....	Very low waves	2	8.00
Dark brown	7	23.33	Low waves	13	52.00
Brown	0	....	Deep waves	6	24.00
Reddish brown	4	13.33	Curly-frizzly	4	16.00
Light brown	0	....	Woolly	0	....
Red	0	....	Total	25	100.00
Black and gray	4	13.33			
Dark brown and gray	1	3.33			
Light brown and gray	0	....	Texture	No.	Per cent
Gray	1	3.33	Coarse	9	33.33
White	0	....	Coarse-medium	0	....
			Medium	14	51.85
Total	30	99.98	Medium-fine	0	....
			Fine	4	14.81
			Total	27	99.99

Head hair (quantity)	No.	Per cent	Beard (quantity)	No.	Per cent
− −	0	....	− −	2	8.00
−	1	4.55	−	2	8.00
Average	0	....	Average	0	....
+	4	18.18	+	8	32.00
+ +	17	77.27	+ +	12	48.00
+ + +	0	....	+ + +	1	4.00
Total	22	100.00	Total	25	100.00

Body hair	No.	Per cent
− −	0	....
−	2	11.11
Average	0	....
+	14	77.78
+ +	2	11.11
+ + +	0	....
Total	18	100.00

EYES

Color	No.	Per cent	Iris	No.	Per cent
Black	0	.....	Homogeneous	7	21.21
Dark brown	9	25.71	Rayed	0	.....
Blue-brown	20	57.14	Zoned	26	78.79
Blue-brown	0	.....			
Green-brown	0	.....	Total	33	100.00
Green-brown	6	17.14			
Gray-brown	0	.....	Sclera	No.	Per cent
Blue	0	.....	Clear	23	67.65
Gray	0	.....	Yellow	0	.....
Light brown	0	.....	Speckled	1	2.94
Blue-gray	0	.....	Bloodshot	10	29.41
Blue-green	0	.....	Speckled and bloodshot	0	.....
			Speckled and yellow	0	.....
Total	35	99.99	Yellow and bloodshot	0	.....
			Total	34	100.00

NOSE

Profile	No.	Per cent	Wings	No.	Per cent
Wavy	0	.....	Compressed	7	22.58
Concave	5	14.29	Compressed-medium	1	3.23
Straight	26	74.29	Medium	17	54.84
Convex	1	2.86	Medium-flaring	2	6.45
Concavo-convex	3	8.57	Flaring	2	6.45
			Flaring plus	2	6.45
Total	35	100.01			
			Total	31	100.00

Tip thickness	No.	Per cent	Tip elevation	No.	Per cent
−	1	14.29	Raised	1	50.00
Average	1	14.29	Horizontal	0	.....
+	1	14.29	Depressed	1	50.00
+ +	4	57.14			
			Total	2	100.00
Total	7	100.01			

TEETH

Bite	No.	Per cent	Loss	No.	Per cent
Under	0	.....	None	26	81.25
Edge-to-edge	0	.....	1–4	5	15.63
Slight over	23	79.31	5–8	0	.....
Marked over	6	20.69	9–16	1	3.13
			17–	0	.....
Total	29	100.00	All	0	.....
			Total	32	100.01

Eruption	No.	Per cent	Wear	No.	Per cent
Complete	34	100.00	None	0	.....
Incomplete	0	.....	Slight	1	25.00
			Average	0	.....
Total	34	100.00	+	1	25.00
			+ +	2	50.00
Caries	No.	Per cent	+ + +	0	.....
None	0	.....			
+	0	.....	Total	4	100.00
+ +	3	100.00			
+ + +	0	.....			
Total	3	100.00			

BODY DEVELOPMENT

Musculature	No.	Per cent	Chest	No.	Per cent
Poor	1	2.86	− −	0	.....
Fair	1	2.86	−	0	.....
Average	0	.....	Average	0	.....
Good	27	77.14	+	31	91.18
Excellent	6	17.14	+ +	3	8.82
Total	35	100.00	Total	34	100.00

HEALTH	No.	Per cent	TATTOOING Quantity	No.	Per cent
Poor	1	2.86	None	17	56.67
Fair	0	.....	Some	13	43.33
Average	0	.....	Extensive	0	.....
Good	30	85.71			
Excellent	4	11.43	Total	30	100.00
Total	35	100.00			

RACIAL POSITION OF THE ARABS

Sir Arthur Keith (pp. 75–76) writes: "How does the Arab stand with regard to other races of mankind? On entering into this inquiry we must note the relationship of Arabia to adjacent racial frontiers. The Red Sea separates the great Arabian peninsula from the Hamitic peoples of Africa, many of which, to be sure, have received Arab infusion. Arabia is separated from the mainland of Asia by the Persian Gulf and the Gulf of Oman. This inlet of the Indian Ocean is also a racial frontier separating the Arab from a people not remotely akin to him, people of the Indo-Afghan type. Also, in the north the base of the peninsula abuts on another racial frontier, the southern frontier of the main or purer Caucasian stock. Then away in the east are the peoples of India, who have many other resemblances to the Arab besides a dark brown skin and dark brown or black hair. If we presume that the modern stocks of mankind have been evolved in or near the regions which they now occupy then we ought to find that the Arab has an evolutionary relationship to all surrounding peoples. That is what we have found in the course of our analysis. The Arab shares traits with Hamitic peoples of Africa, with the Dravidian and Indo-Aryan peoples of India, and with the peoples which extend from the gates of India to the Levant. The Arab's facial features are often so Caucasoid in appearance that we may mistake him for a south European but his pigmentation is usually deeper than that seen in south Europeans. Undoubtedly in his composition we recognize many Negroid traits, and traits which link him with Dravidian and with Hamite.

"Now, how are we to account for Arabia's being occupied by people who are mainly Caucasian in their physical make-up and yet possess so many features in common with dark-skinned neighboring races? In seeking to explain these facts there are other circumstances and relationships which have to be considered. Even today a belt of pigmented human races crosses the Old World. At one extreme we have the Negro of Africa, at the other extreme the Negro of the Pacific. India lies midway in this pigmented belt, one which we suspect extended continuously in Pleistocene times from one extremity of the Old World to the other. On this theory the original inhabitants of Arabia were deeply pigmented and akin to the Hamites of Africa on the one hand and to the Dravidians of India on the other. To the north of the black belt there were two other evolutionary centers: the Mongolian, north of the Himalayas, and the Caucasian, north of the upland mountainous plateau which extends westward from the Himalayas across Iran to Asia Minor. That there was an early break-through from the Mongolian center at the eastern end of the Himalayas is manifest; the Mongol stock at different times broke into the black belt and spread out in the Pacific. There was a Caucasian southward migration at the western end of the Himalayas. In Pleistocene times the great Arabian peninsula was a land to tempt adventurous hunters. The peoples of Arabia might thus represent a mixture of darker-skinned Dravidians into which invaders from the southern or Semitic fringe of the Caucasian center had infused their blood. Such a theory explains many of the facts relating to the racial composition and affinities of the inhabitants of Arabia. Or did the evolutionary center of the Caucasian type actually extend into Arabia? . . . .

"Our interest in the ancient inhabitants of Arabia, particularly of the northern plain, has been stimulated by the expectation that we shall yet be able to prove that our modern way of living—our modern civilization—was initiated by a people or peoples living on or near the frontier of northern Arabia. Were the pioneers of civilization really Arabs (Semites)? Or were they of the less deeply pigmented Caucasian stock farther to the north? We have little evidence to sway us either way, but the only real difference I can perceive between the ancient Mesopotamians of Kish (fourth millennium B.C.) and the modern Arabs of central Iraq relates to size of skull and brain. The average cranial capacity of the ancient Mesopotamian or Arab exceeded that of the average modern inhabitant of central Iraq. I expect that it will yet be proved that the Arab of today is the descendant of the men who built the ancient cities and early civilization along the Euphrates and Tigris rivers."

V. THE TRIBES AND SUB-TRIBES OF THE UPPER EUPHRATES

The following statistical data were obtained from reliable sources which prefer to remain anonymous. During the past fifteen years since these data were compiled numerous individual changes have occurred. Many sheikhs have been succeeded by their sons or nephews. The range, as listed under habitat, has tended to decrease wherever pastoral nomadism has been discouraged. Recent information indicates that the number of families, tents, or houses has remained relatively constant. In general, the information belongs to the period from 1920 to 1925. In any specific instance, however, conditions in 1940 may or may not be as outlined, since no contemporary data are available.

Confederation, tribe, or section	Chiefs	Families, tents, or houses	Habitat
ANAIZA[1] (Tribe)		17,700	Triangle based on Lat. 30°, with its center at Jauf, its apex near Alep. Also visited left bank of Euphrates north of Deir-ez-Zor and the Khabur.
Section			
AMARAT	Fahad Beg ibn Hadhdhal	4,500	Eastern portion of Hamad, from Karbala to Deir-ez-Zor. In autumn near Wadi Ubaiyidh between Karbala and Shithatha; more recently sixteen miles north of Ar Rahhaliya, and near Hindiya Canal.
Sub-sections			
Al Jabal	Fahad Beg ibn Hadhdhal	2,000	......
Al Hiblan	Fahad Beg ibn Hadhdhal	400	......
Al Salqah	Murdi al Rafdi	1,000	......
Al Mutarafah	Jarjir al Hunaidis	......	......
Al Nasrah	Chasib al Sahali	......	......
Al Hussani	Mashan ibn Shamran	......	......
Al Bajaidah	Shami ibn Shami	......	......
A Mudhaiyan	Tahir ibn Dakhil	......	......
Al Sanid	Daiyan ibn Sahlan	......	......
Al Shimlan	Ghadhi al Rubadi	......	......
Al Suqur	Dairbi ibn Mujaf	600	......
Al Dahaman	Huwaichim ibn Dhulaur	......	......
Al Musaib	Hasham al Zuwain	......	......
Al Dilamah	Mutlaq ibn Marzuq	......	......
Al Jalal	Taban ibn Khudhari	......	......

[1] Sunni; nomadic; chiefly camel-breeders, but also horse- and sheep-breeders.

Confederation, tribe, or section	Chiefs	Families, tents, or houses	Habitat
Al Dahamshar	Jazza ibn Mijlad	2,500	
Al Zabanah	Jazza ibn Mijlad		
Al Muhallaf	Dhari ibn Dhubaiyan		
Al Suwailmat	Ayid ibn Bakr		
Al Khamishat	Shallash al Aridh		
Al Mukhaiyat	Nasir abu al Rus		
Al Jalaid	Banaidi ibn Jalaud		
Al Salatin	Ibn Kanfadn		
Al Watbah	Nijris Daidas		
Section FADAN	Mujhim ibn Muhaid. Hachim ibn Muhaid.	3,500	Near Euphrates west of Deir-ez-Zor and on the Khabur. Occasionally came downstream with the Amarat, otherwise in the desert from Deir-ez-Zor to Alep.
Sub-sections			
Dhanna Majid	Mazud ibn Quaishish	1,800	
Al Hazalat	Aswad ibn Harij		
Al Jifal			
Al Khashtah	Salman		
Al Malhud			
Al Mukatharah	Majul al Rahit		
Al Hardha	Faris al Saman		
Al Amarah	Sulaiman al Amir		
Al Wulud	Mujhim ibn Muhaid	1,700	
Al Muhaid	Mujhim ibn Muhaid		
Al Ajrah	Ijrais ibn Fadhal		
Al Rus	Jadu ibn Kira		
Al Sari	Jurais ibn Jaad		
Al Shumailat	Wadi ibn Hubaiyan		
Section MUHALLAF	Ibn Majil, Ibn Majid, Ibn Jandal	1,500	With Wulud Ali section
Sub-sections			
Abdullah			
Ashja			
Budur			
Suwalma			
Section RUWALLA	Nuri ibn Shalan (Paramount)	4,000	Hama to Qasr-el-Azraq and down Wadi Sirhan to Jauf. Eastern limit was high ground where Wadi Hauran rises (Jebel Enaze).
Sub-sections			
Kaka			
Muridh	Nuri ibn Shalan	800	
Nusair	Nuri ibn Shalan	1,000	
Kawakibah		400	
Duran		300	
Furjah		500	
Dughman		450	
Manayi		400	
Mashittah		150	

TRIBES AND SUB-TRIBES OF UPPER EUPHRATES

Confederation, tribe, or section	Chiefs	Families, tents, or houses	Habitat
Section			
AL SBAA[1]	Ghadwan ibn Murshid	2,400	Between Homs and Hama on west, Rasafa on east, to near Alep on north. *Winter.*—With Amarat and Fadan to Kulban el Mat and Wadi Hauran near Rutba.
Sub-sections			
Al Butainat	Ghadwan ibn Murshid	1,200	......
Al Ubidah	Barjas ibn Hufaib	1,200	......
Section			
WULUD ALI	Rashid ibn Sumair	1,800	Mathk plain watered by Barrada. *Winter.*—East and southeast of Damascus.
Sub-sections			
Ataifat			
Fuqarah			
Hajjaj			
Hammamid			
Hasanah			
Mashadiqah			
Musalikh			
Saqra			
Tuluh			
AQAIDAT (Akeydat). (Confederation)		1,200	Both banks of the Euphrates from Tibni to Abu Kemal.
AL DIMIM[2]	Saiyah al Jirrah	170	About eight miles downstream from Khan Kalasil (Salihiya), right bank of Euphrates in area known as Qariyat al Musallakha.
Sections			
AL DIMIM	Saiyah al Jirrah	75	Qariyat al Musallakha (see AL DIMIM).
AL IDHAR	Ali al Barjas	10	About six miles downstream from Khan Kalasil, left bank of Euphrates in area known as Al Bahara.
AL AJARJAH	Harrash al Muhammad and Muhammad al Abdullah	85	Immediately downstream from Khan Kalasil, right bank of Euphrates in area known as Kharaitah.
AL BU HARDAN	Sulaiman al Abdu Rahman	60	......

[1] Famous camel-breeders. I have spent several pleasant days in the tents of Rakkan ibn Murshid near Tellul Basatin, west of Rutba and north of Jebel Enaze on the way to Jebel Tinf.

[2] Sunni; semi-nomadic; agricultural and pastoral.

Confederation, tribe, or section	Chiefs	Families, tents, or houses	Habitat
Sections			
AL SUBAIKHAN[1]	Sulaiman al Abdu Rahman	50	In Jazira north of Tell Hajin.
AL BU HARDAN[2]	Manawakh al Khalil	10	Near Tell Hajin about ten miles below Khan Kalasil on left bank of Euphrates in area known as Hajin.
AL HASSUN	Muhammad al Dindil	255	......
Sections			
AL ALI	Muhammad al Dindil	125	Two to eight miles below Abu Kemal on right bank, in tract known as Suwaiya.
AL MUHAMMAD	Asi al Hawwal	100	About four miles above Abu Kemal on left bank, in tract known as Al Susa.
AL HAMUDI	Hatrush al Shallah	30	About fourteen miles above Abu Kemal on right bank, in tract known as Hasarrat.
AL MAJAWADAH	Salih al Ashban	35	About seven miles downstream from Khan Kalasil on right bank of Euphrates, in tract known as Qariyat al Gattah.
AL MARASIMAH	Abud al Hussain al Shuraidah	15	Opposite Abu Kemal on left bank, in tract known as Shijlah.
	Huwaijah al Abud	15	About four miles downstream from Abu Kemal on left bank, in tract known as Al Susa.
AL BU MIRI	......	65	......
Sections			
AL QADRAU	Niza al Hussain	25	Left bank of Euphrates from about eight to eighteen miles below Khan Kalasil in tract known as Shaafah.
AL ISA	Taraf al Diq	30	......
AL TAUMAH	Fadi al Saiyal	10	On right bank of Euphrates about twelve miles below Khan Kalasil in tract known as Shiyal.
AL MUSHAHIDAH	Budaiwi al Hamid	15	Near Abu Kemal on left bank in tract known as Sukariya.
	Barjas al Abid	15	

[1] Pastoral nomads.
[2] All *charid* owners.

Tribes and Sub-Tribes of Upper Euphrates

Confederation, tribe, or section	Chiefs	Families, tents, or houses	Habitat
AL BU QAAN	Amash al Abid	30	Right bank of Euphrates, near Tell Ramadi, about twelve miles below Khan Kalasil in tract known as Ramadi.
AL BU SARAI	Hamud al Shallash and Faiyadh al Nasir	160	Right bank of Euphrates, Tibni to Deir-ez-Zor.
AL SHAITAT	......	115	......
Sections			
AL JADU	Nadham al Salih	35	Immediately downstream from Khan Kalasil on left bank of Euphrates, in tract known as Gharanij.
AL BU ALIYAT	Zalan al Jasim	40	Immediately upstream from Khan Kalasil on left bank of Euphrates, in tract known as Chischiya.
AL KHANFAR	Raju al Hutaitah	40	About four miles upstream from Khan Kalasil on left bank of Euphrates, in tract known as Al bu Hammam.
AL THULTH	Turki al Nijris	250	Both banks of Euphrates between Meyyadin and Khan Kalasil.
Sections			
AL BU HASSAN	Hamid al Hussain al Nijris	....	......
AL QURAN	Munadir	....	......
AL BU RAHAMAH	Qawan al Jabarah	....	......
AL SHUAIT	Bargash al Muhammad	....	......
BAQQARAH[1] (of the Euphrates)	......	1,200	Left bank of Euphrates from Raqqa to Buseira (at mouth of Khabur); also on both banks of Khabur near junction with Euphrates. *Winter.*—From September to April pastoral element of tribe moved into Jazira.
Sections			
AL ABAIDAT	Ajil al Mahmud	....	......
AL ALI	Sulaiman al Hassan	....	......
AL ASHAHIN	Mahmud al Qahit	....	......
AL BU BADRAN	Salih abu Jarad and Dhahim al Mulla Abid	....	......
AL BU GHANIM	Salih ibn Hassan	....	......
AL BU HAMDAN	Wawi ibn Amtair	....	......
AL KHANJAR	Satam al Muhammad	....	......

[1] Agricultural and pastoral; sheep-breeders.

Confederation, tribe, or section	Chiefs	Families, tents, or houses	Habitat
AL BU MISH	Ataiyit al Hassan	....	......
AL BU MUSA	Dhiyab ibn Bishrah	....	......
AL BU MAISH	Hassan Hamad al Awad	....	......
AL NABBIZAH	Chibbin al Muhammad	....	......
DULAIM[1]	Ali ibn Sulaiman ibn Bakr (Paramount)	19,015 tents	Both banks of Euphrates from Al Qaim to five miles downstream from Al Falluja on left bank, and to Imam Hamza on right bank. Also on Saqlawiya and Aziziya canals. *Winter.*—From September to April pastoral sections migrated to Jazira and Shamiya.
Section			
AL BU ALWAN[2]	Muhanna al Muhammad al Salih and Jasim al Muhammad	510 tents	Right bank of Euphrates from six miles upstream from Ramadi to four miles downstream. Also on right bank four miles upstream from Al Falluja. *Winter.*—Approximately half the section moved to Jazira or Shamiya for winter grazing.
Sub-sections			
Al bu Araf	Muhanna al Muhammad al Salih	....	......
Al bu Ghadir	Faraj al Zauhar	....	......
Al bu Ghurrah	Jasim al Muhammad	....	Right bank four miles upstream from Al Falluja.
Section			
AL BU DHIYAB[3]	Mushhin ibn Hardan	1,700 tents	Left bank of Euphrates from five miles upstream to six miles downstream from Ramadi.
Sub-sections			
Al bu Hamad al Dhiyab	Fadam ibn Muhammad	1,400	......
Al bu Aithah	Fadam ibn Muhammad	....	......
Al bu Ali al Jasim	Mutlaq al Hamzah	....	......
Al bu Hazim	Rushaiyid al Ahmad	....	......
Al Mulahimah	Jasim al Muhammad	....	......
Al Qartan	Abu al Hussain	....	......

[1] Sunni; semi-nomadic; agricultural and pastoral.

[2] Cultivators and sheep-breeders; also own donkeys and act as carriers.

[3] Cultivators, with a few sheep-breeders, who migrate into Jazira during winter.

TRIBES AND SUB-TRIBES OF UPPER EUPHRATES

Confederation, tribe, or section	Chiefs	Families, tents, or houses	Habitat
Al bu Saqr	Chachan ibn Sahu	....	......
Al bu Saudah	Shaham al Hardan	....	......
Al bu Tamah	Sulaim al Hamad	....	......
Al bu Ubaid	Naman al Khalaf	....	......
Al bu Muhammad al Dhiyab	Shaukah ibn Mutlaq	300	......
Al bu Hantush	......	....	......
Al bu Jadan	......	....	......
Al bu Khalifah	......	....	......

Section

AL BU FAHAD[1]	Abdul Muhsin al Farhan	1,700 tents	Right bank of Euphrates from Ramadi to fifteen miles downstream. One small section on left bank downstream from Ramadi.

Sub-sections

Al bu Arab	Qudaiyan al Humaid	....	......
Al bu Faiyadah	Unaizi al Mukhlif	....	......
Al bu Musa	Ali al Nasar	....	......
Al bu Raihan	Faris al Muhammad	....	......
Al bu Salih al Ali	Mutlaq al Darach	....	......
Al bu Taha	Suaiyid al Ali	....	......
Al bu Ujur	Ali al Saad	....	......

Section

AL BU ISA	Harat al Jasim	2,500 tents	Right bank of Euphrates from Al Falluja to sixteen miles downstream. Also cultivated portion of land on Saqlawiya Canal, granted by Jumailah section.

Sub-sections

Al bu Hatim	Aqab al Shuwaidikh	....	......
Al bu Hawa	Ali al Suwait	....	......
Al bu Huraiwat	Farhan al Dhahir	....	......
Al bu Khamis	Dalal al Ali and Fahad al Shahadhah	....	......
Al bu Muhammad al Jasim	Abdas al Ibad	....	......
Al bu Muhanna	Muhammad al Dhahir	....	......
Al bu Quraiti	Matar al Murais	....	......
Al bu Salih	Abd al Khalaf	....	......

Section

JUMAILAH	Abbas al Jassan[2]	1,275 tents	Both banks of Euphrates from Al Falluja to four miles downstream; six miles north of Al Falluja on Saqlawiya Canal.

[1] Cultivators and sheep-breeders.

[2] One of the first to settle on the Saqlawiya Canal.

ANTHROPOLOGY OF IRAQ

Confederation, tribe, or section	Chiefs	Families, tents, or houses	Habitat
Sub-sections			
Al bu Ausaj	Muhaimid al Hajwal	....	......
Al Dukhaiyil	Ali al Khanfar	....	......
Al bu Haddad	Sumair al Fadhil	....	......
Al bu Jasim	Mashkur al Khalaf	....	......
Al bu Muqallad	*Hajji* Jasim	....	......
Al bu Ramlah	Ali al Abbas	....	......
Section			
AL BU KHALIFAH	Khurbit al Jasim	600 tents	Right bank of Euphrates, ten miles downstream from Ramadi to two miles upstream from Dhibban. Also on left bank eighteen miles east of Ramadi.
Sub-sections			
Al bu Ghazail	None	....	......
Al bu Jabar	None	....	......
Al bu Juhaish	Khurbit al Jasim	....	......
Al bu Madlij	Audah al Latif	....	......
Section			
AL BU MUHAMDAH	Habib al Shallal	1,500 tents	Left bank of Euphrates from Al Falluja ten miles upstream. Right bank from Sinn al Dhibban to four miles downstream.
Sub-sections			
Al bu Aql	Dhaif al Salih	....	......
Al bu Akash	Mukhlif al Saiyah	....	......
Al bu Ashshihah	Mahdi al Salih	....	......
Al bu Azzam	Dhiyab al Ahmad	....	......
Al Baqqarah	Abid al Humaiyish	....	......
Al bu Dhiyab	Mulla Hussain al Muhammad and Abdullah al Haif	....	......
Al Falahat	Faiyadh al Jasim	....	......
Al bu Khamis	Sulaiman al Muhammad	....	......
Al Musalihah	Daham al Uwaiyid	....	......
Al bu Quraifa	Hammadi al Faiyadh	....	......
Al Rad	Ali al Ahmad	....	......
Al bu Shahab	Ali al Ibrahim	....	......
Section			
AL BU NIMR [1]	Shaukah ibn Mutlaq	800 tents	Left bank of Euphrates, fifteen miles upstream to seven miles downstream from Hit. Nomadic sections move to country between Ana and Wadi Thahthar for winter grazing.

[1] Chiefly sheep-breeders but also agriculturists.

TRIBES AND SUB-TRIBES OF UPPER EUPHRATES

Confederation, tribe, or section	Chiefs	Families, tents, or houses	Habitat
Sub-sections			
Al bu Farraj	Jadi al Salih	....	......
Al bu Hamad al Hussain	Fahad al Hilal	....	......
Al bu Huntush	Rashid al Salih	....	......
Al bu Hassan	Nijris ibn Qaud	....	......
Al bu Hilal	Rudaini ibn Hilal	....	......
Al bu Jadan	Turki ibn Faris	....	......
Al bu Mani	Hardan al Shindi	....	......
Al bu Mujbil	Farhan al Jadi	....	......
Al bu Samalah	Abid al Fallah	....	......
Al bu Saqr	Shimran al Dhahir	....	......
Al bu Shaban	Farhan al Jadi	....	......
Al bu Sumaidi	Audah al Farhan	....	......
Al bu Tuwaisat	Baddar	....	......
Section			
AL BU RUDAINI[1]	Ali ibn Sulaiman ibn Bakr	3,430	Right bank of Euphrates from Ramadi to Al Qaim; nomadic sections move to Jazira for winter grazing.
Sub-sections			
Al bu Assaf	Ali ibn Sulaiman ibn Bakr, Farhan al Qata, and Huwair al Thamir	600	Right bank of Euphrates twelve miles upstream from Ramadi; small part of this section on left bank. Nomadic.
Al bu Halabsah	Abdullah al Muhammad	420	Detached section settled on Saqlawiya Canal between Al Falluja and Khan Nuqta on north bank.
Al bu Hazim[2]	Shergh ibn Shallaib	100	Beside Aziziya Canal.
Al bu Hussain al Ali[2]	Hamid al Abid	100	Beside Aziziya Canal.
Al bu Kulaib	Radhi al Sulaiman	420 tents	With Hussain al Ali on Aziziya Canal and at Abu Jir.
Al bu Mahal	Aftan al Sharqi	1,270 tents	Right bank of Euphrates five miles upstream to ten miles downstream from Al Qaim.
Al bu Abd	Aftan al Sharqi	....	......
Al bu Taiyib	Hussain al Izbah	....	......
Al bu Tuaimah	Lutaiyif al Fadhil	....	......

[1] Cultivators and sheep-breeders.

[2] Sunni; settled and semi-nomadic; cultivators and sheep-breeders. Sheep in desert south to west of Habbaniya Lake. Market town Ramadi. Followed Ali ibn Sulaiman.

100   ANTHROPOLOGY OF IRAQ

Confederation, tribe, or section	Chiefs	Families, tents, or houses	Habitat
Al bu Miri[1]	Kurdush al Lahaimus	140	Beside Aziziya Canal.
Al bu Matrad (Al bu Jabir)	Sharqi al Shallash	240 tents	Beside Aziziya Canal opposite Marai.
Al bu Fahad[1]	Ibrahim	140	Beside Aziziya Canal opposite Marai.
Section			
AL BU JAGHAIFAH[2]	Haif al Ali	3,500 tents	Left bank of Euphrates between Al Qaim and Ana, and in Jazira.
Sub-sections			
Al bu Ajaj	Battah al Salamah		
Al bu Ali	Ubaid al Hishah		
Al bu Duhail	Ali al Ayash		
Al bu Khalaf	Quran al Sattam		
ZOBA sections[3] which now follow the DULAIM		2,000	
LUHAIB	Juwad al Assaf		On north bank of Saqlawiya Canal, adjoining Halabsah section of Dulaim, seven and one-half miles west of Aqarquf.
SHUWARTAN	Ibrahim al Saba		On north bank of Saqlawiya Canal, west of Tell Ibrahim.
BANI ZAID	Quaiyid al Faiyadh		South bank of Saqlawiya Canal opposite Shuwartan.
KHURUSHIYIN	Shunaitir al Jasim		Between Madhiya and Saqlawiya Canals. Also on north bank of Saqlawiya Canal between Shuwartan and Tell Ibrahim.
QARA-GHUL[4]	Jarrah al Khalaf		On south bank of Saqlawiya Canal northwest of Khan Nuqta.
AL SAADAN	Ursan al Ali		On north bank of Madhiya Canal at junction with Saqlawiya Canal.

[1] Settled and semi-nomadic; cultivators and sheep-breeders. Sheep in the desert south to west of Habbaniya Lake. Market town Ramadi. Followed Ali ibn Sulaiman.

[2] Nomadic; sheep breeders, with little cultivation. They grazed their flocks in Jazira, sometimes moving as far east as the Tigris. Not Dulaimis, but followed Ali ibn Sulaiman. When speaking of the Dulaim collectively they were included.

[3] Sunni; sedentary; cultivators chiefly on the Saqlawiya Canal. Possessed a small number of flocks.

[4] A detached colony from the Qara-Ghul located on left bank of Euphrates six miles downstream from Imam Hamza. Not of Zoba origin, the main Qara-Ghul have always been an independent tribe.

TRIBES AND SUB-TRIBES OF UPPER EUPHRATES

Confederation, tribe, or section	Chiefs	Families, tents, or houses	Habitat
SHITI	Ali al Muslih	....	On south bank of Madhiya Canal at junction with Saqlawiya Canal.
SUBAIHAT	Mulla Munawir al Mulla Faiyadh	....	On south bank of Saqlawiya Canal eight miles east of Al Falluja.
SUMAILAT	Abbas al Hussain	....	Between Madhiya and Qurmah canals, six miles north of Khan Nuqta.
BANI KUBAIS[1]	Farraj ibn Abdullah	400	Kubaisa town.
Sections			
Bait DARIAH	Farraj ibn Abdullah	....	......
AL BU HAIDAH	Muhammad ibn Farraj	....	......
Bait Hajji ISA	Karim ibn *Hajji* Najm	....	......
AL BU HAMAD	Muhammad Ahmad	....	......
Bait MATHLUTHAH	Kasar ibn Ali	....	......
SHADDID and FARRAJ ALLAH	Andah and Lishlash	....	......
AR RAHHALIYA[1]			
(Townsmen)	Chaad	175	Ar Rahhaliya town.
Sections			
BAIQAT[2]	Muhammad al Ataimi	....	......
AL HARUB[3]	Fahad al Abbas	....	......
AL BU SALMAN[4]	Abdul Muhsin as *Sayyid*	....	......

Zoba Sections Which Have Become Independent Tribes

CHITADAH	Dhirb al Sulaiman	420 tents	Between Ridhwaniya Canal and left bank of Euphrates eight to sixteen miles from canal head.
Sections			
AL AZZAH	Jasim al Muhaimid	....	......
AL BARGHUTH	Dhirb al Sulaiman	....	......
AL HUMAID	Addai al Chali	....	......
AL KHAMMAS	Nawwar al Shahwan	....	......
AL QUMZAN	Sharmukh al Thunaiyan	....	......
AL RADHI	Dhaba al Ammar	....	......
AL SUMAIL	Mahbul al Unair	....	......
AL ZUBAR	Najm al Mughamis	....	......

[1] Sunni; sedentary; agriculturists and merchants. Very little cultivation, except for extensive date groves.

[2] Origin Kubaisa.

[3] Origin Hedjaz.

[4] Origin Mosul.

Confederation, tribe, or section	Chiefs	Families, tents, or houses	Habitat
FADDAGHAH	Ursah ibn Dhaidan	300	Left bank of Euphrates at head of Yusufiya Canal.
Sections			
AL DUGHAIYIM	Ali ibn Abdul Muhsin	....	......
AL BU MUFARRAJ	Muhammad ibn Umar	....	......
AL NABIT	No chief	....	......
AL NASSAR	Mutlaq ibn Fajri	....	......
HAIWAT	No chief; Dhari ibn Dhahir was formerly chief	170 tents	On Abu Ghuraib Canal from Khan Nuqta to ten miles southwest.
Sections			
AL ANNAS	Ali al Rajjah	....	......
FALLUJIYIN	Muhammad al Shabib	....	......
FAIYADAH	Muhammad al Madhhur	....	......
AL GURAIBAWIYIN	Nair al Habib	....	......
AL HULAIYIL	Zaidan al Khudaiyir	....	......
AL BU KHALIL	Dalaf al Khalil	....	......
HITAWIYIN	Ali al Muslit	200 tents	Left bank of Euphrates from four to seven miles downstream from Al Falluja, south of Abu Ghuraib Canal.
SHAAR	Sultan ibn Hussain	270 tents	On western edge of Aqarquf, six miles north-northeast of Khan Nuqta.
DULAIM QARTAN[1]	Abdullah ibn Mulla Ahmad Abtan Hassan ibn Salim	420 tents	Left bank of Euphrates; Imam Hamza, to one mile north of Mufraz Post. Also a small section on new Yusufiya Canal.

[1] Sunni; settled cultivators; Dulaim by origin, but eventually became a Zoba section.

APPENDIX A: THE POPULATION OF IRAQ

In order to present the recent population figures these data were obtained from Major C. J. Edmonds, in Baghdad, to whom I am most grateful for generous assistance.

Prior to recording the registered and unregistered population up to the end of November, 1935, it seems desirable to quote excerpts from a review on Sir Ernest Dowson's paper (see Appendix B) by Sir A. T. Wilson, who writes:

"The total population of Iraq in 1930 is given as 2,824,000, a figure which corresponds very closely with the very rough census of 1919, which estimated the population excluding Sulaimaniya at 2,695,000. Sulaimaniya is credited in 1930 with 94,000, so that on this basis the total figure in 1930 is almost exactly the same as for 1919.

"The total area within the frontiers of Iraq today is 453,500 square kilos; that of Iraq in 1920, before great acres of the western state were added to the borders of the infant state, was about 300,000 square kilos. Sir E. Dowson estimates the region of cultivable land within the Rainfall Zone at 41,000 square kilos and that within the Irrigation Zone at 51,000 square kilos, representing 9 per cent and 11 per cent of the total surface of the country respectively. Only one-fifth to one-tenth of these zones is actually cultivated in any given year. The mean density of the highly mobile rural population per square kilo in the cultivated region works out at 19 in the Rainfall and 35 in the Irrigation Zone, a very low proportion in each case."

The tables supplied by Major Edmonds will be found on pages 104 and 105:

STATEMENT OF REGISTERED AND UNREGISTERED POPULATIONS TO THE END OF NOVEMBER, 1935

	Unregistered Townsfolk			Unregistered Tribesmen			Newly Registered Townsfolk and Tribesmen		
	Females	Males	Total	Females	Males	Total	Females	Males	Total
Baghdad	305	270	575	4420	5530	9950	939	1002	1941
Mosul	1365	1765	3130	8775	8721	17496	99	107	206
Erbil	265	255	520	2135	2115	4250	11	23	34
Sulaimaniya	14100	15900	30000	10090	13310	23400	29	25	54
Kirkuk	6530	7200	13730	14626	17205	31831	86	128	214
Diyala	610	390	1000	3805	2375	6180	187	283	470
Kut	550	400	950	2600	2400	5000	22	41	63
Amara	135	130	265	10455	10680	21135			
Basra	450	400	850				201	129	330
Muntafiq	95	100	195	2830	3420	6250	7	13	20
Ad Diwaniya	20	40	60	108277	39626	147903	226	268	494
Hilla	320	230	550	1100	1050	2150	168	198	366
Karbala	328	421	749	658	502	1160	8	15	23
Dulaim	259	210	469	8295	8240	16535	5	6	11
Total	25332	27711	53043	178066	115174	293240	1988	2238	4226

	Females	Males	Total
Unregistered tribesmen according to reports from administrative officials	178066	115174	293240
Unregistered townsfolk according to report of administrative officials	25332	27711	53043
Total	203398	142885	346283

STATEMENT OF REGISTERED AND UNREGISTERED POPULATIONS TO THE END OF NOVEMBER, 1935

	REGISTERED TOWNSFOLK			TRIBESMEN REGISTERED			GRAND TOTAL OF ALL COLUMNS		
	Females	Males	Total	Females	Males	Total	Females	Males	Total
Baghdad	182422	193827	376249	54421	56274	110695	242507	256903	499410
Mosul	163280	164332	327612	49545	55015	104560	223064	229940	453004
Erbil	23159	24521	47680	62921	65266	128187	88491	92180	180671
Sulaimaniya	59163	66740	125903	2352	2495	4847	85734	98470	184204
Kirkuk	84091	82900	166991	5587	5281	10868	110920	112714	223634
Diyala	66756	68908	135674	34475	38101	72576	105843	110057	215900
Kut	16861	17155	34016	48070	50101	98171	68103	70097	138200
Amara	22030	21462	43492	94334	105282	199616	126954	187554	264508
Basra	108362	112533	220895	32087	32150	64237	141100	145212	286312
Muntafiq	16932	15734	32666	97205	95654	192859	117069	114921	231990
Ad Diwaniya	25484	23885	49369	108574	110431	219005	242581	174250	416831
Hilla	32049	31295	63344	70245	75011	145256	103882	107784	211666
Karbala	46011	43874	89885	15203	17270	32473	62208	62082	124290
Dulaim	17989	20150	38139	35511	39171	74682	62059	67777	129836
Total	864599	887316	1751915	710530	747502	1458032	1780515	1779941	3560456

	Females	Males	Total
Tribes registered	710530	747502	1458032
Townsfolk	864599	887316	1751915
Registered by administrative councils	1988	2238	4226
Total unregistered tribesmen and townsfolk	1577117	1637056	3214173
	203398	142885	346283
Grand total	1780515	1779941	3560456

APPENDIX B: LAND TENURE IN IRAQ

BY

Ernest Dowson[1]

About four-fifths of Iraq consists of unproductive or slightly productive desert, steppe, marsh, and hill masses. The productive core is divided broadly into two regions within which cultivation is practised regularly. The northern region is fed by rainfall supplemented by perennial streams rising in the mountains and, to a limited extent, by lift from the rivers. The southern region largely depends upon irrigation, supplied by canals drawn from the river system, following winter rainfall. The former may be appropriately called the "Rainfall Zone," and the latter the "Irrigation Zone." Actually the southward extension of the northern rainfall varies annually, while the date gardens of the Basra *Liwa* constitute a distinct (and of course economically a very important) study. But this does not affect the correctness of the broad picture presented.

Reference to Table I will show that the region of cultivated and cultivable land within the Rainfall Zone covers approximately 41,000 square kilometers, and that within the Irrigation Zone about 51,000

TABLE I.—Approximate Classification of Land Surface (1930)
(*Expressed in sq. km.*)

			Cultivated Region			
	Total Area All Land	Hill Mass	Rainfall Zone		Irrigation Zone	
Liwa			(?) 2% hill mass	Plains	Canal-fed territory*	Machine-fed territory†
Mosul.......	45,800	9,350	190	14,580	......	......
Erbil........	16,600	7,620	150	7,010	......	......
Sulaimaniya..	9,500	6,400	130	2,420	......	......
Kirkuk.......	20,800	......	...	12,020	......	......
Diyala.......	16,200	......	...	710	2,760	90
Baghdad......	22,100	......	...	......	1,710	2,270
Dulaim.......	124,500	......	...	......	920	630
Karbala......	21,200	......	...	......	660	......
Hilla.........	8,100	......	...	......	4,570	330
Kut..........	16,400	......	...	......	4,680	3,860
Ad Diwaniya..	83,000	......	...	......	3,770	2,180
Muntafiq.....	38,700	......	...	......	4,440	270
Amara........	19,700	......	...	......	5,670	1,010
Basra.........	10,900	......	...	......	610†	60
Totals......	453,500	23,370	470	36,740	29,790	10,700

* Territories so classified are at present very incompletely irrigated.
† Tidally watered date gardens.

[1] These notes are quoted from pages 11 et seq. of "An Inquiry into Land Tenure and Related Questions. Proposals for the Initiation of Reform" by Ernest Dowson, K.B.E., formerly Surveyor-General of Egypt, and later successively Under-Secretary of State for Finance, and Financial Adviser to the Egyptian government. This report was printed for the 'Iraq government.

TABLE I.—APPROXIMATE CLASSIFICATION OF LAND SURFACE (1930)—continued
(Expressed in sq. km.)

Liwa	ADDITIONAL POTENTIALLY CULTIVABLE TERRITORY	TOTAL CULTIVATED AND CULTIVABLE REGION	TERRITORY CONTAINING TAPU HOLDINGS	SIZE OF HOLDINGS IN *Misharas* *			
				1–100	101–500	501–1000	1001+
Mosul.......	270	15,040	7,870	.....	...	...	...
Erbil........	.....	7,160	2,420	7,418	728	500	...
Sulaimaniya.	.....	2,550	2,280	.....	...	...	...
Kirkuk......	3,240	15,260	6,280	.....	...	...	...
Diyala......	260	3,820	3,410	4,092	...	546	...
Baghdad.....	890	4,870	1,500	162	220	120	360
Dulaim......	20	1,570	920	2,344	109	121	3
Karbala.....	20	680	620	.....	...	...	...
Hilla........	1,630	6,530	2,380	452	364	98	82
Kut.........	2,170	10,710	2,580	.....	...	...	...
Ad Diwaniya.	5,520	11,470	2,270	8,378	...	155	69
Muntafiq....	370	5,080	6,260	.....	...	...	...
Amara.......	.....	6,680	110	.....	10	5	50
Basra........	110	780	1,190	.....	...	...	...
Totals.....	14,500	92,200	40,090	.....	...	...	...

* Where information was available. Such information must not be presumed to be exhaustive. A *mishara* equals 2,500 square meters or thereabouts.

square kilometers. These figures represent 9 per cent and 11 per cent respectively of the total land surface of the country. The cultivation is preponderantly of an extensive character. Only a fraction of these zones, possibly from a fifth to a tenth, appears to be actually cultivated in any given year. So that land is available for a very great development of agriculture, when other factors are favorable.

The information given in the table regarding the size of holdings is derived from fiscal returns. Although the classification of these holdings by area cannot be expected to be accurate, the figures possibly give some indication of the frequency of larger and smaller holdings in the districts actually concerned. The latter must not be taken to coincide with the cultivated areas of the *Liwas* themselves.

AGRICULTURAL POPULATION

In any general study of the land tenure of a country it is desirable to know the numbers and distribution of the agricultural population. Thus it would greatly assist in planning development in 'Iraq if trustworthy information were available as to the rural population, their main occupations (e.g. cultivation, with type of crops, stock-keeping, fishing, reed-cutting, the numbers of the sedentary, semi-nomadic and truly nomadic population, etc.). The figures need not be closely accurate, but they should necessarily be sufficiently reliable relative approximations to allow dependable deductions to be drawn from them. But although continuous and painstaking efforts were made to satisfy this need by the *Liwa* authorities, it has to be admitted that no basis exists for arriving at figures that can be utilised with any confidence. Table II contains the best estimates that the *Liwa* authorities were able to furnish; and the Census Department was not in a position to give me any better material. I include the Table, because it at least represents the best local opinion of the position under the various heads cited. It should perhaps be noted more particularly that it was found impossible to separate in any systematic way those engaged in urban, from

TABLE II.—APPROXIMATE POPULATION (1930)
(*Expressed in thousands*)

Liwa	TOTAL ESTIMATED POPULATION	THREE PRINCIPAL TOWNS	NOMADIC TRIBAL SECTIONS	RURAL POPULATION Settled	RURAL POPULATION Tribal	RURAL POPULATION Total	MEAN DENSITY*
Mosul	320	79	45	176	20	196	22
Erbil	106	...	3	47	56	103	15
Sulaimaniya	94	...	15	51	28	79	37
Kirkuk	160	...	19	63	78	141	13
Diyala	240	...	1	79	160	239	67
Baghdad	388	219	2	93	74	167	98
Dulaim	147	...	59	39	49	88	95
Karbala	90	...	2	83	5	88	136
Hilla	103	...	...	30	73	103	21
Kut	170	...	...	60	110	170	20
Ad Diwaniya	238	...	58	79	101	180	40
Muntafiq	340	...	20	25	295	320	72
Amara	238	...	...	36	202	238	36
Basra	190	46	10	34	100	134	284
Totals	2,824	344	234	895	1,351	2,246	36†

* Mean density per sq. km. cultivated region.
† Mean density total 78,000 sq. km. cultivated region.

those engaged in rural pursuits. Finally, only the populations of Baghdad, Basra, and Mosul were excluded from the latter. This explains the misleading density figure for the Karbala *Liwa*. It should also be noted that the figures given for the agricultural population include all those engaged in rural occupations, other than genuine nomads.

So far as the figures can be accepted the mean density of the rural population per square kilometer of the cultivated region works out at about nineteen in the four *Liwas* of the Rainfall Zone, and at about thirty-five in the five most typical *Liwas* of the Irrigation Zone. These figures are very low, especially for the potentially fertile irrigable lands of the latter; but there is no reason to think that they err on this side. However, it will be appreciated that dependable statistics of the agricultural population are needed to enable development of the country to be pursued to the greatest advantage and with the greatest economy.

The extreme mobility of the majority of the population, especially throughout the Irrigation Zone, is an important factor in the consideration of development schemes of all sorts, having regard to the general sparsity of the population and the limited resources of the country. If work can be concentrated on such schemes in a few of the most suitable areas, and facilities can be given to the population to colonize them as opportunity and occasion justify, much more rapid and satisfactory results will be obtained for the same effort and expenditure in a given time than if attempts have to be made to carry out development in a scattered and incomplete manner throughout the whole country at once.

APPENDIX C: GENERAL HEALTH OF THE KISH ARABS[1]

In order to obtain data on the health of the Arabs of the Kish area, each individual who was studied anthropometrically was questioned, particularly as to whether he was susceptible or immune to attacks of fever. In many instances the individual was afraid to admit to sickness since this might reflect against his being taken as a workman on the excavations. An Arab, too, is inordinately proud of his strength and endurance and ashamed of sickness and its resultant weakness.

The will of Allah accounts for sickness or health, sorrow or happiness, poverty or wealth. Consequently, the Arab believes there is little use in working for or against the divine will when to follow the latter course must only be to court disaster and final disappointment.

Thus, the Arab suffers from a particularly virulent form of malaria because he makes little or no effort to eliminate the many pools of stagnant water that lie, especially in the winter, within a few miles of the villages.

Paroxysms of chill and violent shivering followed by a rapidly rising temperature and pulse count are symptomatic of the fever. The body is soon bathed in a copious sweat and the patient begins to feel more comfortable. Headache and nausea are frequently felt. Eyes become tired, often bloodshot, and the patient feels depressed.

Few Arabs die of malaria, but the general lassitude and debility caused by the disease lower their resistance against fevers of a more malignant nature, which often prove fatal.

The only remedy used against malaria is quinine (local Arabic *kanaqina*), which can be purchased in the *suq* at Hilla. Since the Arabs, however, do not believe in European prophylactic measures, they use this only as a cure.

During the winter season of 1927–28 Mr. Eric Schroeder[2] dispensed medicine every evening before sunset. Doses of quinine were much in demand and it was observed that the patients pre-

---

[1] These notes were based on data obtained while the writer was attached as physical anthropologist to the Field Museum–Oxford University Joint Expedition to Kish during 1927–28 and were written during the latter part of 1928. (See also Field, 1935a.)

[2] Now Curator of Near Eastern Art, Fogg Art Museum, Cambridge, Massachusetts.

ferred prescriptions in the form of pills to liquid medicines. From fifteen to twenty grains of quinine lowered the temperature within a short time and reduced the attacks of shivering, although in some cases smaller doses at regular intervals over a period of twenty-four hours were required.[1]

The attack of fever generally lasted from three to five days, and because of the high temperature and nausea, left a general weakness, particularly in the lower limbs.

The fever statistics obtained among the Arabs are appended herewith, although the figures should not be taken as representative of the entire group living in the Kish area today.

Fever	No.
No attacks	11
Attacks for three to five days	10
Occasional attacks	15
Frequent attacks	5
Yearly attacks	6
Attack for one month (1927)	1
Attack for four months (1925)	1
Attack for one year (1926)	1
Attack for two years continuously	1

The majority of the eleven individuals who reported that they were not subject to attacks of fever admitted that they had had occasional attacks during their childhood and youth. It may be that such individuals develop a localized partial immunity to the malarial parasite. One man, not listed among these eleven, claimed that he was fevered frequently before marriage but not afterwards.

When I was at Jemdet Nasr in March, 1928, I had an attack of giddiness accompanied by partial blindness, racking pains, and shivering fits. Twenty grains of quinine and a complete rest ended this attack within twenty-four hours. I was confident that my illness had been due primarily to bad water and resolved to investigate the matter.

Jemdet Nasr, eighteen miles northeast of Kish, lies about midway between the Tigris and Euphrates rivers. The irrigation canals do not come within ten miles of Jemdet Nasr, but following the spring rains a large neighboring catchment basin (*kessereh*) partially fills with water. Because of this supply of water it was possible for Mr. Louis Charles Watelin[2] to conduct the excavations at Jemdet

---

[1] As a prophylaxis against malaria, travelers in Iraq should take five grains of quinine every third day before sunrise or after sunset, but never during the heat of the day.

[2] Director of the Field Museum–Oxford University Joint Expedition to Kish, Iraq, from 1929 to 1933. Mr. Watelin died in July, 1934.

Nasr. During the heat of the day many of the workmen drank nearly two gallons apiece of this water, which was brought in tanks by automobile truck from the *kessereh* to the camp and to the excavations. In the early morning our own water jars were filled and upon inquiry I found that the Arab truck driver and native assistants had filled the tanks from the same part of the *kessereh* in which the Beduin women were washing their feet and their clothes. In spite of the fact that all drinking water was boiled, this undoubtedly accounted for my sickness. Arrangements were made for obtaining water from a different part of the *kessereh;* consequently, there was no more illness.

Smallpox.—It is interesting to note that the first accurate and reliable account of smallpox was given by Rhazes, an Arabian physician, who lived in the ninth century.

Among the individuals recorded there were nineteen persons who had suffered from smallpox (*jidri*). Ten persons admitted being affected during childhood; the remainder suffered the disease during adult life. One man recalled having had an attack of smallpox at the age of ten. The scars or pockmarks were always visible on the face and could readily be distinguished from any other local disease. One individual (No. 197) had pockmarks on the inside of his right forearm.

Apart from the virulence of the disease and its attendant high rate of mortality, the principal effect of confluent smallpox on the face is inflammation or ulceration of the eyes, often resulting in partial or total blindness. Vaccination was unknown in the Kish area. Fortunately, however, the malady affected only a small proportion of the population.

Eyes.—Because there was no qualified medical service available within these little camps, it was not surprising that contagious diseases, such as trachoma and granular conjunctivitis, were passed from person to person.

Although many suffered from various diseases of the eyes, the eyesight was relatively good. The prevalence of blowing sand and the almost entire absence of washing, combined with the natural glare of the sun sharply reflected from the light-colored alluvial plain, tended to cause inflammation of the eyes and occasional cases of follicular conjunctivitis.

During the summer, as in Egypt, one could see small children with sore and inflamed eyes surrounded by numerous flies, which they did not trouble to drive from their faces. It seems certain that much of the eye trouble was thus derived from some slight infection during childhood which was left unattended and never disinfected.

The available statistics from my observations include the following data:

Eyes	No.
Both eyes blind	3*
Right eye blind	5
Left eye blind	2
Cataracts in either eye	8
Right eye bad	2
Left eye bad	2
Both eyes bad	3
Slightly cross-eyed	2

* One man aged 70.

Headaches were the common complaint among men, women, and children. When unaccompanied by fever these were caused primarily by the intense glare of the sun, which undoubtedly affects the eyesight.

I had no opportunity to study the women, but their frequent complaints of headache and pains in the eyes were an indication that the various diseases of the eyes were also prevalent among them.

Ears.—Only one individual, Hamoiser el-Abid (No. 50) was observed to have an infected left ear which might have developed into a mastoid infection.

Teeth.—According to Frazer (vol. 9, p. 181), among the heathen Arabs, when a boy's tooth fell out, he used to take it between his finger and thumb and throw it towards the sun, saying, "Give me a better for it." After that his teeth were sure to grow straight and close and strong. "The sun," says Tharafah, "gave the lad from his own nursery-ground a tooth like a hailstone, white and polished." Thus the reason for throwing old teeth towards the sun would seem to have been a notion that the sun sends hail, from which it naturally follows that it can send a man a tooth as smooth and white and hard as a hailstone.

Two individuals (Nos. 10 and 40) had remarkably good teeth which were not only strong but also clean and in perfect condition. Very bad teeth were noted in five Arabs, including Hashim Hradhun (No. 65), who was only nineteen years of age. He stated that his father had extremely poor teeth and much dental decay. Hassan el-Murjan (No. 130) also had an extreme case of dental decay which had caused absorption of the gums. This was to be expected, however, since he was about seventy years of age.

One individual (No. 135) had poor teeth, which had grown at every conceivable angle in both jaws. A few persons had large and prominent front incisor teeth, but this condition was rare. The only broken tooth recorded was that of No. 209, whose upper right first incisor was broken off in the middle and gave the owner considerable dull throbbing pain. Since Hassan el-Abud (No. 1) smoked many native cigarettes daily, his teeth were badly stained with a brown film.

TEETH

Wear	No.	Loss	No.
Normal	168	None	165
Slight	63	One	46
Plus	33	Two	35
Double plus	19	Three	7
Triple plus	2	Four	6
		Five	5
		Six	3
Caries	No.	Ten	1
None	189	Ten or more	9
Slight	44	Sixteen	1
Double plus	29	Thirty-one	1
Triple plus	21	Thirty-two	1
		Lower jaw	76
		Upper jaw	41
Bite	No.		
Edge-to-edge	16		
Slight over	238		
Marked over	100		
Under	4		

Since there were many objections to opening the mouth and holding it thus for a time sufficient to obtain accurate numerical results, the number of teeth indicated as lost must be taken as far from correct. The other figures, however, are useful in determining the general dental condition of the people.

Skin Infections and Scarring.—The "Baghdad boil"[1] begins like a pimple and quickly increases in size until it forms a hard red lump in the skin. The discharge continues for several months and the boil generally leaves a large, ugly scar. Three individuals were observed with these scars, two on the left cheeks and the other on the left upper lip near the nasal orifice.

No. 284 had a scar behind the right ear which was not, however, the result of inflammation of the mastoid process. Of the three other individuals who had facial scars, No. 43 had a mole scar on the left side of the nose, No. 60 a scar on glabella, and No. 170 had a small circular scar just above and on the right side of glabella.

[1] Cf. Schlimmer (pp. 81–91), and Field, 1939a, p. 693.

Deformation.—The heads of the Arab children are in no way bound or tied down so as to produce artificial cranial deformation, even from an involuntary cause.

Sayyid Abid el-Hassan (No. 398) had a serious cut on the upper lip which made a slight deformation.

One individual's left ear was slightly punctured near the lobe, and the right ear of Alway-an-Nuar (No. 125), was peculiarly flexed in the apical region.

Two subjects (Nos. 230 and 286) had deformed right hands, and in the former case, that of a middle-aged man, the lower arm was also affected causing the flesh on the upper arm surrounding the right humerus to become pinched and withered.

Respiratory Diseases.—Throat diseases were rarely observed, but Hadawi il-Mehenna (No. 443) was always hoarse and often complained of a sore throat.

There was a remarkable, although by no means total, absence of influenza and any inflammatory affection of the nasal mucous membranes among these people. This, again, was probably due to a local immunity caused by an adaptation to the environmental changes of climate throughout a succession of generations living under more or less similar conditions.

Tuberculosis.—According to a Health Officer stationed in Baghdad, tuberculosis was a prevalent disease. When I visited Kish in June, 1928, Juad, brother of Sheikh Atiyeh, and one of the armed sentries in camp, begged me on his knees to save his life with European medicine, but he was beyond the power of medical aid. Such also was the case of one of the servants, Majid, aged twenty-two, who had a continuous racking cough.

In several of the village encampments, men with hollow chests and deeply sunken eyes would beg for medicine to cure their coughs and pains. I suspected that many of these were tubercular.

Ventral Disorders.—Owing to the restricted diet of dates (*tamr*) and unleavened cakes (*chupattis*), and the quantity of tea (*chai*) and coffee (*kahwa*) imbibed, ventral disorders were common. One individual (No. 16) admitted that the drinking of coffee caused nausea, and that the blowing of the east wind[1] brought a similar complaint. Another man said that he had had fever and vomiting attacks during the cholera outbreak in 1927.

[1] Cf. Field, 1939a, p. 566.

During the burning heat of the summer, diarrhea was prevalent among infants and accounted for the high infant death rate in the village encampments.

Jaundice was never observed among the Arabs, but one individual (No. 26) said that several years ago his skin "turned yellow in color" and that he was incapacitated for several weeks. Another individual stated that his left wrist was branded as a cure for this disease.

While stones in the bladder were said to be common affections, I never heard of a single case of appendicitis. The operation for the removal of the appendix was totally unknown.

Venereal Diseases.—In a discussion of the probable relationship between syphilis, bejel and yaws, Hudson (1939, pp. 1840-45) states that "my statistics, covering thousands of cases, show at least 60 per cent of those who reach adult life have passed through this stage [bejel, an eruption in the mouth or on the body, lasting about one year] in childhood and are therefore syphilitic."

According to Harrison (p. 318) there appears to be a localized immunity[1] to the disease, as tertiary syphilis, including locomotor ataxia and paresis, is extremely uncommon, despite the prevalence of primary and secondary manifestations of the disease.

We found only one apparent case of syphilis in Iraq. It was at the end of our trip to the Tigris River (see Field, 1935a, map, p. 84), during the latter part of June, 1928. Soon after dawn one morning I set out in a seven-passenger touring car with Mr. Showket as photographer and interpreter, a mechanic, and five men equipped with shovels, ropes, wire-netting, and food and water for several days.

In order to cross the irrigation canals we followed the Jemdet Nasr track and at the north end of Tell Barguthiat we turned in a northeasterly direction and continued toward the Tigris River.

There was no track or route of any kind but after driving for several hours over the hard, rough, alluvial plain we saw the black tents of Sheikh *Hajji* Hunta's encampment, which stood near the right bank of the Tigris. The Sheikh, a venerable old man, who passed many hours in prayer, received us warmly and bade us remain as his guests until the following morning.

---

[1] Among the Al bu Muhammad Arabs living in the Hor al Hawiza to the east of Amara, we found many individuals with either bejel or syphilis, although no cases of the advanced stages of the disease were observed. Two Government doctors were engaged in treating about two hundred cases daily. The most successful treatment was with intravenous injections of bismuth.

After preliminary arrangements I began to measure and photograph the men sitting around the Sheikh's tent and to reward them with Arab cigarettes. The large feast at noon, combined with a shade temperature of 118° F., delayed my anthropometric work for three hours.

In the evening they brought me the sick and suffering of all ages, and I prescribed for each out of my medicine chests. There were many complaints of aches in the head, eyes, and stomach, and I observed several cases showing the symptoms of rheumatoid arthritis. In these individuals, all past middle age, the knee joints were affected and the small joints of the fingers were stiff and altered in shape. Finally, one man, who gave his approximate age as sixty-five, came into the tent and begged on his knees for medical attention. He unwrapped his headcloth and bent his head down toward me. Near the bregma there was a large gummy tumor, unprotected from the filthy head-dress to which it adhered. The iris was greatly inflamed and the patient complained of partial blindness. He appeared to have an advanced case of syphilis. The risk of spreading the infection among the members of the tribe was considerable, if not certain. Yet the Sheikh refused to permit a doctor to visit his camp. Instead, he turned to me and said that he would send the diseased old man to water the camels at a desert well, and that he would not be allowed to return.

Such treatment is not dictated generally, for the Arabs know the value of mercury, if only for secondary lesions; primary and tertiary stages of syphilis are not recognized as the same disease (cf. Harrison, p. 310). Mercury is inhaled through tobacco smoke. Although it produces horrible salivation, it seems to clear up secondary lesions quite effectively. If this medicated tobacco is shaken in water it will yield a considerable amount of finely divided mercury.

Other Types of Native Treatment.—According to Harrison (p. 309) the Arabs use branding (cf. *kawi* or *chawi*) to treat all kinds of complaints. The principle is counter-irritation, and the practice is often beneficial. In pleurisy the application of a hot iron acts as a powerful and, from a medical viewpoint, valuable counter-irritant.

For purposes of hemostasis the Arabs have learned to make incisions with a red-hot knife. Since amputation of the hand was the customary punishment for theft, it was the most common major surgical operation. The stump was dipped in boiling oil to check the hemorrhage, as was the general practice in the Middle Ages in Europe.

Blood-letting was practiced by a few individuals. No. 647 had scars on his right cheek where his mother had tried to relieve head pains by gashing his cheek with a razor. No. 32 had a round scar on the left side of the chin and lined scars on the right temple, from which blood had been taken to eradicate and cure frequent headaches.

The study of anatomy was unknown[1] among the Arabs, and human dissection was regarded with horror. But the treatment of fractures was important because they were often the result of gunshot, which affected the soft tissues surrounding the wounds. No effort was made to reduce a fracture but an excellent substitute for splints was applied in the following manner (cf. Harrison, p. 311). The patient was laid on the sand, and small stakes were driven into the ground along the sides of the fractured bone, which was held in place by means of cords. A tent was erected over the injured person to protect him from the intense rays of the sun. The patient remained in this position for several months until the natural processes of healing had knitted the broken bone. Since the fractures were not reduced the positions of the joined bone fragments were often remarkable, but after a period of complete immobility the great majority of fractures were united.

Remedies.—There were many quack remedies for sale in every small market and wandering dealers passed through each town, village, and near-by encampment armed with miraculous powders, draughts, and charms against all forms of sickness. These medicines often contained simple and innocent constituents purchased in the bazaar or market a few hours earlier. The women also believed in the curative properties of various herbs which were prepared and administered by them to the various members of their households (cf. Hooper and Field).

Attitude toward Medical Treatment.—Throughout this entire region a doctor was unknown. As a matter of fact, if a strange doctor were to visit a small encampment he would be prohibited from seeing the sick people because of their extreme superstition. The Arabs preferred to remain in their huts, suffering in silence. The only exception we found was when Mr. Schroeder and I were asked to visit the village of Sheikh *Hajji* Miniehil in an attempt to save the life of the newly born son of one of the workmen. We rode on horses to the village. Equipped with our medicine chest we entered

---

[1] This statement refers to Arabs of central Iraq. On the other hand the excellent medical work of the graduates of the Royal College of Medicine in Baghdad and other medical centers in Iraq has now (in 1939) begun to change the picture.

a tiny mud hut with a low entrance. It was filled with smoke from an oil lamp, and there were about twenty people crowding around the mother and baby, who were on the floor. We ordered everyone out of the hut and attempted to take the baby's temperature under the arm. After some time we managed to quiet the mother and examine the baby, who was feverish and evidently in considerable pain. We prescribed a quarter of a cascara sagrada tablet morning and evening, and left, saying that Allah is omniscient and omnipotent. In this way we removed from ourselves all responsibility for the baby's death, which seemed almost certain. However, the baby lived, and as a result our medical fame went abroad far and wide.

Constitution.—There were few obese Arabs, although some corpulent persons were always to be seen in the Hilla bazaar. The usual thinness was due primarily to the struggle for existence and to the lack of fattening foods during childhood and adolescence. Mitteb (No. 13) was a small, frail-looking young man who wore a string around his wrist so that he could measure if he were growing thinner or fatter. He was subject to frequent attacks of fever and headaches and the resultant debility. Abid-en-Nasser (No. 192) was disproportionately large, and his general overgrowth, particularly in his hands and feet, suggested an unbalanced metabolism, possibly due to the abnormal functioning of his endocrine glands. Since there was considerable enlargement and overdevelopment of the hands and feet as well as a pronounced extension of the supraorbital crest, this case suggested acromegaly.

I do not believe that the Arabs were in general as sensitive to pain as Europeans. This may have been due to their inherent belief in the power, might, and wisdom of Allah, and their innate stoicism. One unique characteristic was displayed by their cruelty to wild life but disproportionate fondness for their own domesticated animals.

The men and women had tremendous physical endurance, which was largely due to the hard struggles for existence from early childhood, which the weaker do not survive. They were remarkably good walkers and runners but they had little strength in their arms and legs for lifting or pushing weights.

The women, who were tireless workers, aged rapidly, so that they appeared worn out and wrinkled soon after they reached twenty years of age.

The men were naturally lazy, and, judging from general opinion, appeared to be more subject to attacks of fever than the women.

The workmen at the excavations were under continuous supervision, which was most necessary. They were incapable of working at high speed for more than a few minutes, but when allowed to work at their own speed they could continue to excavate daily for eight and a half hours (cf. Field, 1929b).

Development of Public Welfare.—Local health authorities are making every effort to guard against the spread of virulent diseases brought about through pilgrimages and inadequate medical care.

Formerly, cholera often spread from India to Europe, carried by individuals among the vast throngs of pilgrims who visited the sacred shrines of Iraq, Syria, and Mecca. A Mohammedan who died on the road to or from a pilgrimage became a martyr to his faith. Thus, individuals were inspired to continue the pilgrimage in the face of sickness, even to death. Usually the pilgrims were poor Mohammedans, who carried no luggage except money in a small bag or leathern wallet. The conditions under which they were forced to travel were by no means conducive to cleanliness, and since they were united in the common desire to worship in Mecca, they would befriend each other on any pretext. Thus, the danger of the introduction into Europe of diseases such as cholera, plague, and smallpox was an ever-present one, since among the many thousands of pilgrims who visited these shrines each year, there were many individuals who carried the diseases, and who came in contact with travelers en route to European ports (cf. Clemow).

The danger of the spread of disease increased when pilgrims from India and Persia (Iran) began to travel by the thousands every year through Baghdad and Damascus to Haifa and by sea to Jidda, the port of Mecca on the Red Sea. The sea route, which had been in vogue for centuries, became almost entirely superseded by the trans-desert automobile services.

At present the Iraq Medical Health Officers at Baghdad and Ramadi inspect all passengers and detain any suspected cases of contagious diseases. From March to October, in the year 1892, Asiatic cholera spread from India all through Europe to the United States, leaving in its wake a trail of victims. During the summer of 1928 there was an outbreak of plague in India and Persia, and it was necessary for every traveler to be inoculated against the *Bacillus pestis* before entering or leaving Iraq. Each passport carried an Iraq Health Service quarantine pass, giving the name of the person and stating that he or she "proceeding out of Iraq is found on examination to be free from infectious disease. Quarantine measures taken: Inoculated against plague." The date was appended to each form.

During the past decade the Director General of Health and the faculty and graduates of the Royal College of Medicine in Baghdad have entirely reorganized medical care and prevention of disease throughout Iraq; the hospitals in Baghdad, Mosul, Basra, Kirkuk, and Amara have made rapid strides in the dissemination of medical practice throughout the country. Furthermore, the staff doctors of the Iraq Petroleum Company, not only near the trans-desert pipe-line stations but also at Kirkuk, have taught tens of thousands of their native workmen to appreciate the benefits of medical care and the elements of preventive medicine.

The wisdom of the general health policy of the government of Iraq will be reflected in the better health of their future settled and nomadic citizens.

APPENDIX D: ANTHROPOMETRIC DATA FROM ROYAL HOSPITAL, BAGHDAD

BY

DR. B. H. RASSAM[1]

INTRODUCTION

The raw data were recorded on 497 individuals during the period beginning February 3 and ending June 30, 1932.

The following information and measurements were recorded on each individual: name, age, sex, nationality, religion, tribe, town, head length, and head breadth.

All individuals nineteen years of age and under have been grouped as children.

In preparing these data for publication, the figures have been reclassified so that twenty groups result. The cephalic indices and the statistical summaries were calculated at Harvard by Dr. Carl C. Seltzer and Miss Elizabeth Reniff.

INDIVIDUALS MEASURED BY DR. B. H. RASSAM (497)

No.	Localities
148	Arabs from Baghdad (pp. 123–124).
39	Arabs from Ad Diwaniya (1), Al Mahmudiya (2), Amara (3), Basra (5), Diala (1), Ezza (1), Hilla (2), Karbala (1), Karrada (1), Khanaqin (1), Kut (1), Mendali (3), Mosul (10), Ramadi (1), Rawa (1), Samarra (2), Shafii (1), Shahraban (1), and Tikrit (1) (pp. 124–125).
47	Arab females from Baghdad (p. 125).
18	Arab females from Kut al Hai (1), Hilla (3), Mosul (10), Samarra (1), Shergat (2), and Tikrit (1) (pp. 125–126).
4	Arab children from Baghdad (p. 126).
8	Arabs of Sheikh Saad (1), Beni Saad (5), and Dulaim tribes (2) (p. 126).
33	Arab children of Al Mahmudiya (1), Beni Saad (30), Chefil (1), and An Najaf tribes (1) (p. 126).
7	Beduins from Mosul *Liwa* (p. 127).
49	Kurds from Erbil (4), Kirkuk (26), Khanaqin (2), Mosul (5), and Sulaimaniya (12) (p. 127).
4	Kurd females from Kirkuk (3) and Erbil (1) (p. 127).
20	Christians from Baghdad (p. 128).
39	Christians from Mosul (36) and Tell Kaif (3) (p. 128).
5	Christian females from Mosul (p. 129).
6	Christian females from Baghdad (p. 129).
19	Jews from Baghdad (17), Erbil (1), and Kirkuk (1) (p. 129).
7	Jewesses from Baghdad (p. 129).

[1] Graduate of the Royal College of Medicine, Baghdad, and member of the Medical Staff of the Royal Hospital, Baghdad. My deep gratitude to Dr. Rassam for placing the original records at my disposal must be recorded (H.F.).

No.		Localities
33	....	Kurds from Tehran (13), Irani Tabriz (13), Waly (1), Pestako (3), Hussain Kuli Khan (1), Ali Sharwan (1), and Kermanshah (1) (p. 130).
3	....	Irani Kurd females from Tabriz (p. 130).
4	....	Irani Christians from Urmia (3) and Tabriz (1) (p. 130).
4	....	Turks from Van (2) and Istanbul (1), and one Christian female from an unidentified locality (p. 130).

148 Arabs (Baghdad)

No.	Age	G.O.L.	G.B.	C.I.	No.	Age	G.O.L.	G.B.	C.I.
3790	20	185	140	75.7	3841	32	185	150	81.1
3791	20	183	143	78.4	3842	32	178	145	81.5
3792	20	185	155	86.1	3843	32	175	140	80.0
3793	20	175	145	82.9	3844	32	180	140	77.8
3794	20	183	148	80.9	3845	33	170	140	82.4
3795	21	183	148	80.9	3846	35	183	145	79.2
3796	22	185	140	75.7	3847	35	180	135	75.0
3797	22	180	145	80.6	3848	35	185	150	81.1
3798	22	175	150	85.7	3849	35	183	150	82.0
3799	22	180	145	80.6	3850	35	180	143	79.4
3800	23	180	145	80.6	3851	35	185	150	81.1
3801	24	185	160	86.5	3852	35	190	150	78.9
3802	24	183	160	87.4	3853	35	180	145	80.6
3803	24	175	145	82.9	3854	35	188	143	76.1
3804	25	180	135	75.0	3855	35	180	145	80.6
3805	25	170	130	76.5	3856	35	188	145	77.1
3806	25	185	135	73.0	3857	35	180	140	77.8
3807	25	178	150	84.3	3858	35	168	145	86.3
3808	25	180	150	83.3	3859	35	180	140	77.8
3809	25	180	150	83.3	3860	35	175	140	80.0
3810	26	190	148	77.9	3861	35	175	140	80.0
3811	26	180	145	80.6	3862	35	185	140	75.7
3812	26	175	145	82.9	3863	35	190	145	76.3
3813	26	180	150	83.3	3864	35	180	150	83.3
3814	26	175	140	80.0	3865	35	178	140	78.7
3815	27	183	158	86.3	3866	35	180	148	82.2
3816	27	180	145	80.6	3867	35	178	140	78.7
3817	27	180	140	77.8	3868	35	190	145	76.3
3818	27	180	145	80.6	3869	35	168	140	83.3
3819	28	180	145	80.6	3870	35	180	143	79.4
3820	28	183	153	83.6	3871	36	190	140	73.7
3821	28	180	143	79.4	3872	36	175	145	82.9
3822	28	180	140	77.8	3873	36	175	140	80.0
3823	28	180	143	79.4	3874	36	175	145	82.9
3824	28	180	148	82.2	3875	36	170	140	82.4
3825	29	183	145	79.2	3876	38	190	150	78.9
3826	30	183	140	76.5	3877	38	175	140	80.0
3827	30	175	140	80.0	3878	38	175	143	81.7
3828	30	183	145	79.2	3879	38	180	145	80.6
3829	30	185	160	86.5	3880	39	180	143	79.4
3830	30	180	150	83.3	3881	40	173	135	78.0
3831	30	190	150	78.9	3882	40	185	143	77.3
3832	30	185	143	77.3	3883	40	180	150	83.3
3833	30	185	150	81.1	3884	40	190	145	76.3
3834	30	178	145	81.5	3885	40	170	145	85.3
3835	30	170	143	84.1	3886	40	183	140	76.5
3836	30	185	140	75.7	3887	40	190	150	78.9
3837	32	185	145	78.4	3888	40	178	143	80.3
3838	32	173	145	83.8	3889	40	178	140	78.7
3839	32	180	143	79.4	3890	40	185	145	78.4
3840	32	185	148	80.0	3891	40	180	140	77.8

148 Arabs (Baghdad)—continued

No.	Age	G.O.L.	G.B.	C.I.	No.	Age	G.O.L.	G.B.	C.I.
3892	40	183	140	76.5	3915	45	178	140	78.7
3893	40	180	145	80.6	3916	45	178	140	78.7
3894	40	180	145	80.6	3917	45	185	140	75.7
3895	40	180	143	79.4	3918	46	180	143	79.4
3896	40	180	143	79.4	3919	46	170	140	82.4
3897	42	183	150	82.0	3920	48	185	145	78.4
3898	42	188	150	79.8	3921	50	190	130	68.4
3899	42	180	140	77.8	3922	50	185	140	75.7
3900	42	185	145	78.4	3923	50	180	140	77.8
3901	42	173	140	80.9	3924	50	175	140	80.0
3902	42	178	148	83.1	3925	50	180	145	80.6
3903	42	180	150	83.3	3926	50	190	145	76.3
3904	45	180	145	80.6	3927	50	175	140	80.0
3905	45	180	145	80.6	3928	50	180	143	79.4
3906	45	188	150	79.8	3929	55	180	150	83.3
3907	45	178	145	81.5	3930	55	180	148	82.2
3908	45	178	148	83.1	3931	56	178	143	80.3
3909	45	180	140	77.8	3932	58	180	148	82.2
3910	45	183	145	79.2	3933	60	183	145	79.2
3911	45	180	145	80.6	3934	60	175	140	80.0
3912	45	190	148	77.9	3935	63	170	140	82.4
3913	45	180	143	79.4	3936	65	180	140	77.8
3914	45	178	150	84.3	3937	70	175	140	80.0

Measurements and Indices of 148 Arabs (Baghdad)

Measurements	No.	Range	Mean	S.D.	C.V.
Age..............	148	20–74	37.65 ± 0.57	10.20 ± 0.40	27.09 ± 1.06
Head length......	148	167–196	180.42 ± 0.28	5.13 ± 0.20	2.84 ± 0.11
Head breadth.....	148	129–161	143.83 ± 0.29	5.31 ± 0.21	3.69 ± 0.14
Indices					
Cephalic.........	148	68–88	79.71 ± 0.18	3.24 ± 0.13	4.06 ± 0.16

Thirty-Nine Arabs (Nineteen Towns)

No.	Town	Age	G.O.L.	G.B.	C.I.	No.	Town	Age	G.O.L.	G.B.	C.I.
3992	Ad Diwaniya	32	193	140	72.5	4010	Kut	40	190	150	78.9
3993	Al Mahmudiya	30	183	133	72.7	4011	Mendali	25	180	148	82.2
						4012	Mendali	28	180	145	80.6
3994	Al Mahmudiya	27	190	140	73.7	4013	Mendali	45	180	143	79.4
						4014	Mosul	34	180	140	77.8
3995	Amara	26	175	150	85.7	4015	Mosul	35	183	145	79.2
3996	Amara	29	185	140	75.7	4016	Mosul	40	180	148	82.2
3997	Amara	30	180	140	77.8	4017	Mosul	38	175	140	80.0
3998	Basra	20	168	140	83.3	4018	Mosul	42	190	145	76.3
3999	Basra	22	180	148	82.2	4019	Mosul	45	180	140	77.8
4000	Basra	28	170	143	84.1	4020	Mosul	45	183	145	79.2
4001	Basra	35	188	160	85.1	4021	Mosul	50	185	140	75.7
4002	Basra	48	183	150	82.0	4022	Mosul	52	178	150	84.3
4003	Diala	45	190	135	71.1	4023	Mosul	54	190	148	77.9
4004	Ezza	35	170	130	76.5	4024	Ramadi	30	183	140	76.5
4005	Hilla	30	178	138	77.5	4025	Rawa	24	185	160	86.5
4006	Hilla	45	180	150	83.3	4026	Samarra	30	195	135	69.2
4007	Karbala	32	185	145	78.4	4027	Samarra	35	188	135	71.8
4008	Karrada	30	188	135	71.8	4028	Shafii	30	188	140	74.5
4009	Khanaqin	25	180	145	80.6	4029	Shahraban	22	190	150	78.9
						4030	Tikrit	25*	178	138	77.5

* Age uncertain.

Measurements and Indices of Thirty-nine Arabs (Iraq)

Measurements	No.	Range	Mean	S.D.	C.V.
Age	39	20–59	35.20±0.98	9.05±0.69	25.71±1.96
Head length	39	167–196	182.61±0.66	6.09±0.47	2.79±0.21
Head breadth	39	129–161	143.23±0.73	6.72±0.51	4.69±0.36
Indices					
Cephalic	39	68–88	78.24±0.47	4.32±0.33	5.52±0.42

Forty-seven Arab Females (Baghdad)

No.	Age	G.O.L.	G.B.	C.I.	No.	Age	G.O.L.	G.B.	C.I.
3938	22	170	140	82.4	3962	38	165	145	87.9
3939	22	175	140	80.0	3963	38	175	145	82.9
3940	25	175	140	80.0	3964	38	180	155	86.1
3941	25	170	145	85.3	3965	40	175	143	81.7
3942	25	180	135	75.0	3966	40	180	138	76.7
3943	26	175	150	85.7	3967	40	190	148	77.9
3944	27	168	143	85.1	3968	40	170	140	82.4
3945	28	180	145	80.6	3969	40	170	140	82.4
3946	29	180	145	80.6	3970	40	175	140	80.0
3947	30	175	140	80.0	3971	40	168	140	83.3
3948	30	180	150	83.3	3972	40	173	140	80.9
3949	30	170	148	87.1	3973	40	170	140	82.4
3950	30	175	143	81.7	3974	42	175	145	82.9
3951	30	170	140	82.4	3975	42	170	140	82.4
3952	31	180	140	77.8	3976	45	170	148	87.1
3953	33	185	145	78.4	3977	45	175	145	82.9
3954	35	165	140	84.8	3978	45	165	140	84.8
3955	35	170	143	84.1	3979	46	173	140	80.9
3956	35	170	145	85.3	3980	46	165	140	84.8
3957	35	180	145	80.6	3981	50	170	140	82.4
3958	35	168	140	83.3	3982	20	180	150	83.3
3959	35	178	140	78.7	3983	20	165	145	87.9
3960	35	180	150	83.3	3984	20	180	140	77.7
3961	36	170	140	82.4					

Measurements and Indices of Forty-seven Arab Females (Baghdad)

Measurements	No.	Range	Mean	S.D.	C.V.
Age	47	20–54	35.70±0.77	7.85±0.55	21.99±1.53
Head length	47	164–190	174.00±0.54	5.49±0.38	3.16±0.22
Head breadth	47	135–155	142.57±0.43	4.35±0.30	3.05±0.21
Indices					
Cephalic	47	74–88	82.20±0.28	2.82±0.20	3.43±0.24

Eighteen Arab Females (Six Towns)

No.	Town	Age	G.O.L.	G.B.	C.I.	No.	Town	Age	G.O.L.	G.B.	C.I.
4031	Kut al Hai	30	180	148	82.2	4040	Mosul	42	178	140	78.7
4032	Hilla	25	175	140	80.0	4041	Mosul	42	170	143	84.1
4033	Hilla	30	180	140	77.8	4042	Mosul	45	170	145	85.3
4034	Hilla	30	170	140	82.4	4043	Mosul	48	180	145	80.6
4035	Mosul	30	175	145	82.9	4044	Mosul	55	170	140	82.4
4036	Mosul	32	178	143	80.3	4045	Samarra	40	185	148	80.0
4037	Mosul	35	170	145	85.3	4046	Shergat	35	175	140	80.0
4038	Mosul	36	170	140	82.4	4047	Shergat	50	170	143	84.1
4039	Mosul	40	173	140	80.9	4048	Tikrit	45	170	140	82.4

Measurements and Indices of Eighteen Arab Females (Six Towns)

Measurements	No.	Range	Mean	S.D.	C.V.
Age	18	25–59	39.80±1.25	7.85±0.88	19.72±2.22
Head length	18	170–187	174.66±0.69	4.32±0.49	2.47±0.28
Head breadth	18	138–149	141.82±0.52	3.24±0.36	2.28±0.26
Indices					
Cephalic	18	77–85	81.51±0.29	1.80±0.20	2.21±0.25

Four Arab Children (Baghdad)

No.	Age	G.O.L.	G.B.	C.I.
3985	14	190	145	76.3
3986	14	180	140	77.8
3987	15	170	140	82.3
3988	16	183	148	80.8
Averages	14.8	180.8	143.3	79.3

Eight Tribal Arabs (Iraq)

No.	Tribe	Age	G.O.L.	G.B.	C.I.
4082	Sheikh Saad	36	200	148	74.0
4083	Beni Saad	30	170	140	82.4
4084	Beni Saad	30	190	140	73.7
4085	Beni Saad	40	195	140	71.8
4086	Beni Saad	46	185	140	75.7
4087	Beni Saad	50	185	145	78.4
4088	Dulaim	34	193	145	75.1
4089	Dulaim	60	183	145	79.2
Averages		40.8	187.6	142.9	76.3

Thirty-three Arab Children of Various Tribes (Iraq)

No.	Town	Age	G.O.L.	G.B.	C.I.	No.	Town	Age	G.O.L.	G.B.	C.I.
4049	Al Mahmudiya	12	175	135	77.1	4065	Beni Saad	12–14	175	140	80.0
4050	Beni Saad	6	170	125	73.5	4066	Beni Saad	12–14	180	145	80.6
4051	Beni Saad	6	175	135	77.1	4067	Beni Saad	12–14	170	130	76.5
4052	Beni Saad	6	180	120	66.7	4068	Beni Saad	12–14	180	135	75.0
4053	Beni Saad	7	180	140	77.8	4069	Beni Saad	12–14	180	148	82.2
4054	Beni Saad	8	180	140	77.8	4070	Beni Saad	12–14	178	130	73.0
4055	Beni Saad	8	180	128	71.1	4071	Beni Saad	12–14	170	130	76.5
4056	Beni Saad	10	175	145	82.9	4072	Beni Saad	12–14	185	140	75.7
4057	Beni Saad	10	175	148	84.6	4073	Beni Saad	12–14	175	125	71.4
4058	Beni Saad	12–14	188	145	77.1	4074	Beni Saad	12–14	170	130	76.5
4059	Beni Saad	12–14	175	135	77.1	4075	Beni Saad	14	190	140	73.7
4060	Beni Saad	12–14	188	138	73.4	4076	Chefil	15	190	140	73.7
4061	Beni Saad	12–14	175	130	74.3	4077	Beni Saad	15	180	138	76.7
4062	Beni Saad	12–14	185	140	75.7	4078	Beni Saad	16	188	148	78.7
4063	Beni Saad	12–14	170	140	82.4	4079	Beni Saad	18	190	140	73.7
4064	Beni Saad	12–14	165	130	78.8	4080	Beni Saad	19	190	140	73.7
						4081	An Najaf	18	183	140	76.5

Averages of the above figures would be valueless since the ages range from six to nineteen. Under the town heading, tribal names, such as Beni Saad, have been included. Presumably these Arab children belong to semi-nomadic groups, which can not be classed either as Beduins or town-dwellers.

SEVEN BEDUINS (MOSUL Liwa)

No.	Liwa	Age	G.O.L.	G.B.	C.I.
4090	Mosul	30	188	145	77.1
4091	Mosul	32	183	140	76.5
4092	Mosul	35	185	145	78.4
4093	Mosul	35	185	140	75.7
4094	Mosul	35	180	140	77.8
4095	Mosul	35	180	143	79.4
4096	Mosul	40	175	140	80.0
	Averages..	34.6	182.3	141.9	77.8

FORTY-NINE KURDS (FIVE TOWNS)

No.	Town	Age	G.O.L.	G.B.	C.I.	No.	Town	Age	G.O.L.	G.B.	C.I.
4193	Erbil	32	190	140	73.7	4218	Kirkuk	42	170	150	88.2
4194	Erbil	35	170	150	88.2	4219	Kirkuk	45	175	140	80.0
4195	Erbil	45	180	150	83.3	4220	Kirkuk	55	180	140	77.8
4196	Erbil	50	183	148	80.9	4221	Kirkuk	55	170	153	90.0
4197	Kirkuk	20	180	138	76.7	4222	Kirkuk	60	180	140	77.8
4198	Kirkuk	25	188	153	81.3	4223	Khanaqin	32	183	150	82.0
4199	Kirkuk	26	180	158	87.8	4224	Mosul	35	178	150	84.3
4200	Kirkuk	28	175	140	80.0	4225	Mosul	35	178	140	78.7
4201	Kirkuk	28	188	145	77.1	4226	Mosul	36	173	143	82.7
4202	Kirkuk	30	170	140	82.4	4227	Mosul	40	173	145	83.8
4203	Kirkuk	30	188	143	76.1	4228	Mosul	42	175	143	81.7
4204	Kirkuk	30	180	150	83.3	4229	Sulaimaniya	25	185	150	81.1
4205	Kirkuk	32	178	145	81.5	4230	Sulaimaniya	25	178	148	83.1
4206	Kirkuk	32	185	145	78.4	4231	Sulaimaniya	30	183	150	82.0
4207	Kirkuk	32	175	145	82.9	4232	Sulaimaniya	40	188	153	81.4
4208	Kirkuk	35	185	145	78.4	4233	Sulaimaniya	40	180	150	83.3
4209	Kirkuk	35	180	145	80.5	4234	Sulaimaniya	45	180	155	86.1
4210	Kirkuk	36	185	153	82.7	4235	Sulaimaniya	46	175	155	88.6
4211	Kirkuk	40	185	148	80.0	4236	Sulaimaniya	50	180	153	85.0
4212	Kirkuk	40	180	150	83.3	4237	Sulaimaniya	50	180	155	86.1
4213	Kirkuk	40	185	148	80.0	4238	Sulaimaniya	60	173	145	83.8
4214	Kirkuk	40	163	143	87.7	4239	Sulaimaniya	60	180	140	77.8
4215	Kirkuk	40	180	150	83.3	4240	Sulaimaniya	65	170	145	85.3
4216	Kirkuk	40	185	145	78.4	4241*	Khanaqin	17	173	135	78.0
4217	Kirkuk	40	178	143	80.4						

* No. 4241 (age 17) was omitted from the averages.

MEASUREMENTS AND INDICES OF FORTY-EIGHT KURDS (IRAQ)

Measurements	No.	Range	Mean	S.D.	C.V.
Age................	48	20–69	40.55±1.03	10.60±0.73	25.42±1.75
Head length........	48	164–190	179.43±0.57	5.82±0.40	3.24±0.22
Head breadth.......	48	138–158	146.95±0.53	5.43±0.37	3.70±0.25
Indices					
Cephalic...........	48	74–91	82.05±0.34	3.48±0.24	4.24±0.29

FOUR FEMALE KURDS (KIRKUK AND ERBIL)

No.	Town	Age	G.O.L.	G.B.	C.I.
4242	Kirkuk	25	185	143	77.3
4243	Kirkuk	28	188	145	77.1
4244	Erbil	40	175	150	85.7
4245	Kirkuk	40	183	140	76.5
	Averages..	33.2	182.8	144.5	79.2

Twenty Christians (Baghdad)

No.	Age	G.O.L.	G.B.	C.I.	No.	Age	G.O.L.	G.B.	C.I.
4123	20	190	148	77.9	4133	30	173	140	80.9
4124	20	190	150	78.9	4134	35	188	150	79.8
4125	20	180	150	83.3	4135	35	185	140	75.7
4126	24	180	145	80.6	4136	36	175	140	80.0
4127	26	178	145	81.5	4137	36	180	150	83.3
4128	27	185	130	70.3	4138	40	183	150	82.0
4129	28	178	148	83.1	4139	42	188	148	78.7
4130	30	190	150	78.9	4140	45	180	140	77.8
4131	30	190	150	78.9	4141	45	175	140	80.0
4132	30	180	145	80.6	4142	55	180	140	77.8

Measurements and Indices of Twenty Christians (Baghdad)

Measurements	No.	Range	Mean	S.D.	C.V.
Age	20	20–64	34.00±1.42	9.40±1.00	27.65±2.95
Head length	20	173–190	182.25±0.82	5.43±0.58	2.98±0.32
Head breadth	20	129–155	145.00±0.90	6.00±0.64	4.14±0.44
Indices					
Cephalic	20	68–88	79.50±0.51	3.36±0.36	4.23±0.45

Thirty-nine Christians (Mosul and Tell Kaif)

No.	Town	Age	G.O.L.	G.B.	C.I.	No.	Town	Age	G.O.L.	G.B.	C.I.
4149	Mosul	20	190	150	78.9	4169	Mosul	40	173	155	89.5
4150	Mosul	22	180	150	83.3	4170	Mosul	40	180	150	83.3
4151	Mosul	25	185	148	80.0	4171	Mosul	40	175	140	80.0
4152	Mosul	25	180	150	83.3	4172	Mosul	40	175	145	82.9
4153	Mosul	25	180	143	79.4	4173	Mosul	40	180	140	77.8
4154	Mosul	26	185	150	81.1	4174	Mosul	40	178	148	83.1
4155	Mosul	28	173	140	80.9	4175	Mosul	45	180	143	79.2
4156	Mosul	29	178	143	80.3	4176	Mosul	45	180	145	80.6
4157	Mosul	30	178	143	80.3	4177	Mosul	45	175	140	80.0
4158	Mosul	30	185	153	82.7	4178	Mosul	45	175	150	85.7
4159	Mosul	32	180	150	83.3	4179	Mosul	46	180	153	85.0
4160	Mosul	32	188	150	80.0	4180	Mosul	50	180	150	83.3
4161	Mosul	33	175	140	80.0	4181	Mosul	60	180	145	80.6
4162	Mosul	35	180	145	80.6	4182	Mosul	65	185	148	80.0
4163	Mosul	35	180	145	80.6	4183	Tell Kaif	35	180	150	83.3
4164	Mosul	35	180	143	79.4	4184	Tell Kaif	60	183	140	76.5
4165	Mosul	35	190	150	78.9	4185	Tell Kaif	65	180	150	83.3
4166	Mosul	36	180	145	80.6	4186*	Mosul	12	165	138	83.6
4167	Mosul	36	185	145	78.4	4187*	Mosul	18	170	140	82.4
4168	Mosul	38	180	143	79.4						

* Nos. 4186 and 4187 have been omitted from the averages—ages 12 and 18.

Measurements and Indices of Thirty-seven Christians (Iraq)

Measurements	No.	Range	Mean	S.D.	C.V.
Age	37	20–69	39.45±1.28	11.50±0.90	29.15±2.29
Head length	37	173–190	180.24±0.49	4.38±0.34	2.43±0.19
Head breadth	37	138–155	146.47±0.55	4.92±0.39	3.36±0.26
Indices					
Cephalic	37	74–91	81.48±0.32	2.91±0.23	3.57±0.28

Anthropometric Data: Royal Hospital

Five Christian Females (Mosul)

No.	Age	G.O.L.	G.B.	C.I.
4188	35	175	143	81.7
4189	35	175	140	80.0
4190	40	173	150	86.7
4191	42	170	150	88.2
4192	18	168	148	88.1
Averages..	34	172.2	146.2	84.9

Six Christian Females (Baghdad)

No.	Age	G.O.L.	G.B.	C.I.
4143	30	175	140	80.0
4144	35	170	143	84.1
4145	40	170	143	84.1
4146	40	180	140	77.8
4147	45	175	140	80.0
4148	46	180	145	80.6
Averages..	39.3	175	141.8	81.1

Nineteen Jews (Baghdad, Erbil, and Kirkuk)

No.	Town	Age	G.O.L.	G.B.	C.I.	No.	Town	Age	G.O.L.	G.B.	C.I.
4097	Baghdad	23	170	140	82.3	4107	Baghdad	40	180	145	80.6
4098	Baghdad	25	180	143	79.4	4108	Baghdad	45	180	150	83.3
4099	Baghdad	26	183	148	80.9	4109	Baghdad	45	170	143	84.1
4100	Baghdad	30	180	140	77.8	4110	Baghdad	50	175	140	80.0
4101	Baghdad	32	178	145	81.5	4111	Baghdad	55	180	145	80.6
4102	Baghdad	32	185	145	78.4	4112	Erbil	26	180	140	77.8
4103	Baghdad	35	183	143	78.1	4113	Kirkuk	32	180	140	77.8
4104	Baghdad	35	170	140	82.4	4114*	Baghdad	12	183	135	73.8
4105	Baghdad	38	170	140	82.4	4115*	Baghdad	18	178	143	80.3
4106	Baghdad	40	175	140	80.0						

* Nos. 4114 and 4115 have been omitted from the averages—ages 12 and 18.

Measurements and Indices of Seventeen Jews (Iraq)

Measurements	No.	Range	Mean	S.D.	C.V.
Age..............	17	20–59	37.00±1.54	9.40±1.09	25.41±2.94
Head length.........	17	170–187	177.72±0.76	4.65±0.54	2.62±0.30
Head breadth.........	17	138–152	142.18±0.59	2.63±0.42	2.55±0.29
Indices					
Cephalic............	17	77–85	80.28±0.32	1.92±0.22	2.39±0.28

Seven Jewesses (Baghdad)

No.	Age	G.O.L.	G.B.	C.I.
4116	20	190	148	77.9
4117	25	170	140	82.4
4118	25	180	135	75.0
4119	35	170	140	82.4
4120	40	170	145	85.3
4121	40	165	140	84.8
4122	48	165	143	86.7
Averages..	33.2	172.9	141.6	82.1

Thirty-three Kurds (Iran)

No.	Town	Age	G.O.L.	G.B.	C.I.	No.	Town	Age	G.O.L.	G.B.	C.L.
4247	Tehran	25	168	140	83.3	4265	Tabriz	38	180	150	83.3
4248	Tehran	26	168	140	83.3	4266	Tabriz	38	180	148	82.2
4249	Tehran	28	190	150	78.9	4267	Tabriz	40	180	143	79.4
4250	Tehran	32	165	140	84.8	4268	Tabriz	40	178	140	78.7
4251	Tehran	35	178	148	83.1	4269	Tabriz	45	185	148	80.0
4252	Tehran	36	175	148	84.6	4270	Tabriz	48	183	150	82.0
4253	Tehran	36	178	145	81.5	4271	Tabriz	50	180	143	79.4
4254	Tehran	42	178	150	84.3	4272	Waly	25	183	140	76.5
4255	Tehran	45	190	150	78.9	4273	Pestako	25	180	150	83.3
4256	Tehran	45	183	150	82.0	4274	Pestako	30	180	135	75.0
4257	Tehran	50	180	148	82.2	4275	Pestako	30	183	140	76.5
4258	Tehran	50	180	148	82.2	4276	Hussain				
4259	Tabriz	30	178	148	83.1		Kuli Khan	32	190	150	78.9
4260	Tabriz	35	170	140	82.4	4277	AliSharwan	35	188	138	73.4
4261	Tabriz	35	183	168	91.8	4278	Kerman-				
4262	Tabriz	35	183	150	82.0		shah	42	183	140	76.5
4263	Tabriz	35	180	148	82.2	4279*	Tehran	16	180	143	79.4
4264	Tabriz	38	190	150	78.9						

* No. 4279 (age 16) was omitted from the averages.

Measurements and Indices of Thirty-two Kurds (Iran)

Measurements	No.	Range	Mean	S.D.	C.V.
Age	32	25–54	37.95±.89	7.45±.63	19.63±1.66
Head length	32	164–193	180.18±.74	6.24±.53	3.46±0.29
Head breadth	32	135–170	146.23±.78	6.57±.55	4.49±0.38
Indices					
Cephalic	32	71–94	80.52±.47	3.96±.33	2.92±0.41

Three Kurd Females (Iran)

No.	Town	Age	G.O.L.	G.B.	C.I.
4280	Tabriz	40	178	145	81.5
4281	Tabriz	40	170	145	85.3
4282	Tabriz	45	180	145	80.6
	Averages	41.7	176	145	82.5

Four Christians (Iran)

No.	Town	Age	G.O.L.	G.B.	C.I.
4283	Tabriz	34	183	140	76.5
4284	Urmia	26	180	150	83.3
4285	Urmia	30	178	155	87.1
4286	Urmia	19	180	153	85.0
	Averages	27.3	180.2	149.5	82.9

Four Turks (Turkey)

No.	Town	Age	G.O.L.	G.B.	C.I.
4287	Van	28	180	148	82.2
4288	Van	45	180	153	85.0
4289	Istanbul	35	185	160	86.5
4246*	?	36	170	140	82.4
	Averages	36	181.7	153.7	84.6

* No. 4246, a Christian woman, was omitted from the averages.

APPENDIX E: INDIVIDUALS MEASURED IN ROYAL HOSPITAL, BAGHDAD

BY

WINIFRED SMEATON[1]

INTRODUCTION

During the period from November, 1934, to February, 1935, thirty-two males and forty-one females were measured in the Royal Hospital, Baghdad, where Dr. Shaib Shawkat facilitated the work in every possible manner. Eleven girls in the Central School for Girls in Baghdad also were measured during the winter of 1932-33.

In order to present these anthropometric data so that they will be comparable to other statistics from Iraq the results are presented according to the Harvard and Keith systems.

It must, however, be borne in mind that random sampling in a centrally located hospital does not yield valid anthropological deductions, particularly where the sample is small in number. For this reason the number of individuals, not the percentages, has been used in the following text.

On the other hand, every additional individual measured and observed throws some degree of light on the racial composition of the peoples of Iraq.

THIRTY-THREE MALES EXAMINED IN ROYAL HOSPITAL, BAGHDAD

Introduction.—Among these individuals twenty-three men were placed in an Arab group; the remainder were left as separate entities.

TWENTY-THREE ARABS FROM VARIOUS TOWNS

No.	Tribe	Locality	No.	Tribe	Locality
4398	....	An Nasiriya	4476	Chaab	Near Baghdad
4464	....	Mosul	4477	Daaya	Near Aziziya
4465	....	Diyala	4478	Ugrair	Hammam Ali
4466	....	Baghdad	4479	....	Shatra
4467	....	Baghdad	4480	Al bu Sultan	Near Mahmudiya
4469	....	Baghdad	4481	Baalwan	Born in Ramadi
4470	....	Baghdad	4482	Umairi	Near Baquba
4471	....	Near Aziziya	4483	....	Baasaf near Al Falluja
4472	Al bu Sultan	Hamza	4484	Nefafsha	Near Aziziya
4473	....	Ana	4485	Al bu Sultan	Near Latifiya
4474	....	Diltawa	4486	Jenabi	Near Yusufiya
4475	Ambergujah	Near Baquba			

[1] A member of the Field Museum Anthropological Expedition to the Near East from April to July, 1934.

Demography.—There was an identical number of sons and daughters. The size of the families appears to have been small, although No. 4466 reported one son living and many dead and one daughter living and many dead.

DEMOGRAPHY

Sons	No.	Per cent	Daughters	No.	Per cent
None	2	16.67	None	2	16.67
1	3	25.00	1	3	25.00
2	2	16.67	2	1	8.33
3–4	4	33.33	3–4	4	33.33
5–6	1	8.33	5–6	2	16.67
7 or more	0	.....	7 or more	0	.....
Total	12	100.00	Total	12	100.00

Age.—The mean was 38.30 with a range of 18–64. Our group shows a very wide distribution, with thirteen men under 40 and ten over this age.

AGE DISTRIBUTION

Age	No.	Per cent	Age	No.	Per cent
18–19	2	8.70	45–49	4	17.39
20–24	3	13.04	50–54	2	8.70
25–29	5	21.74	55–59	1	4.35
30–34	0	.....	60–64	3	13.04
35–39	3	13.04	65–69	0	.....
40–44	0	.....	70–x	0	.....
			Total	23	100.00

MORPHOLOGICAL CHARACTERS OF TWENTY-THREE ARABS

Skin.—Nos. 4472 and 4477 possessed dark and Nos. 4481 and 4485 very dark skins. With the exception of the latter these were listed as having Negro blood.

Hair.—The color was either dark brown or black, sometimes tinged with gray.

HAIR

Color	No.	Per cent	Texture	No.	Per cent
Black	0	.....	Coarse	6	46.15
Very dark brown	0	.....	Coarse-medium	0	.....
Dark brown	4	30.77	Medium	5	38.46
Brown	0	.....	Medium-fine	0	.....
Reddish brown	0	.....	Fine	2	15.38
Light brown	0	.....			
Red	0	.....	Total	13	99.99
Black and gray	6	46.15			
Dark brown and gray	2	15.38			
Light brown and gray	0	.....			
Gray	1	7.69			
White	0	.....			
Total	13	99.99			

Three men (Nos. 4464, 4467, and 4473) had low wavy hair, and six had coarse, five medium, and two fine, hair.

Seven men (Nos. 4398, 4466, 4471, 4472, 4474, 4475, and 4486) wore mustaches, Nos. 4466 and 4472 being black and No. 4398 brown. Seven individuals (Nos. 4466, 4474, 4477, 4479, 4481, 4482, and 4483) had shaven heads.

Eyes.—While nineteen individuals had dark brown eyes, one individual had black, one green-brown, and two gray-brown eyes. The sclera were bloodshot (14), yellow (3), clear (3), or yellow and bloodshot (2). The iris was homogeneous in Nos. 4467 and 4472, rayed in No. 4471, and zoned in Nos. 4464 and 4473. Seven men (Nos. 4456, 4458, 4464, 4474, 4481, 4482, and 4484) had blue-ringed irises, possibly arcus senilis. No. 4467 had a dark rim around his iris, and Nos. 4477 and 4480 had Negroid eyes.

EYES

Color	No.	Per cent	Sclera	No.	Per cent
Black	1	4.35	Clear	3	13.64
Dark brown	19	82.61	Yellow	3	13.64
Blue-brown	0	....	Speckled	0	....
Blue-brown	0	....	Bloodshot	14	63.64
Green-brown	1	4.35	Speckled and bloodshot	0	....
Green-brown	0	....	Speckled and yellow	0	....
Gray-brown	2	8.70	Yellow and bloodshot	2	9.09
Blue	0	....			
Gray	0	....	Total	22	100.01
Light brown	0	....			
Blue-gray	0	....			
Blue-green	0	....			
Total	23	100.01			

Nose.—The profile was convex (11), straight (6), concave (4), or concavo-convex (2). The alae were medium (10), flaring (9), or compressed (4). In thickness the nasal tip was thin (No. 4467), slightly more than average (Nos. 4466, 4471, and 4473), and double plus (Nos. 4465 and 4472). Nos. 4479 and 4482 had high nasal bridges. Fourteen individuals had depressed and four elevated nasal tips. The septum was either straight (13) or convex (10); the inclination was upwards in eighteen cases and downwards in only four individuals.

The following observations on the nose were recorded: No. 4470, marked nasion depression and high, aquiline angle; No. 4472, very flat and broad; No. 4476, small; No. 4477, short and broad nose and eyes were chief indications of Negroid blood; No. 4478, broad; No. 4479, very aquiline; No. 4480, Negroid; and No. 4483, small.

Nose

Profile	No.	Per cent	Wings	No.	Per cent
Wavy	0	.....	Compressed	4	17.39
Concave	4	17.39	Compressed-medium	0	.....
Straight	6	26.09	Medium	10	43.48
Convex	11	47.83	Medium-flaring	0	.....
Concavo-convex	2	8.70	Flaring	9	39.13
			Flaring plus	0	.....
Total	23	100.01	Total	23	100.00

Tip elevation	No.	Per cent	Septum	No.	Per cent
Elevated	4	22.22	Straight	13	56.52
Horizontal	0	.....	Convex	10	43.48
Depressed	14	77.78			
			Total	23	100.00
Total	18	100.00			

Septum inclination	No.	Per cent
Up	18	81.82
Down	4	18.18
Total	22	100.00

Teeth.—The occlusion was recorded as marked-over (8), slight-over (7), edge-to-edge (6), and under bite (1). The small number of teeth lost indicates a relatively healthy oral condition. Nos. 4464, 4477, 4483, and 4486 had good and No. 4476 excellent teeth. Wear was slightly more than average in six cases (Nos. 4372, 4379, 4380, 4384, 4385, and 4398) and double plus in Nos. 4465, 4471, 4481, and 4482. Eruption was recorded as incomplete in Nos. 4469, 4470, 4478, and 4483, and complete in Nos. 4467, 4471, and 4472.

The following observations were recorded on the teeth: stained, Nos. 4470, 4473, 4474, 4477, 4480–4482, and 4484; tartar deposit, Nos. 4398, 4469, 4478, and 4479; broken, Nos. 4471 (2), and No. 4470, lower first molars; good and strong, Nos. 4479 and 4483; fairly clean, No. 4485; crooked but strong and white, No. 4486; three gold-capped, No. 4398; and all premolars and molars lost, No. 4475.

Teeth

Bite	No.	Per cent	Loss	No.	Per cent
Under	1	4.55	None	6	33.33
Edge-to-edge	6	27.27	1–4	6	33.33
Slight over	7	31.82	5–8	1	5.56
Marked over	8	36.36	9–16	3	16.67
			17–	2	11.11
Total	22	100.00	All	0	.....
			Total	18	100.00

Prognathism.—Alveolar prognathism was observed in Nos. 4469, 4473, and 4479.

Lips.—Eversion was recorded as slightly more than average in Nos. 4472, 4480 (everted lower lip), and 4481 and double plus in No. 4479. Nos. 4472, 4477, 4480, and 4481 appeared to have some Negroid blood.

Physical Appearance.—Nos. 4398 and 4467 were pale. No. 4479 was very thin. No. 4486 had bad posture.

Pathological Cases.—No. 4476 bore smallpox scars. No. 4398 had scalp disease, probably favus. Nos. 4466 and 4483 were blind in the left eye, and both possessed a filmed right eye. No. 4474 had both eyes filmed but could see dimly.

No. 4483 had a sprained elbow and swollen forearm which was bent around, the result of a fall six weeks before he came to the hospital.

CAUTERIZATION (*Chawi*)

No.
4465: Both forearms.
4471: Right arm, both legs.
4472: Left wrist and both forearms.
4473: Both legs, right arm "for pain after fever."
4474: Above right knee.
4475: Right leg.
4476: Both forearms.
4478: Both forearms.
4481: Belly (10) for "tubercular lesion which didn't heal."
4483: Elbow, "to relieve sprained elbow."
4484: Both arms.
4485: Both arms.
4486: Above ankle.

Tattooing.—Sixteen men bore simple tattooed designs, but no individual was extensively tattooed.

TATTOOING

	No.	Per cent
None	4	**20.00**
Some	16	**80.00**
Extensive	0	.....
Total	20	**100.00**

MEASUREMENTS AND INDICES OF TWENTY-THREE ARABS

Stature.—The mean was 167.58, range 155.0–175.0. The threefold Harvard classificatory system places eleven men as medium (160.6–169.4), eight as tall (169.5–x), and only two as short (x–160.5). According to the fourfold Keith system thirteen men were medium (160.0–169.9), six tall (170.0–179.9), and two short (x–159.9). No individual was in the very tall (180.0–x) group.

STATURE*

Harvard system	No.	Per cent	Keith system	No.	Per cent
Short (x–160.5)	2	9.52	Short (x–159.9)	2	9.52
Medium (160.6–169.4)	11	52.38	Medium (160.0–169.9)	13	61.90
Tall (169.5–x)	8	38.10	Tall (170.0–179.9)	6	28.57
			Very tall (180.0–x)	0	.....
Total	21	100.00	Total	21	99.99

*Nos. 4464 and 4486 omitted.

Sitting Height.—The mean was 86.14, range 81.0–92.0. The trunk length was long (85.0–89.9) or medium (80.0–84.9). One man had a very long (90.0–x) trunk. No individuals were in the short (75.0–79.9) or very short (x–74.9) categories.

SITTING HEIGHT (Trunk Length)

Group	Range	No.	Per cent
Very short	x–74.9	0	.....
Short	75.0–79.9	0	.....
Medium	80.0–84.9	6	28.57
Long	85.0–89.9	14	66.67
Very long	90.0–x	1	4.76
Total		21	100.00

Head Measurements and Indices.—Fifteen Arabs had wide (140–149) heads, six had very wide (150–x) heads, and two had narrow (130–139) heads. No man had a very narrow (x–129) head. Seventeen men had narrow (100–109) foreheads. Although there were no individuals in the very wide (120–x) category, there were three individuals in both the wide (110–119) or very narrow (x–99) classifications.

The Harvard threefold system places fourteen men as dolichocephals (x–76.5) and nine as mesocephals (76.6–82.5). There were no brachycephals (82.6–x). According to the Keith system sixteen men were mesocephals (75.1–79.9) and seven were dolichocephals (70.1–75.0). No individual was in the ultradolichocephalic (x–70.0), brachycephalic (80.0–84.9), or ultrabrachycephalic (85.0–x) divisions.

HEAD BREADTH

Group	Range	No.	Per cent
Very narrow	x–129	0	.....
Narrow	130–139	2	8.70
Wide	140–149	15	65.22
Very wide	150–x	6	26.09
Total		23	100.01

MINIMUM FRONTAL DIAMETER

Group	Range	No.	Per cent
Very narrow	x–99	3	13.04
Narrow	100–109	17	73.91
Wide	110–119	3	13.04
Very wide	120–x	0	.....
Total		23	99.99

CEPHALIC INDEX

Harvard system	No.	Per cent	Keith system	No.	Per cent
Dolichocephalic (x–76.5)	14	60.87	Ultradolichocephalic (x–70.0)	0	.....
Mesocephalic (76.6–82.5)	9	39.13	Dolichocephalic (70.1–75.0)	7	30.43
Brachycephalic (82.6–x)	0	.....	Mesocephalic (75.1–79.9)	16	69.57
Total	23	100.00	Brachycephalic (80.0–84.9)	0	.....
			Ultrabrachycephalic (85.0–x)	0	.....
			Total	23	100.00

Head Form.—No. 4473 had a flat area near bregma. No. 4480 had a flattened area above the occipital region.

Facial Measurements and Indices.—The upper facial height was either medium long (70–75) or long (76–x). Three men had medium short (64–69) upper faces. No Arab was in the short (x–63) group. The mean was 75.90, range 65–89.

The total facial height was either medium long (120–129) or medium short (110–119). Two men had long (130–x) faces and one a short (x–109) face. The mean was 120.40, range 105–134. No. 4474 was omitted.

The facial index was either mesoprosopic (84.6–89.4) or leptoprosopic (89.5–x). Two Arabs were euryprosopic (x–84.5).

The mean upper facial index was 55.88, range 49–63. The mean facial index was 89.50, range 80–99.

No. 4477 had small features. Nos. 4480 and 4481 had well-developed supraorbital crests.

FACIAL MEASUREMENTS

Upper facial height	No.	Per cent	Total facial height	No.	Per cent
Short (x–63)	0	.....	Short (x–109)	1	4.55
Medium short (64–69)	3	13.04	Medium short (110–119)	9	40.91
Medium long (70–75)	10	43.48	Medium long (120–129)	10	45.45
Long (76–x)	10	43.48	Long (130–x)	2	9.09
Total	23	100.00	Total	22	100.00

TOTAL FACIAL INDEX*

Group	No.	Per cent
Euryprosopic (x–84.5)	2	9.09
Mesoprosopic (84.6–89.4)	10	45.45
Leptoprosopic (89.5–x)	10	45.45
Total	22	99.99

*No. 4474 omitted.

Nasal Measurements and Indices.—Twenty men had medium (50–59) and three short (x–49) nasal heights. No individual was in the long (60–x) class. The mean was 53.14, range 40–59. Eleven Arabs had medium wide (36–41), nine medium narrow (30–35), and two wide (42–x) noses. No man had a very narrow (x–29) nose. The mean was 36.77, range 31–42. No. 4476 was omitted.

Eleven men were mesorrhine (67.5–83.4), ten were leptorrhine (x–67.4), and one platyrrhine (83.5–x). The mean was 68.94, range 56–87.

NASAL MEASUREMENTS

Nasal height	No.	Per cent	Nasal width*	No.	Per cent
Short (x–49)	3	13.04	Very narrow (x–29)	0	....
Medium (50–59)	20	86.96	Medium narrow (30–35)	9	40.91
Long (60–x)	0	....	Medium wide (36–41)	11	50.00
Total	23	100.00	Wide (42–x)	2	9.09
			Total	22	100.00

NASAL INDEX*

Group	No.	Per cent
Leptorrhine (x–67.4)	10	45.45
Mesorrhine (67.5–83.4)	11	50.00
Platyrrhine (83.5–x)	1	4.55
Total	22	100.00

*No. 4476 omitted.

INDIVIDUALS OMITTED FROM STATISTICAL ANALYSES

Since the remainder of the males measured in the Royal Hospital, Baghdad, belonged to various racial stocks and different religious groups, no statistical analyses could be made, although the measurements and indices for the ten Arabs have been calculated merely for comparative purposes.

Provenance.—No. 4456, Chaldean from Tell Kaif; No. 4457, Afghan from Herat (12 years before); No. 4458, Armenian from Istanbul; No. 4459, Armenian from Van; No. 4460, Turkoman from Tuz Khurmatli near Kirkuk; No. 4461, Turkoman from Kirkuk;

MEASUREMENTS AND INDICES OF BAGHDAD ROYAL HOSPITAL MALES

Measurements	No.	Range	Mean	S.D.	C.V.
Age	23	18–64	38.30±2.09	14.85±1.48	38.77±3.86
Stature	21	155–175	167.58±0.76	5.19±0.54	3.10±0.32
Sitting height	21	81–92	86.14±0.38	2.55±0.27	2.96±0.31
Head length	23	176–211	193.56±0.95	6.72±0.67	3.47±0.35
Head breadth	23	138–155	146.29±0.59	4.23±0.42	2.89±0.29
Minimum frontal diameter	23	97–116	104.58±0.62	4.40±0.44	4.21±0.42
Bizygomatic diameter	23	125–149	136.15±0.77	5.45±0.54	4.00±0.40
Bigonial diameter	22	90–117	101.70±0.86	6.00±0.61	5.90±0.60
Total facial height	22	105–134	120.40±0.93	6.50±0.66	5.40±0.55
Upper facial height	23	65–89	75.90±0.82	5.85±0.58	7.71±0.77
Nasal height	23	40–59	53.14±0.60	4.24±0.42	7.98±0.79
Nasal breadth	22	31–42	36.77±0.42	2.94±0.30	8.00±0.81
Ear length	23	56–79	65.50±0.79	5.64±0.56	8.61±0.86
Ear breadth	23	26–40	34.44±0.43	3.03±0.30	8.80±0.88
Indices					
Relative sitting height	21	48–53	51.46±0.14	0.98±0.10	1.90±0.20
Cephalic	23	68–82	75.39±0.40	2.82±0.28	3.74±0.37
Fronto-parietal	23	66–77	71.44±0.43	3.03±0.30	4.24±0.42
Zygo-frontal	23	72–83	76.98±0.42	2.96±0.29	3.85±0.38
Zygo-gonial	22	69–83	75.04±0.47	3.30±0.34	4.40±0.45
Total facial	22	80–99	89.50±0.60	4.20±0.43	4.69±0.48
Upper facial	23	49–63	55.88±0.46	3.24±0.32	5.80±0.58
Nasal	22	56–87	68.94±0.84	5.84±0.59	8.47±0.86
Ear	23	41–64	52.58±0.76	5.40±0.54	10.27±1.02

No. 4462, Assyrian from Shemsaddin tribe, now resident at Erbil; No. 4463, Turk from Istanbul; No. 4468, Arab from Baghdad; No. 4487, Arab from between Baghdad and Diltawa.

No. 4456, obviously a non-Arab type, had a high, vaulted forehead (straight up), a flat area rather high on the head and "terrible" teeth with deposits. He had *chawi* scars on his right knee and his right arm.

No. 4457, a Mongoloid type, had a high, sloping vault, narrow head, face and features with a medium epicanthic fold, large eye pupils, and most of the lower teeth replaced by bridgework made in Khurasan. He had a *chawi* on his right thigh.

No. 4458 had a very straight nose, "not at all the Armenian type of nose." His teeth were stained.

No. 4459 had a flat occiput, small nose, and teeth stained but strong-looking.

No. 4460 had a bad deposit on the teeth and scurf on the scalp.

No. 4461 had a high, sloping vault and a prominent strong chin. His left hand was paralyzed, the fingers bent and immovable, due to injuries received while working for the Iraq Petroleum Company.

Measurements and Indices of Baghdad Royal Hospital Males

Measurements	No.	Mean	S.D.	C.V.
Stature	31	167.71±.62	5.09±.44	3.04±.26
Sitting height	31	85.89±.39	3.21±.27	3.74±.31
Head length	32	191.94±.90	7.60±.64	3.96±.33
Head breadth	31	147.55±.58	4.78±.41	3.24±.28
Minimum frontal diameter	32	104.59±.45	3.77±.32	3.60±.31
Bizygomatic breadth	32	137.22±.63	5.30±.45	3.86±.33
Bigonial breadth	30	102.00±.70	5.72±.50	5.61±.49
Total facial height	30	121.97±.80	6.52±.57	5.35±.47
Upper facial height	31	76.23±.62	5.09±.44	6.68±.58
Nasal height	32	53.38±.52	4.34±.37	8.13±.69
Nasal breadth	32	36.47±.32	2.68±.23	7.35±.63
Ear length	32	66.50±.67	5.62±.47	8.45±.71
Ear breadth	32	35.09±.39	3.26±.27	9.29±.77
Indices				
Relative sitting height	31	51.16±.19	1.60±.14	3.13±.27
Cephalic	31	77.00±.51	4.20±.36	5.45±.47
Fronto-parietal	31	70.87±.38	3.11±.27	4.39±.38
Zygo-frontal	32	76.28±.32	2.67±.23	3.50±.30
Zygo-gonial	30	74.57±.39	3.15±.27	4.22±.36
Total facial	30	89.15±.49	3.98±.35	4.46±.39
Upper facial	31	55.58±.38	3.13±.27	5.63±.49
Nasal	32	68.83±.66	5.54±.47	8.05±.68
Ear	31	53.15±.64	5.29±.45	9.95±.85

No. 4462 had a flat, broad occiput, large nose, some deposit on his teeth, and some smallpox scars; although his eyes appeared normal he had been blinded by a 24-foot fall from a housetop.

No. 4463 had a flat occiput, a nose broad throughout its entire length, some deposit on his teeth, and "vision all right," although he was blind in the left eye and the right appeared filmed.

No. 4468 had lost all his teeth ten years ago and had cancer of the tongue. He had *chawi* scars on his left foot "to relieve pain," and below his left knee.

No. 4487 had a high, sloping vault. His teeth were stained; the lower incisors and canines were present. He had "cancer" on the right arm in three places.

Anthropometric Data: Royal Hospital

Measurements of Baghdad Royal Hospital Males

No.	Age	Stature	SH	L	B	B'	J	go-go	GH	G'H	NH	NB
4398	29	1567	836	189	149	107	140	110	119	74	52	34
4456*	55	1690	850	187	153	101	143	109	(136)†	(81)†	57	38
4457*	30	1777	892	202	140	105	137	100	113	72	45	35
4458*	48	1670	936	185	152	107	138	107	128	81	58	35
4459*	35	1613	815	175	158	107	146	...	126	78	53	35
4460*	33	1606	829	196	153	106	142	100	127	78	51	33
4461*	25	1684	893	180	148	108	140	106	124	72	52	34
4462*	30	1653	862	186	154	107	146	110	124	79	59	34
4463*	16	1579	808	183	151	102	132	93	113	70	50	37
4464	47	....	....	191	150	105	141	100	128	82	59	37
4465	53	1699	882	195	151	113	143	104	123	79	59	42
4466	58	1620	859	191	145	101	133	106	(114)‡	(73)‡	53	38
4467	35	1690	887	193	146	104	135	102	127	80	58	40
4468*	78	1647	799	197	...§	107	140	102	(124)§	(77)§	57	39
4469	18	1727	890	181	139	103	129	99	119	74	53	34
4470	21	1588	830	197	148	98	137	105	126	73	53	35
4471	60	1737	872	191	150	99	137	96	117	73	52	34
4472	38	1731	871	204	144	98	131	106	124	73	52	35
4473	45	1677	881	191	140	101	127	91	123	80	57	33
4474	60	1734	874	209	147	112	141	97	....	88	58	38
4475	60	1695	868	192	150	109	146	114	124	81	57	42
4476	28	1655	853	192	153	101	133	103	117	73	50	(35)¶
4477	25	1693	826	192	146	102	135	94	114	69	47	37
4478	23	1721	917	198	145	110	137	...	130	80	54	38
4479	25	1678	854	200	147	109	140	110	129	76	55	39

* Omitted throughout averages.
† Questionable—"Teeth poor."
‡ Questionable.
§ "Seems swollen on both sides above ears;" aged; all teeth gone ten years ago.

Indices of Baghdad Royal Hospital Males

No.	EL	EB	RSH	B/L	B'/B	GH/J	G'H/J	NB/NH	EB/EL	go-go/J	B'/J
4398	66	33	53.4	78.8	71.8	85.0	52.9	65.4	50.0	78.6	76.4
4456	68	36	50.3	81.8	66.0	95.1	56.6	66.7	52.9	76.2	70.6
4457	63	35	50.2	69.3	75.0	82.5	52.6	77.8	55.6	73.0	76.6
4458	65	37	56.0	82.2	70.4	92.8	58.7	60.3	56.9	77.5	77.5
4459	66	36	50.5	90.3	67.7	86.3	53.4	66.0	54.5	....	73.3
4460	65	30	51.6	78.1	69.3	89.4	54.9	64.7	46.2	70.4	74.6
4461	70	34	53.0	82.2	73.0	88.6	51.4	65.4	48.6	75.7	77.1
4462	68	42	52.1	82.8	69.5	84.9	54.1	57.6	61.8	75.3	73.3
4463	61	35	51.2	82.5	67.5	85.6	53.0	74.0	57.4	70.5	77.3
4464	66	37	....	78.5	70.0	90.8	58.2	62.7	56.1	70.9	74.5
4465	71	34	51.9	77.4	74.8	86.0	55.2	71.2	47.9	72.7	79.0
4466	71	36	53.0	75.9	69.7	85.7	54.9	71.7	50.7	79.7	75.9
4467	71	33	52.5	75.6	71.2	94.1	59.3	69.0	46.5	75.6	77.0
4468	79	35	48.5	....	....	88.6	55.0	68.4	44.3	72.9	76.4
4469	58	36	51.5	76.8	74.1	92.2	57.4	64.2	62.1	76.7	79.8
4470	66	38	52.3	75.1	66.2	92.0	53.3	66.0	57.6	76.6	71.5
4471	66	28	50.2	78.5	66.0	85.4	53.3	65.4	42.4	70.1	72.3
4472	57	35	50.3	70.6	68.1	94.7	55.7	67.3	61.4	80.9	74.8
4473	63	36	52.5	73.3	72.1	96.9	63.0	57.9	57.1	71.7	79.5
4474	76	38	50.4	70.3	76.2	....	62.4	65.5	50.0	68.8	79.4
4475	72	38	51.2	78.1	72.7	84.9	55.5	73.7	47.2	78.1	74.7
4476	59¶	29¶	51.5	79.7	66.0	88.0	54.8	....¶	49.2	77.4	75.9
4477	62	34	48.8	76.0	69.9	84.4	51.1	78.7	54.8	69.6	75.6
4478	66	37	53.3	73.2	75.9	94.9	58.4	70.4	56.1	....	80.3
4479	67	38	50.9	73.5	74.1	92.1	54.3	70.9	56.7	78.6	77.9

¶ Small nose, smallpox scars affect nasal measurement; right ear was measured.

Measurements of Baghdad Royal Hospital Males

No.	Age	Stature	SH	L	B	B'	J	go-go	GH	G'H	NH	NB
4480	18	1691	853	194	140	106	130	98	112	69	45	38
4481	45	1736	885	198	145	105	140	105	132	88	59	41
4482	48	1626	868	191	146	102	136	101	121	72	51	33
4483	20	1616	832	176	138	100	127	95	108	66	42	32
4484	50	1637	838	188	148	104	138	99	116	75	52	33
4485	28	1636	825	193	145	107	133	96	121	75	52	37
4486	35	(1677)†	814	199	151	103	134	100	119	76	54	38
4487*	70	1779	870	196	151	105	144	112	...	..	54	38

* Omitted in statistical series.   † Measurement uncertain because of bad posture.

Indices of Baghdad Royal Hospital Males

No.	EL	EB	RSH	B/L	B'/B	GH/J	G'H/J	NB/NH	EB/EL	go-go/J	B'/J
4480	63	33	50.4	72.2	75.7	86.2	53.1	84.4	52.4	75.4	81.5
4481	76	36	51.0	73.2	72.4	94.3	62.9	69.5	47.4	75.0	75.0
4482	65	34	53.4	76.4	69.9	89.0	52.9	64.7	52.3	74.3	75.0
4483	57	32	51.5	78.4	72.5	85.0	52.0	76.2	56.1	74.8	78.7
4484	69	31	51.2	78.7	70.3	84.1	54.3	63.5	44.9	71.7	75.4
4485	60	32	50.4	75.1	73.8	91.0	56.4	71.2	53.3	72.2	80.5
4486	65	38	(48.5)†	75.9	68.2	88.8	56.7	70.4	58.5	74.6	76.9
4487	77	44	48.9	77.0	69.5	....	....	70.4	57.1	77.8	72.9

Morphological Characters of Baghdad Royal Hospital Males

	HAIR			EYES			NOSE	
No.	Form	Texture	Color	Color	Sclera	Iris	Profile	Wings
4398	...	......	......	black	clear	...	conv	medium
4456	...	medium	dk br	dk br	blood	ray	conv	medium
4457*	...	......	dk br	dk br	......	...	str	flar
4458	...	coarse	blk, gray	dk br	blood	...	str	medium
4459	l w	med-fine	black	dk br	yellow	...	str	comp
4460	...	medium	blk, gray	gr-br	blood	...	conv	comp
4461*	...	coarse	black	dk br	blood	...	conv	medium
4462†	...	......	black	dk br	blood	...	str	comp
4463*	...	......	......	dk br	clear	...	str	flar
4464	l w	fine	dk br	dk br	blood	zon	conv	medium
4465	...	......	br, gray	gray-br	blood	...	conv	flar
4466*	...	......	......	dk br	blood	...	conv	medium
4467	l w	coarse	dk br	dk br	blood	hom	conv	medium
4468	l w	coarse	gray	dk br	blood	...	conc	flar
4469‡	...	fine	dk br	dk br	clear	...	conc	medium
4470	...	medium	black	dk br	yellow	...	conv	medium
4471	...	......	br, gray	gray-br	blood	ray	conv	comp
4472	...	......	......	dk br	blood	hom	conc	flar
4473‡	l w	medium	blk, gray	gr-br	blood	zon	c-c	comp
4474*	...	......	gray	dk br	blood	...	str	medium
4475	...	......	blk, gray	dk br	blood	...	str	flar
4476‡	...	coarse	black	dk br	blood	...	conc	medium
4477*	...	medium	blk, gray	dk br	yellow	...	c-c	flar
4478‡	...	coarse	black	dk br	yellow	...	str	flar
4479*	...	......	black	dk br	blood	...	conv	medium
4480 ¶	...	......	black	dk br	yell, blood	...	conc	flar
4481*	...	coarse	blk, gray	dk br	yell, blood	...	conv	flar
4482*	...	......	blk, gray	dk br	blood	...	conv	comp
4483*	...	coarse	blk	dk br	......	...	str	medium
4484	...	medium	blk, gray	dk br	blood	...	conv	comp
4485	...	medium	black	dk br	blood	...	str	flar
4486	...	coarse	dk br	dk br	clear	...	str	flar
4487 ¶	...	medium	gray	gray-br	blood	...	conv	flar

* Shaved.   † Baldness plus.   ‡ Hair very short.   ¶ Hair short.

Fifty-two Females Measured in Royal Hospital, Baghdad

Introduction.—Within this series there are twenty women, who can be grouped together. The remainder must be left as separate entities.

Notes.—Since these individuals may at some future time be included in larger series from the same areas it is desirable to record the tribal information.

Among the Arab series of twenty women Nos. 4506–4508, 4510 (mother from Basra), 4511, and 4512 were from Baghdad; No. 4513, from An Najaf; No. 4514, from Mahmudiya; No. 4515, from Baquba; No. 4516, from Hiyaliya tribe near Baghdad; No. 4517, from Al bu Muhammad tribe east of Amara; No. 4518, from Ajili tribe near Karrada; No. 4519, from Tai tribe near Baquba; No. 4520, from Karrada; No. 4522, from Muadhdham; No. 4523, non-tribal from Samarra; No. 4524, from Al-Umara (?tribe) near Mahmudiya; No. 4526, from Rabia tribe near Kut; and No. 4527, from Hufaiya tribe near Hilla.

Twenty Arab Women from Various Towns in Iraq

Demography.—In this group of Arab women there was a slight female preponderance—fifteen daughters to eleven sons.

Demography

Sons	No.	Per cent	Daughters	No.	Per cent
None	5	31.25	None	1	6.25
1	1	6.25	1	5	31.25
2	3	18.75	2	4	25.00
3–4	7	43.75	3–4	6	37.50
5–6	0	.....	5–6	0	.....
7 or more	0	.....	7 or more	0	.....
Total	16	100.00	Total	16	100.00

Age.—Three-quarters of the group were between 20–34 years of age. The mean was 30.50, range 20–59.

Age Distribution

Age	No.	Per cent	Age	No.	Per cent
18–19	0	.....	45–49	1	5.00
20–24	7	35.00	50–54	0	.....
25–29	4	20.00	55–59	1	5.00
30–34	4	20.00	60–64	0	.....
35–39	2	10.00	65–69	0	.....
40–44	1	5.00	70–x	0	.....
			Total	20	100.00

Morphological Characters of Twenty Arab Women

Skin.—The color was dark in Nos. 4517 and 4524.

Hair.—The color shaded from dark brown to black, with about an equal number in each division. The majority (17) had low wavy hair; the other two individuals, deep waves. In about half of the group the texture was medium, with an almost equal number in the coarse and fine categories. No. 4524 had a shaven head.

Hair

Color	No.	Per cent	Form	No.	Per cent
Black	9	45.00	Straight	0	....
Very dark brown	0	....	Very low waves	0	....
Dark brown	8	40.00	Low waves	17	89.47
Brown	0	....	Deep waves	2	10.53
Reddish brown	0	....	Curly-frizzly	0	....
Light brown	0	....	Woolly	0	....
Red	0	....			
Black and gray	1	5.00	Total	19	100.00
Dark brown and gray	1	5.00			
Light brown and gray	0	....	Texture	No.	Per cent
Gray	1	5.00	Coarse	4	21.05
White	0	....	Coarse-medium	0	....
			Medium	10	52.63
Total	20	100.00	Medium-fine	0	....
			Fine	5	26.32
			Total	19	100.00

Eyes.—The color was dark brown (10) or black (6). No. 4508, omitted from the following table on color, had eyes of green-gray flecked with brown. No. 4510 had light brown eyes. About three-quarters of the group possessed clear sclera, the remainder being bloodshot. No. 4514 had a homogeneous iris. No. 4526 had small eyes, which she kept only partly open. No. 4507 was recorded with a blue ringed iris, probably arcus senilis. Nos. 4510 and 4511 had filmed eyes, and No. 4523 bluish filmed eyes.

Eyes

Color	No.	Per cent	Sclera	No.	Per cent
Black	6	33.33	Clear	15	78.95
Dark brown	10	55.56	Yellow	0	....
Blue-brown	0	....	Speckled	0	....
Blue-brown	0	....	Bloodshot	4	21.05
Green-brown	1	5.56	Speckled and bloodshot	0	....
Green-brown	0	....	Speckled and yellow	0	....
Gray-brown	0	....	Yellow and bloodshot	0	....
Blue	0	....			
Gray	0	....	Total	19	100.00
Light brown	1	5.56			
Blue-gray	0	....			
Blue-green	0	....			
Total	18	100.01			

ANTHROPOMETRIC DATA: ROYAL HOSPITAL

Nose.—The profile was either convex (8) concave (7) or straight (5). The alae were medium (11), the remainder tending to be more flaring (5) than compressed (3). No. 4523 had a high nasal bridge and Nos. 4507 and 4513 broad nasal bridges. The septum was either straight (6) or convex (6). Three-quarters of the group possessed a nasal septum with an upward inclination. The nasal tip was either depressed (12) or elevated (5). Nos. 4514, 4515, and 4526 had slightly thicker than average nasal tips, but in No. 4507 the fleshy part of the nose was thin.

Nos. 4513 and 4518 possessed small noses, No. 4514 a broad, No. 4515 a short and broad, and No. 4507 a narrow nose except in the bridge.

NOSE

Profile	No.	Per cent	Wings	No.	Per cent
Wavy	0	.....	Compressed	1	5.26
Concave	7	35.00	Compressed-medium	2	10.53
Straight	5	25.00	Medium	11	57.89
Convex	8	40.00	Medium-flaring	1	5.26
Concavo-convex	0	.....	Flaring	4	21.05
			Flaring plus	0	.....
Total	20	100.00	Total	19	99.99

Septum	No.	Per cent	Tip thickness	No.	Per cent
Straight	6	50.00	− −	0	.....
Convex	6	50.00	−	1	25.00
			Average	0	.....
Total	12	100.00	+	3	75.00
			+ +	0	.....
			Total	4	100.00

Septum inclination	No.	Per cent	Tip elevation	No.	Per cent
Up	14	73.68	Elevated	5	29.41
Down	5	26.32	Horizontal	0	.....
			Depressed	12	70.59
Total	19	100.00	Total	17	100.00

DESCRIPTION OF NASAL SEPTUM

No.	Septum	Inclination	Elevation	No.	Septum	Inclination	Elevation
4506	convex	up	depressed	4517	.....	.....	.....
4507	convex	up	depressed	4518	.....	up	.....
4508	straight	down	depressed	4519	.....	down	depressed
4510	straight	up	depressed	4520	.....	up	elevated
4511	straight	up	depressed	4522	.....	up	elevated
4512	straight	up	elevated	4523	.....	down	depressed
4513	convex	up	depressed	4524	straight	up	elevated
4514	convex	up	depressed	4526	convex	up	.....
4515	convex	up	elevated	4527	.....	down	depressed
4516	straight	down	depressed	4528	.....	up	depressed

Teeth.—The majority (14) possessed a marked-over bite and two women had an edge-to-edge bite. Only three women had normal slight-over occlusion. As the group is relatively young the number of teeth lost is high, indicating poor dental condition among these town-dwellers. Wear was slight on the teeth of No. 4522, slightly more than average on Nos. 4518 and 4526 and double plus on No. 4512. Nos. 4506, 4518, and 4520 possessed complete eruption, No. 4516 incomplete.

TEETH

Bite	No.	Per cent	Loss	No.	Per cent
Under	0	.....	None	2	15.38
Edge-to-edge	2	10.53	1–4	5	38.46
Slight over	3	15.79	5–8	1	7.69
Marked over	14	73.68	9–16	4	30.77
			17–	0	.....
Total	19	100.00	All	1	7.69
			Total	13	99.99

NOTES ON DENTITION

No.	Description
4506	Teeth unusually white but a slight deposit.
4507	Bad deposits on teeth.
4508	Teeth stained. Six or seven lost ("one for each pregnancy").
4509	Teeth rather yellow.
4511	Upper incisors pulled out.
4512	Teeth stained.
4513	Yellow deposit on teeth. Two teeth broken.
4514	Teeth stained.
4515	Teeth stained; three broken off.
4516	White, strong teeth.
4520	Some deposit on teeth.
4524	Not much deposit on teeth.
4527	Excellent teeth.

Prognathism.—Nos. 4508, 4512, 4520, 4522, and 4524 had alveolar prognathism.

Malars.—Nos. 4507 and 4512 had slightly more than average lateral projection of the malars.

Tattooing.—The majority of the women recorded were tattooed, seven extensively.

Tattooing	No.	Per cent
None	5	25.00
Some	6	30.00
Extensive	7	45.00
Total	18	100.00

SPECIAL OBSERVATIONS

No.	Description
4506	Bad scars on the right side of the nose as a result of a "Baghdad boil"; bad goiter.
4507	Handsome.
4509	Large dark scars of "Baghdad boil" between eyes and on forehead.
4511	Thin, pleasant face; hair cut for mourning.

No.	Description
4513	Growth like small tumor in large navel; abdomen distended; at hospital for a genito-urinary operation.
4514	Broad face; nose broad throughout.
4516	Smallpox scars; tattooed on both wrists, on back of right wrist specifically to relieve pain. Breath foul.
4517	Pretty; gonorrheal complications in eyes of month-old baby.
4518	Suffering from bilharziasis. Had *chawi* scars on right ankle, leg, back, belly, and under breast "to relieve pain."
4519	Looks like a mummy, extremely thin; stomach greatly distended by water.
4522	Smallpox scars.
4523	Bad posture; does not look like an Arab woman.
4524	Tattooed, specifically on the belly to "relieve pain."
4527	Hair clean; very pretty.
4528	Part Negro.

MEASUREMENTS AND INDICES OF TWENTY ARAB WOMEN

In grouping the women, special divisions of stature and sitting height have been assigned by Dr. Hooton, since these are the two measurements in which there is a marked sexual difference.[1]

Stature.—The majority (15) were medium (149.0–159.0); there were no very short (x–139.0) and no very tall (170.0–x) individuals. The mean was 154.50, range 143.0–169.0.

STATURE

Harvard system	Range	No.	Per cent
Very short	x–139.0	0	.....
Short	140.0–148.0	3	15.00
Medium	149.0–159.0	15	75.00
Tall	160.0–169.0	2	10.00
Very tall	170.0–x	0	.....
Total		20	100.00

Sitting Height.—The majority (12) were medium (74.0–78.9) in trunk length but seven women possessed long (79.0–83.9) trunks. No individual was very short (x–68.9) or very long (84.0–x) in trunk length. This increase in sitting height does not appear in the stature so that these seven women must tend to have shorter legs. The mean was 79.00, range 72.0–86.0.

SITTING HEIGHT (Trunk Length)

Group	Range	No.	Per cent
Very short	x–68.9	0	.....
Short	69.0–73.9	1	5.00
Medium	74.0–78.9	12	60.00
Long	79.0–83.9	7	35.00
Very long	84.0–x	0	.....
Total		20	100.00

[1] If the females are grouped according to the male classifications the result is as follows: nineteen short (x–160.5), one medium (160.6–169.4), and no tall (169.5–x) individuals.

Head Measurements.—The head breadth (mean 141.0, range 129.0–152.0) was wide (140.0–149.0) or narrow (130.0–139.0). No women had very narrow (x–129.0) heads, but two were in the very wide (150.0–x) category. The minimum frontal diameter (mean 99.50, range 89.0–112.0) was either very narrow (x–99.0) or narrow (100.0–109.0). The cephalic index (mean 77.85, range 71.0–88.0) according to the Harvard classificatory system was either dolichocephalic (9) or mesocephalic (8), but there were three women in the brachycephalic (82.6–x) group. The Keith fivefold divisions show a different arrangement: nine mesocephals (75.1–79.9), five dolichocephals (70.1–75.0), five brachycephals (80.0–84.9), and one ultrabrachycephal (85.0–x).

HEAD BREADTH

Group	Range	No.	Per cent
Very narrow	x–129.0	0	.....
Narrow	130.0–139.0	8	40.00
Wide	140.0–149.0	10	50.00
Very wide	150.0–x	2	10.00
Total		20	100.00

MINIMUM FRONTAL DIAMETER

Group	Range	No.	Per cent
Very narrow	x–99.0	10	50.00
Narrow	100.0–109.0	10	50.00
Wide	110.0–119.0	0	.....
Very wide	120.0–x	0	.....
Total		20	100.00

CEPHALIC INDEX

Harvard system	No.	Per cent	Keith system	No.	Per cent
Dolichocephalic (x–76.5)	9	45.00	Ultradolichocephalic (x–70.0)	0	.....
Mesocephalic (76.6–82.5)	8	40.00	Dolichocephalic (70.1–75.0)	5	25.00
Brachycephalic (82.6–x)	3	15.00	Mesocephalic (75.1–79.9)	9	45.00
Total	20	100.00	Brachycephalic (80.0–84.9)	5	25.00
			Ultrabrachycephalic (85.0–x)	1	5.00
			Total	20	100.00

Facial Measurements and Indices.—The upper facial height (mean 69.00, range 60.0–84.0) was medium short (11) or medium long (7), but there was one woman in the short (x–63.0) and one in the long (76.0–x) categories. The total facial height (mean 111.00, range 100.0–124.0) was medium short (12) or short (7). Despite the number of individuals with medium long upper faces only one woman was in the medium long (120.0–129.0) group for total facial height. The total facial index (mean 87.25, range 80.0–94.0) was either mesoprosopic (12), leptoprosopic (5), or euryprosopic (3).

FACIAL MEASUREMENTS

Upper facial height	No.	Per cent	Total facial height	No.	Per cent
Short (x–63.0)	1	5.00	Short (x–109.0)	7	35.00
Medium short (64.0–69.0)	11	55.00	Medium short (110.0–119.0)	12	60.00
Medium long (70.0–75.0)	7	35.00	Medium long (120.0–129.0)	1	5.00
Long (76.0–x)	1	5.00	Long (130–x)	0	.....
Total	20	100.00	Total	20	100.00

TOTAL FACIAL INDEX

Group	Range	No.	Per cent
Euryprosopic	x–84.5	3	15.00
Mesoprosopic	84.6–89.4	12	60.00
Leptoprosopic	89.5–x	5	25.00
Total		20	100.00

Nasal Measurements and Indices.—In fourteen individuals the nose was short (x–49.0) and in six it was medium (50.0–59.0). Eighteen individuals had medium narrow and two very narrow nasal widths. Eleven individuals were leptorrhine (x–67.4) and nine mesorrhine (67.5–83.4). There were no long (60–x), no medium wide or wide (36–x), and no platyrrhine (83.5–x) noses in the group.

NASAL MEASUREMENTS

Nasal height	No.	Per cent	Nasal width	No.	Per cent
Short (x–49)	14	70.00	Very narrow (x–29)	2	10.00
Medium (50–59)	6	30.00	Medium narrow (30–35)	18	90.00
Long (60–x)	0	.....	Medium wide (36–41)	0	.....
Total	20	100.00	Wide (42–x)	0	.....
			Total	20	100.00

NASAL INDEX

Group	Range	No.	Per cent
Leptorrhine	x–67.4	11	55.00
Mesorrhine	67.5–83.4	9	45.00
Platyrrhine	83.5–x	0	.....
Total		20	100.00

Measurements and Indices of Females in Baghdad Royal Hospital

Measurements	No.	Range	Mean	S.D.	C.V.
Age	20	20–59	30.50±1.41	9.35±1.00	30.66±3.27
Stature	20	143–169	154.50±0.76	5.07±0.54	3.28±0.35
Sitting height	20	72–86	79.00±0.53	3.54±0.38	4.48±0.48
Head length	20	170–193	180.75±0.80	5.28±0.56	2.92±0.31
Head breadth	20	129–152	141.10±0.78	5.19±0.55	3.68±0.39
Minimum frontal diameter	20	89–112	99.50±0.66	4.36±0.46	4.38±0.47
Bizygomatic diameter	20	120–139	127.50±0.67	4.45±0.47	3.49±0.37
Bigonial diameter	20	86–105	94.70±0.52	3.48±0.37	3.67±0.39
Total facial height	20	100–124	111.00±0.74	4.90±0.52	4.41±0.47
Upper facial height	20	60–84	69.00±0.65	4.30±0.46	6.23±0.66
Nasal height	20	40–59	47.50±0.49	3.24±0.35	6.82±0.73
Nasal breadth	20	28–36	32.60±0.27	1.80±0.19	5.52±0.59
Ear length	19	52–71	60.22±0.60	3.88±0.42	6.44±0.70
Ear breadth	20	29–40	32.70±0.38	2.49±0.27	7.61±0.81
Indices					
Relative sitting height	20	48–55	51.20±0.24	1.58±0.17	3.09±0.34
Cephalic	20	71–88	77.85±0.62	4.08±0.44	5.24±0.56
Fronto-parietal	20	66–80	71.05±0.46	3.03±0.32	4.26±0.45
Zygo-frontal	20	72–83	78.10±0.44	2.92±0.31	3.74±0.40
Zygo-gonial	20	69–80	74.35±0.48	3.21±0.34	4.32±0.46
Total facial	20	80–94	87.25±0.51	3.35±0.36	3.84±0.41
Upper facial	20	49–63	54.35±0.44	2.91±0.31	5.35±0.57
Nasal	20	56–79	67.70±0.82	5.44±0.58	8.04±0.86
Ear	19	45–68	54.06±0.75	4.84±0.53	8.95±0.98

Individuals Omitted from Statistical Series

The following information refers to women who can not be grouped into a series.

Nos. 4488–4490 were Turkomans from Kirkuk; Nos. 4491 and 4492 were Jewesses from Baghdad; No. 4493 was a Jewess from Diarbekr; Nos. 4494–4496 were Kurds from Erbil, Sulaimaniya, and Dohuk, respectively; Nos. 4497 and 4498 were Chaldeans from Tell Kaif and Al Qosh, respectively; Nos. 4499–4501 were Assyrians from Darbank (Iran), Tiyari tribe, and Peshabur tribe near Zakho, respectively; No. 4502 was an Irani from Tehran; Nos. 4503 and 4504 were Armenians from Alep and Istanbul, respectively; No. 4505 was a Syrian from Tripoli; No. 4509 was an Arab, aged 16, from Baghdad; No. 4521 was a Dulaimi from near the Diyala; No. 4525 was an Arab, aged 15 (*Sayyida*), from Karbala; and No. 4528 was an Arab with Negro blood, from Baghdad.

No.	Description
4488	Medium epicanthic eye fold; small nose, nasal bridge low; teeth evenly spaced.
4489	Section of hair cut on top of head "to relieve pain in neck"; smallpox scars; some deposit on teeth.
4490	Operation on eyes, which are filmed; vision poor; central incisors and other teeth missing.
4492	Thin.
4493	Flat occiput; small nose.
4494	Alveolar prognathism; two lower molars missing.

No.	Description
4495	Flat occipital area toward left side; scar above left brow, where she was hit by a knife hurled by her husband; soot (*sukham*) was applied to heal the wound; much deposit on teeth; compound fracture of wrist, after operation still hurt, was tattooed on left wrist and on back of hand, but "it still hurt."
4496	Teeth stained, two lower incisors covered with gold; hair dyed with some preparation giving the same effect as henna.
4497	Breathed with difficulty; looked older than probable age.
4498	Maximum point of head low, flatter above; few teeth left, chiefly in front; three miscarriages.
4500	Upper molars gold-plated.
4501	In hospital for sake of child.
4502	Yellowish skin; teeth stained.
4503	Small nose; bad deposits on teeth, several broken off.
4504	Teeth slightly yellow; short nose.
4505	Several teeth broken and missing; *chawi* on left forearm.
4509	Small round scar on back of left hand where a piece of flesh was cut out to cure internal pain; large dark scars of Baghdad boil between eyes and on forehead; teeth slightly yellow.
4525	Negro admixture; small nose; hair matted, full of lice; one eye lost.

OBSERVATIONS RECORDED ON NASAL SEPTUM

No.	Septum	Inclination	Elevation	No.	Septum	Inclination	Elevation
4488	straight	up	elevated	4498	convex	down	depressed
4489	straight	up	....	4499	convex	down	....
4490	convex	down	elevated	4500	convex	up	depressed
4491	......	down	depressed	4501	convex	up	depressed
4492	straight	....	depressed	4502	......	up	elevated
4493	convex	down	depressed	4503	straight	up	depressed
4494	......	down	depressed	4504	straight	down	depressed
4495	straight	up	depressed	4505	convex	up	elevated
4496	convex	up	elevated	4509	straight	up	depressed
4497	convex	up	depressed	4521	......	up	depressed
				4525	straight	up	....

Prognathism.—Nos. 4494, 4503, and 4525 had slight alveolar prognathism.

Eyes.—No. 4488 had blue-ringed and No. 4501 gray-ringed eyes.

ELEVEN GIRLS EXAMINED IN ROYAL HOSPITAL, BAGHDAD

Provenance.—No. 4530, Arab and Turkish from Baghdad; No. 4531, Arab from Baghdad; No. 4532, Arab from Mosul; No. 4533, Arab from Baghdad, father from Kurdistan; No. 4534, father Turk and Kurd, mother from Iran and the Caucasus; No. 4535, Arab from Baghdad; No. 4536, Arab from Baghdad, father from Kirkuk; Nos. 4537 and 4538, Arabs from Baghdad; No. 4539, Arab from Baghdad, ancestors on both sides from Mosul; and No. 4540, Chaldean from Al Qosh.

No. 4539 belonged to a Christian family, all of whom had light blue or green eyes. According to this informant her Moslem friends possessed darker, curlier hair than those of the Christian group.

With the exception of Nos. 4530 and 4540, these girls were measured and examined in the Central School for Girls, Baghdad.

Morphological Observations on Eleven Girls

Skin.—Nos. 4534 and 4538 had darker than average skin color.

Hair.—No. 4535 had applied peroxide, so the hair was reddish, with lighter parts on the surface. No. 4538 had a line of hair from eyebrows to hairline. Her arms were unusually hirsute.

Physiognomy.—No. 4538 had a low brow.

Nose.—No. 4532 had very round nostrils. No. 4534 had a broad nose. The nasion depression was almost absent in No. 4535. It was difficult to locate the subnasion point of No. 4536 as the nasal tip overhung.

Teeth.—No. 4533 possessed good teeth.

Lips.—No. 4537 had slightly higher than average integumental thickness.

Negroid.—No. 4534 appeared to have slight Negroid admixture.

Pathology.—No. 4531 had a boil scar on the right side of her nose, which invalidated measurement of the nasal breadth. No. 4533 had scars from Baghdad boils. No. 4535 had smallpox scars, which invalidated the measurement of the nasal breadth.

When the forty-one adult females are grouped into one series the following table results:

Measurements and Indices of Females in the Baghdad Royal Hospital

Measurements	No.	Mean	S.D.	C.V.
Stature	41	152.39±.70	6.65±.50	4.36±.33
Sitting height	41	78.29±.42	4.02±.30	5.13±.38
Head length	41	178.51±.67	6.38±.48	3.57±.27
Head breadth	41	142.59±.60	5.73±.43	4.02±.30
Minimum frontal diameter	41	101.20±.46	4.40±.33	4.35±.33
Bizygomatic breadth	41	128.56±.51	4.84±.36	3.76±.28
Bigonial breadth	41	95.34±.48	4.55±.34	4.71±.36
Total facial height	41	109.80±.47	4.48±.33	4.08±.30
Upper facial height	41	68.29±.43	4.04±.30	5.92±.44
Nasal height	41	47.63±.36	3.44±.26	7.22±.55
Nasal breadth	41	32.51±.28	2.64±.20	8.12±.62
Ear length	41	60.54±.43	4.05±.30	6.69±.50
Ear breadth	41	32.68±.27	2.59±.19	7.93±.58
Indices				
Relative sitting height	41	51.37±.15	1.41±.11	2.74±.21
Cephalic	41	80.00±.47	4.44±.33	5.55±.41
Fronto-parietal	41	71.04±.33	3.18±.24	4.48±.34
Zygo-frontal	41	78.76±.28	2.67±.20	3.39±.25
Zygo-gonial	41	74.23±.35	3.37±.25	4.54±.34
Total facial	41	85.49±.37	3.55±.26	4.15±.30
Upper facial	41	53.16±.34	3.19±.24	6.00±.45
Nasal	40	68.04±.64	6.03±.45	8.86±.66
Ear	41	54.21±.56	5.28±.39	9.74±.72

Anthropometric Data: Royal Hospital

Measurements of Females in the Baghdad Royal Hospital

No.	Age	Stature	SH	L	B	B'	J	go-go	GH	G'H	NH	NB
4488*	18	1485	772	177	146	103	130	92	107	66	42	32
4489*	23	1650	862	183	143	102	132	96	114	65	48	33
4490*	50	1506	784	182	143	114	134	109	116	71	49	44
4491*	18	1466	732	175	143	100	122	92	112	67	50	30
4492*	32	1580	804	189	147	99	132	105	109	69	49	29
4493*	21	1415	694	172	140	98	126	93	103	62	46	30
4494*	25	1553	799	179	151	106	135	98	109	65	45	31
4495*	30	1577	824	165	147	103	133	99	116	71	55	30
4496*	18	1411	751	172	154	104	132	96	110	65	46	36
4497*	28	1447	759	163	135	98	123	88	104	67	49	30
4498*	58	1583	770	180	157	104	136	101	(108)†	(69)†	50	35
4499*	38	1610	837	186	150	109	138	93	112	73	55	35
4500*	25	1538	833	171	141	102	130	96	104	60	43	33
4501*	20	1560	825	174	139	104	130	97	111	69	47	37
4502*	17	1385	732	174	146	106	129	89	103	65	42	32
4503*	50	1471	736	168	152	104	136	104	111	71	49	31
4504*	30	1470	766	175	143	104	134	93	111	70	45	32
4505*	35	1450	774	180	142	95	123	93	108	70	50	32
4506	22	1547	761	180	139	98	130	97	111	67	46	(31)§
4507	35	1542	838	187	137	101	122	96	111	69	48	32
4508	30	1526	785	172	141	97	129	94	113	69	50	32
4509*	16	1553	771	181	141	105	124	90	110	69	49	32
4510	55	1550	774	187	145	96	124	91	(112)†	(73)†	51	32
4511	25	1491	804	174	144	103	126	90	(112)¶	(66)¶	47	34
4512	25	1604	821	182	141	103	127	90	111	72	50	35

Indices of Females in the Baghdad Royal Hospital

No.	EL	EB	RSH	B/L	B'/B	GH/J	G'H/J	NB/NH	EB/EL	go-go/J	B'/J
4488*	62	32	52.0	82.5	70.5	82.3	50.8	76.2	51.6	70.8	79.2
4489*	63	33	52.2	78.1	71.3	86.4	49.2	68.8	52.4	72.7	77.3
4490*	68	35	52.1	78.6	79.7	86.6	53.0	89.8	51.5	81.3	85.1
4491*	56	28	49.9	81.7	69.9	91.8	54.9	60.0	50.0	75.4	82.0
4492*	65	35	50.9	77.8	67.3	82.6	52.3	59.2	53.8	79.5	75.0
4493*	58	29	49.0	81.4	70.0	81.7	49.2	65.2	50.0	73.8	77.8
4494*	60	33	51.4	84.4	70.2	80.7	48.1	68.9	55.0	72.6	78.5
4495*	65	36	52.3	89.1	70.1	87.2	53.4	54.5	55.4	74.4	77.4
4496*	54	35	53.2	89.5	67.5	83.3	49.2	78.3	64.8	72.7	78.8
4497*	55	32	52.5	82.8	72.6	84.6	54.5	61.2	58.2	71.5	79.7
4498*	62	36	48.6	87.2	66.2	79.4	50.7	70.0	58.1	74.3	76.5
4499*	64	33	52.0	80.6	72.7	81.2	52.9	63.6	51.6	67.4	79.0
4500*	(60)‡	25	54.2	82.5	72.3	80.0	46.2	76.7	41.7	73.8	78.5
4501*	(60	31	52.9	79.9	74.8	85.4	53.1	78.7	51.7	74.6	80.0
4502*	62)‡	33	52.9	83.9	72.6	79.8	50.4	76.2	53.2	69.0	82.2
4503*	66	33	50.0	90.5	68.4	81.6	52.2	63.3	50.0	76.5	76.5
4504*	62	33	52.1	81.7	72.7	82.8	52.2	71.0	53.2	69.4	77.6
4505*	63	31	53.4	78.9	66.9	87.8	56.9	64.0	49.2	75.6	77.2
4506	62	36	49.2	77.2	70.5	85.4	51.5	67.4	58.1	74.6	75.4
4507	58	34	54.3	73.3	73.7	91.0	56.6	66.7	58.6	78.7	82.8
4508	63	34	51.4	82.0	68.8	87.6	53.5	64.0	54.0	72.9	75.2
4509*	56	30	49.6	77.9	74.5	88.7	55.6	65.3	53.6	72.6	84.7
4510	66	36	49.9	77.5	66.2	90.3	58.9	62.7	54.5	73.4	77.4
4511	56	38	53.9	82.8	71.5	88.9	52.4	72.3	67.9	71.4	81.7
4512	57	34	51.2	77.5	73.0	87.4	56.7	70.0	59.6	70.9	81.1

* Omitted from averages. † Edentulous. ‡ Stretched.
§ Questionable. ¶ Questionable; upper incisors pulled out.

Measurements of Females in the Baghdad Royal Hospital

No.	Age	Stature	SH	L	B	B'	J	go-go	GH	G'H	NH	NB
4513	23	1455	740	179	150	105	129	97	117	73	49	31
4514	33	1582	837	185	141	103	135	101	109	73	51	34
4515	26	1554	786	186	142	99	127	92	110	65	45	33
4516	20	1547	808	180	131	100	125	95	103	62	46	33
4517	22	1532	748	180	140	99	126	94	110	69	50	33
4518	30	1522	770	183	137	96	122	89	109	67	44	32
4519	43	1573	805	183	138	100	124	96	105	70	47	29
4520	23	1427	757	172	147	99	127	101	108	67	43	34
4521*	23	1560	781	185	135	100	122	98	104	67	44	31
4522	33	1561	785	184	150	104	134	92	119	73	47	34
4523	48	1554	785	178	135	98	130	102	120	81	57	33
4524	38	1590	831	180	145	95	129	90	116	73	47	29
4525*	15	1389	696	171	134	95	120	93	102	59	41	30
4526	29	1460	730	174	136	91	120	96	104	64	45	31
4527	20	1541	782	191	140	109	135	96	113	69	49	31
4528	22	1666	865	180	138	98	129	95	105	68	47	35

Indices of Females in the Baghdad Royal Hospital

No.	EL	EB	RSH	B/L	B'/B	GH/J	G'H/J	NB/NH	EB/EL	go-go/J	B'/J
4513	58	33	50.9	83.8	70.0	90.7	56.6	63.6	56.9	75.2	81.4
4514	62	31	52.9	76.2	73.0	80.7	54.1	66.7	50.0	74.8	76.3
4515	53	29	50.6	76.3	69.7	86.6	51.2	73.3	54.7	72.4	78.0
4516	58	31	52.2	72.8	76.3	82.4	49.6	71.7	53.4	76.0	80.0
4517	59	33	48.8	77.8	70.7	87.3	54.8	66.0	55.9	74.6	78.6
4518	62	30	50.6	74.9	70.1	89.3	54.9	72.7	48.4	73.0	78.7
4519	61	33	51.2	75.4	72.5	84.7	56.5	61.7	54.1	77.4	80.6
4520	59	30	53.0	85.5	67.3	85.0	52.8	79.1	50.8	80.0	78.0
4521*	51	36	50.1	73.0	74.1	85.2	54.9	70.5	70.6	80.3	82.0
4522	58	30	50.3	81.5	69.3	88.8	54.5	72.3	51.7	68.7	77.6
4523	67	32	50.5	75.8	72.6	92.3	62.3	57.9	47.8	78.5	75.4
4524	66	36	52.3	80.6	65.5	89.9	56.6	61.7	54.5	69.8	73.6
4525*	58	35	50.1	78.4	70.9	85.0	49.2	73.2	60.3	77.5	79.2
4526	(61)†	32	50.0	78.2	66.9	86.7	53.3	68.9	52.5	80.0	75.8
4527	58	31	50.7	73.3	77.9	83.7	51.1	63.3	53.4	71.1	80.7
4528	68	33	51.9	76.7	71.0	81.4	52.7	74.5	48.5	73.6	76.0

*Omitted from the averages. † Stretched.

ANTHROPOMETRIC DATA: ROYAL HOSPITAL

MORPHOLOGICAL CHARACTERS OF FEMALES IN THE BAGHDAD ROYAL HOSPITAL

	HAIR			EYES			NOSE	
No.	Form	Texture	Color	Color	Sclera	Iris	Profile	Wings
4488	v l w	medium	dk br	dk br	clear	....	str	medium
4489	l w	medium	black	dk br	clear	....	conv	medium
4490	....	fine	white	dk br	blood	....	conv	flar+
4491	l w	medium	dk br	black	clear	....	conv	comp
4492	v l w	coarse	dk br	dk br	clear	....	conv	medium
4493	l w	medium	dk br	dk br	clear	....	c-c	medium
4494	v l w	medium	v dk br	black	yellow-blood	....	str	medium
4495	l w	medium	dk br	dk br	clear	....	conv	cp-m
4496	l w	medium	dk br	bl-gray	blood	hom	conc	flar
4497	l w	medium	dk br	dk br	clear	....	conv	comp
4498	l w	medium	blk, gray	dk br	blood	....	conv	flar
4499	v l w	fine	dk br	gray-gr	blood	....	c-c	....
4500	l w	fine	dk br	green-br	clear	zon	conv	medium
4501	v l w	fine	dk br	dk br	clear	....	c-c	flar
4502	l w	medium	br, gray	black	clear	....	conc	medium
4503*	l w	fine	gray	gray-br	blood	zon	c-c	medium
4504	l w	fine	dk br	green-br	clear	zon	c-c	medium
4505	l w	coarse	v dk br	green-br	clear	zon	conv	medium
4506	l w	fine	dk br	dk br	blood	....	str	medium
4507	l w	medium	blk, gray	black	clear	....	conv	comp
4508	l w	medium	black	gr-gray (flecked)	clear	....	conv	medium
4509	l w	coarse	dk br	dk br	clear	....	conc	medium
4510	l w	fine	gray	lt br	blood	....	conv	m-fl
4511*	l w	medium	dk br	black	blood	....	conv	medium
4512	l w	coarse	black	dk br	clear	....	conc	medium
4513	d w	medium	black	black	clear	....	str	....
4514	l w	fine	dk br	gr-br	clear	hom	conc	medium
4515	l w	medium	dk br	black	clear	....	conv	flar
4516	l w	coarse	black	black	clear	....	str	medium
4517	l w	coarse	black	black	clear	....	str	flar
4518	l w	fine	black	dk br	clear	....	conc	medium
4519	l w†	fine	black	dk br	clear	....	conv	cp-m
4520	l w	medium	dk br	dk br	clear	....	conc	flar
4521‡	....	....	dk br	dk br	clear	....	str	medium
4522	l w	coarse	dk br	dk br	clear	....	conc	medium
4523	l w	medium	br, gray	.....	.....	....	conv	medium
4524‡	....	....	dk br	dk br	clear	....	str	cp-m
4525	....	....	dk br	.....	.....	....	conc	m-fl
4526	l w	medium	black	dk br	blood	....	conc	medium
4527	l w	medium	black	dk br	clear	....	conv	medium
4528	d w	medium	dk br	dk br	clear	....	conc	flar

* Very thin.   † Cut off.   ‡ Shaved.

APPENDIX F: MAMMALS FROM IRAQ

BY

Colin C. Sanborn[1]

In 1934 Dr. Henry Field, leader of the Field Museum Anthropological Expedition to the Near East, and Mr. Richard A. Martin collected zoological specimens in Iraq and Iran.

As a result of their efforts and the subsequent gifts which have come to the Museum from the Near East this list of mammals has been prepared.

Since the zoological material available from this part of the world is limited, there is considerable uncertainty of identification.

The names of collectors have been given in parentheses. Dr. Walter P. Kennedy was a staff member of the Royal College of Medicine in Baghdad. Mr. Austin Eastwood, head of the Cotton Growers Association, maintained a private zoo in Baghdad. Mrs. E. S. Drower, author of several books on Iraqi folklore, has lived in Baghdad for the past fifteen years. Mr. J. H. Dekker, who was in charge of the Iraq Petroleum Company's pipe-line station T-3, died in 1936. Philippus Dinka, an Assyrian, was superintendent of the British Consulate at Diana-Rowandiz in 1934. He is now with the British Oil Development Company near Mosul.

Last, but not least by far, comes Yusuf Lazar, another Assyrian, who was an invaluable zoological collector during 1934. Since that time he has continued to send specimens to the Museum. Dr. Field has supplied the funds necessary for this important work.

The spelling of place names conforms, wherever possible, to the style adopted by the Permanent Committee on Geographical Names of the Royal Geographical Society in London. In addition to the names of places which can be located with ease, such as Baghdad, Basra, Hilla, An Nasiriya, Karbala, Balad Sinjar, Khanaqin, and Sulaimaniya, the following groupings can be made:

Northern Area.—Diana-Baradost, Baadri, Rowandiz, and Guli Ali Bagh.

Southern Area.—Qala Salih, Amara, Chahala, and Halfaya.

We hope that zoological specimens from Southwestern Asia will continue to enrich the study collections in the Museum.

[1] Curator of Mammals at Field Museum.

Hemiechinus auritus Gmelin.

Near Baghdad, female with five young, alcoholic (Kennedy); skull only (Dinkha).

Liponycteris kachhensis magnus Wettstein.

Taphozous magnus Wettstein, Ann. Konigl. Naturhist. Hofmuseums, Wien, 27, p. 466, 1913—Basra.

Taphozous kachhensis babylonicus Thomas, Journ. Bombay Nat. Hist. Soc., 24, p. 58, 1915—Euphrates.

Taphozous kachhensis magnus Cheesman, Journ. Bombay Nat. Hist. Soc., 27, p. 328, 1920.

Liponycteris kachhensis magnus Thomas, Ann. Mag. Nat. Hist., (9), 9, p. 267, 1922.

Baghdad and Tall Tauwa near Baghdad, 4 males, 3 females, April 1–16, 1934 (Field).

This form has been recorded from Lake Tiberias southeastward to the head of the Persian Gulf.

Measurements.—Forearm 79–83.3; second finger, metacarpal 67–71.8; third finger, metacarpal 71.9–76.6, first phalanx 27.5–31.7, second phalanx 30.8–33.9; fourth finger, metacarpal 58.8–63.3, first phalanx 14.3–18.5, second phalanx 8.3–9.3; fifth finger, metacarpal 48.6–51.4, first phalanx 14.7–17.4, second phalanx 7.8–9.6; tibia 31.4–35; ear from meatus 21–24; tragus, height 5.5–6, width 5–6. Skull: male, greatest length 32.2; condylo-basal length from front of canine 27.2; palatal length 8.2; interorbital width 9.4; intertemporal width 5.5; zygomatic width 18.3; mastoid width 16; width of braincase 12; length of upper toothrow 12.7; width across cingula of canines 7.2; width across m^2 11.7; lower toothrow 14.1; mandible 23.3. The males are slightly larger than the females.

Asellia tridens murraiana Anderson.

Baghdad, 13 females, May–August, 1935; 1 male, June 19, 1936 (Lazar).

The forearms on these specimens are so long (51–55.4) that they are referred to this subspecies.

Myotis myotis omari Thomas.

Myotis myotis omari Thomas, Proc. Zool. Soc. Lond., pt. 2, p. 521, 1905.

Diana-Baradost, northeast Iraq, male, female, June 29, 1934.

The only recorded specimens of this form are the type and topotype from Derbent, fifty miles west of Isfahan, Iran, and a female from Telespid, southwestern Iran. The Iraq specimens agree fairly well with the original description.

Measurements (male and female, and skull of male).—Forearm 59.1–60.4; second finger, metacarpal 56–56.6; third finger, metacarpal 55.5–57, first phalanx 19.3–19.6, second phalanx 16–18; fourth finger, metacarpal 54.8–55.9, first phalanx 13.4–14.7, second phalanx 13.6–13.6; fifth finger, metacarpal 53.3–54.1, first phalanx 13–13.8, second phalanx 13.5–11; tibia 26.4–27; calcar 16.3–17.1; ear from meatus 24–25; height of tragus 11–11. Skull: greatest length 22.5; condylo-basal length 20.9; palatal length 9.3; interorbital width 5.1; zygomatic width 13.9; mastoid width 10.1; width of braincase 9.7; upper toothrow 9.1; width across cingula of canines 5.8; width across m^2 9; lower toothrow 9.7; mandible 17.

Pipistrellus kuhli Kuhl.

Baghdad, 33, from April 1, 1934 (Field and Martin) to June 10, 1938 (Lazar); Amara marshes, 1, April 22, 1934 (Field); Sheikh Falih as Saihud's camp, 23, April 27, 1934 (Field); Halfaya, 49, April 28, 29, 1934 (Field); Balad Sinjar, 6, June 4, 1934 (Field); Baadri, 2, June 14, 1934 (Field); Tall Tauwa, near Baghdad, 6, April 1–16, 1934 (Field); An Nasiriya, 6, March 12–24, 1935 (Lazar); Karbala, 4, October 10, 1937 (Lazar); Rustam Farm, near Baghdad, 20, January 9, 1939 (Lazar).

This appears to be the commonest bat in Iraq.

Eptesicus hingstoni Thomas.

Eptesicus hingstoni Thomas, Journ. Bombay Nat. Hist. Soc., **26**, p. 745, 1919 —Iraq (Baghdad).

Baghdad, 1 male, 4 females, April 1–16, 1934, 1 male, January–April, 1935 (Lazar), 1 male, 2 females, May–August, 1935 (Lazar), 1 juv. male, 1 juv. female, June 19, 1936 (Lazar); An Nasiriya, March 12–24, 1935 (Lazar); Karbala, 1 male, October 10, 1937 (Lazar).

Measurements.—Forearm 43.7–47.8; second finger, metacarpal 40.2–45; third finger, metacarpal 41.1–46, first phalanx 12.3–14.5, second phalanx 11.1–12.5; fourth finger, metacarpal 40–45, first phalanx 10.7–11.9, second phalanx 8.9–10.1; fifth finger, metacarpal 38.8–43.1, first phalanx 8.8–9.6, second phalanx 6.4–8.2; tibia 18.5–20.2; calcar 17–19. Skull (female): greatest length 18; condylo-basal length 16.9; palatal length 7.6; interorbital width 3.7; zygomatic width 11.4; mastoid width 8.8; width of braincase 7.8; upper toothrow 6.4; width across cingula of canines 5.4; width across m^2 6.6; lower toothrow 6.9.

Eptesicus walli Thomas.

Eptesicus walli Thomas, Journ. Bombay Nat. Hist. Soc., **26**, p. 746, 1919—Iraq (Basra).

An Nasiriya, 2 males, 7 females, March 12–24, 1935 (Lazar).

This is the second published record of this bat. It appears to be much scarcer than *E. hingstoni* as Lazar has collected it but once in five years.

Measurements.—Forearm 36.3–40.9; third finger, metacarpal 36–37.9, first phalanx 10.4–11.9, second phalanx 10.6–11.7; fourth finger, metacarpal 34.7–37.9, first phalanx 8.6–10.5, second phalanx 8.1–9.8; fifth finger, metacarpal 33.4–36.4, first phalanx 6.5–8.5, second phalanx 6.4–7.6. Tibia 14.6–15.8; ear 12.3–13. Skull of largest female: total length 14.4; condylo-basal length 13.3; palatal length 6.8; interorbital width 3.8; rostral width 5.5; zygomatic width 9.5; mastoid width 7.4; width of braincase 6.6; upper toothrow 5.2; width across canines 4.5; across molars 6.3.

Canis aureus Linnaeus.

Iraq, 1 trade skin, no locality (Field); Diyala, 1 skull (Lazar).

Canis lupus pallipes Sykes.

Seri Hassan Beg Mountains, Rowandiz District, 1 skeleton (Dinkha); Diyala, 1 skeleton (Lazar); Sulaimaniya, 1 skull only (Lazar); Khanaqin, 1 skeleton (Lazar).

Pocock (Proc. Zool. Soc. Lond., 1935) referred the Iraq wolves to this subspecies.

Vulpes vulpes splendens Thomas.

Rowandiz, 1 trade skin, no skull. This very large skin in worn pelage is referred to this form.

Vulpes persica Blanford.

One skin without skull or locality in Iraq (Lazar).

Herpestes persicus Gray.

Baghdad, 1 skull only (Lazar).

Felis chaus Güldenstädt.

Hilla Desert, 1937, 1 skin only (Lazar).

Two skulls without skins from Baghdad (Lazar) appear to be house cats.

Martes foiana Erxleben.

Rowandiz, 1 trade skin without skull (Field).

Lutra lutra Linnaeus.

Qala Salih, near Amara, 1 trade skin without skull (Drower).

Meles meles subsp.

Guli Ali Bagh, 1 male, skin and skeleton, 1937 (Dinkha).

This is the first time this genus has been recorded from Iraq. It probably belongs to either the subspecies *minor* Satunin or *caucasicus* Ognev, but comparative material is not available.

Measurements.—Skull: total length 136.3; condylo-basal length 128; palatal length from in front of incisors 71.8; zygomatic width about 85; mastoid width 63.8; width of braincase 51.9; interorbital width 29.7; intertemporal width 24.3; upper toothrow 43.3; maxillary width 46.6.

Mellivora wilsoni Cheesman.

Mellivora wilsoni Cheesman, Journ. Bombay Nat. Hist. Soc., **27**, p. 335, 1920
—Ram Hormuz, southwestern Iran.

Station T-1, on northern oil pipe-line, western Iraq (Dekker).

This specimen agrees closely with the original description.

Cheesman recorded a specimen from Baksal, Tyb River, which is the only other published record for Iraq.

Measurements.—Greatest length 121.1; condylo-basal length 60.8; interorbital width 33.2; intertemporal width 36; zygomatic width 71.4; mastoid width 67.4; width of braincase 62.5; upper toothrow 33.8; maxillary width 42.

Hyaena hyaena Linnaeus.

Baghdad, skeleton, July 6, 1936 (Eastwood); Baradost, skin and skeleton, 1937 (Dinkha).

These might be the subspecies *zarudnye* but Pocock (Proc. Zool. Soc. Lond., p. 820, 1934) considered this form a probable synonym of *hyaena*.

Measurements.—Total length 229–241; condylo-basal length 214.1–218.7; zygomatic width 152.6; postorbital width 35.9–36.6;

interorbital width 47.8–46.2; maxillary width 86.4–87.5; width at base of canines 51.8–52.5; length of p^4 29.6–30.5.

Sus scrofa attila Thomas.

Chahala, near Amara, male and female, skins with skeletons, 4 young in alcohol, April 23, 1934 (Field). Place: Rhamalla, ten miles from Khanaqin, 4 skulls only, February 28, 1935 (Lazar); Khanaqin, 2 skulls only, November 15, 1935 (Lazar); Baradost, skin and skeleton, 1937 (Dinkha).

Measurements (2 from Chahala; taken on dried skin).—Total length 1450–1430; tail 170 (about)–240. Skull: greatest length 450–370; condylo-basal length 383–343; zygomatic width 166.5–153; interorbital width 95.7–74.7; length of nasals 254–200; width of nasals 36.5–29.4; occipital depth 144–122; length of mandible 330–305; maxillary toothrow c–m^3 167–151; p^1–m^3 133.4–121.4; lower toothrow including canine 173.2–164, excluding canine 143.2–139; third upper molar 42.8 x 26.6–37.6 x 23.2; third lower molar 41.6 x 20.6–41.5 x 17.3; width of internal face of lower canine 23.5–25.

Ursus arctos Linnaeus.

Baghdad, skin and skeleton, 1935 (Eastwood); Baradost, skin and skeleton, 1937 (Dinkha).

Jaculus loftusi Blanford.

Dipus loftusi Blanford, Ann. Mag. Nat. Hist., (4), **16**, p. 312, 1875—Mohammerah (now Khorram Shahr), southwestern Iran.

Near Rutba, Station H-3, female, May 10, 1934.

Measurements.—Total length 270; tail 160; hind foot 58.5. Skull: greatest length 33.6; condylo-basal length 29.5; palatal length 8.2; interorbital width 11.5; zygomatic width 23.4; mastoid width 23.8; width of braincase 18.5; upper molar series 4.9; lower molar series 5.2; mandible 11.7; anterior palatine foramina 4; orbital width 23.4.

Nesokia buxtoni Thomas.

Baghdad, skull only, November 10, 1935 (Dinkha).

Lepus connori Robinson.

Hinaidi Bridge, ten miles southeast of Baghdad, skull only, December 18, 1935 (Lazar); Baghdad, 1 skeleton, 2 skulls only, November 2, 1935 (Dinkha).

Referred to this species on the basis of Cheesman's list.

Gazella sp.

Iraq, 1 skeleton, 2 skins only; Diyala, 2 skulls only, November 20, 1935 (Lazar); Hinaidi Bridge, 2 skulls only, November 26, 1935 (Lazar); Baghdad, skeleton, December 2, 1936 (Lazar); skeleton, September 2, 1936 (Eastwood); 3 skulls only, September 7, 1936 (Eastwood); Karbala, skull only, December 12, 1935 (Eastwood).

There are not enough skins in this collection to determine which species is represented.

Capra aegagrus blythi Lydekker.

Baradost, Rowandiz District, 2 skeletons, February, 1935 (Dinkha); Barzan near Aqra, horns.

APPENDIX G: NOTES ON INSECTS FROM IRAQ

During the Field Museum Anthropological Expedition to the Near East in 1934, we collected a number of insects in Iraq and in Iran.

Since our return to Chicago, Yusuf Lazar, my Assyrian zoological and botanical collector, has sent additions to our collections from the Baghdad area.

Through the kindness of Captain N. W. Riley, Keeper of the Department of Entomology at the British Museum (Natural History) all specimens have been sent to London to be determined.

As a direct result two papers have appeared: "Hemiptera from Iraq, Iran, and Arabia" by W. E. China, and "Orthoptera from Iraq and Iran" by B. P. Uvarov (Field Mus. Nat. Hist., Zool. Ser., vol. 20, Nos. 32 and 33, pp. 439–451).

In a letter from the British Legation, Tehran, dated August 9, 1939, Mr. E. P. Wiltshire writes: "With regard to the material collected by Yusuf [Lazar] in Baghdad for you, I was able to get a glimpse of it recently in London and can confirm that he got a male *Sumeria dipotamica* Tams in Baghdad; also one or two *Nychiodes divergaria* Stgr., which confirms my guess made in my article on the Baghdad Orchard."

Wiltshire has published the following notes (pp. 17, 20) excerpted from the "Entomologist's Record" (Feb. 15, 1939).

"1. *Nychiodes*(?) *divergaria*. Small larvae of this genus were found in XI.37 in numbers at night on apricot trees. Unfortunately I was obliged to take them with me to Tabriz in December, where the winter was longer and severer than Baghdad's. None hibernated successfully, so I cannot be sure of the species' identity, but expect that it will prove to be *divergaria*[1] which I have found not uncommonly in Kurdistan."

"2. Until the life-history of this recently described Notodontid is known, one cannot say to which of the above divisions of the Mesopotamian fauna it pertains, though, to judge from its facies and the situations in which I have taken it, it may well prove to be a reed-feeder. It seems to be most frequent in the delta of the Euphrates and Tigris, but it also occurs up to some height in the Zagros range. In 1938 I captured a female at Basra (25.V.) and a male at Khorram Shahr (Mohammerah) (2.X), both to light near

[1] Confirmed from Field Museum specimens collected in Baghdad by Yusuf Lazar.

the river. I also believe[1] it occurs at Baghdad. Since no description of the female was published by Mr. Tams, I append one hereto:

> "*Sumeria dipotamica*, Tams (Proc. R. Ent. Soc. London (B) 1938). Female Neallotype; Basra, 25.V.1938, in coll. m.
> Antenna: Much more lightly bipectinated than male.
> Expanse: 54 mm., i.e., considerably larger than male.
> In other respects, similar to the male.

"N.B.—The autumnal brood male taken by me at Khorramshahr was only 40 mm. in expanse."

The remainder of the collections, particularly the Lepidoptera, awaits determination.

[1] Confirmed from Field Museum specimens collected in Baghdad by Yusuf Lazar.

APPENDIX H: PLANTS COLLECTED BY THE EXPEDITION

BY

PAUL C. STANDLEY[1]

During 1934, while leader of the Field Museum Anthropological Expedition to the Near East, Dr. Henry Field supervised the collecting of about 10,000 herbarium specimens from Iraq, Iran, Trans-Jordan, Palestine, and Syria. He also collected a number of useful plants and drugs.[2]

The majority of the specimens in the following list were determined at Field Museum, but since the Herbarium does not contain large collections from Southwestern Asia, it was necessary to send series to European experts for determination. Part of the collection was therefore sent to Kew Herbarium where the late Mr. A. R. Horwood identified some of the specimens. A recent letter from Sir Arthur Hill states that as a result of Mr. Horwood's death, followed shortly by Air Raid Precautions, no further work can now be done on this collection.

Prior to 1934 Mr. Evan Guest, who was attached to the Ministry of Agriculture in Baghdad, made a large collection of herbarium specimens in Iraq. During his trips to northern Iraq and Kurdistan, Yusuf Lazar, an Assyrian, accompanied him as a botanical collector. The Guest Collection also awaits identification at Kew.

In 1934, Dr. Field engaged the services of Yusuf Lazar as a botanical and zoological collector. The greater part of the specimens listed are the fruit of his remarkable energy and painstaking devotion in this service in both Iraq and Iran. Since 1935, working as a private collector financed by Dr. Field, he has forwarded to the Museum additional herbarium specimens, mainly from the Baghdad area. In the following list those specimens marked "F & L" were collected during the 1934 Expedition, those with "L" by Yusuf Lazar, 1935-39.

Other undetermined specimens were sent to Dr. Gunnar Samuelsson of the Natural History Museum in Stockholm, and to Professor J. Bornmüller, Weimar. These two specialists have submitted determinations that are included in the following list. Although several hundred numbers still await identification, a provisional list is herewith appended.

[1] Curator of the Herbarium, Field Museum.
[2] See Hooper and Field.

Dr. Rustam Hydar, Director of the Rustam Agricultural Experimental Farm near Baghdad, generously presented to Field Museum a number of varieties of *Gossypium*, *Hordeum*, and *Triticum* which are not listed here, since they were enumerated by Hooper and Field (pp. 122–124, 126–127, 181–183).

The reader is referred to "Useful Plants and Drugs of Iran and Iraq" by David Hooper and Henry Field (Field Mus. Nat. Hist., Bot. Ser., vol. 9, No. 3, pp. 71–241, 1937), and particularly to the list of bibliographical references (pp. 75–78). In addition, the publications of Boissier (1867–84), Schlimmer (1874), Dymock (1885, 1891), Aitchison (1890), Post (1896), Burkill (1909), Bornmüller (1917), Laufer (1919), Gilliat-Smith and Turrill (1930), Guest (1933), Samuelsson (1933 et seq.), and Vavilov (1934 et seq.) should be used as standard references.

The spelling of place names conforms wherever possible to the system adopted by the Permanent Committee on Geographical Names of the Royal Geographical Society in London.

ALPHABETICAL LIST OF PLANTS COLLECTED IN IRAQ

Number	Genus and Species	Locality
L 374	*Acanthophyllum microcephalum* Boiss.	Near Baghdad
F & L 892	*Acanthus longistylis* Freyn.	Jebel Baradost near Diana Rowandiz
F & L 879, 937	*Acer monspessulanum* L.	Jebel Baradost near Diana Rowandiz
F & L 463	*Achillea aleppica* DC.	Muwasul Tiatan Mukzuk Nuwar
F & L 635, 651	*Achillea aleppica* DC.	Jebel Khatchra near Balad Sinjar
F & L 375	*Achillea conferta* DC.	Telegraph pole M90 between Baiji and Mosul
F & L 539	*Achillea conferta* DC.	Mir Khasim between Balad Sinjar and Tall Afar
F & L 474	*Achillea conferta* DC.	Jebel Golat between Ain Tellawi and Balad Sinjar
L 186	*Achillea falcata* L.	Rustam Farm near Baghdad
F & L 519	*Achillea micrantha* M. Bieb.	Between Tall Afar and Balad Sinjar
F & L 573	*Achillea micrantha* M. Bieb.	Tell Es Shur between Tall Afar and Balad Sinjar
L 145	*Achillea micrantha* M. Bieb.	Near Baghdad
F & L 763	*Achillea micrantha* M. Bieb.	Sheikh Adi near Ain Sifni
F & L 307, 472	*Achillea micrantha* M. Bieb.	Jebel Golat between Ain Tellawi and Balad Sinjar
F & L 470	*Achillea micrantha* M. Bieb.	Muwasul Tiatan Mukzuk Nuwar
F & L 535	*Achillea oligocephala* DC.	Mir Khasim between Balad Sinjar and Tall Afar
F & L 479	*Achillea oligocephala* DC.	Jebel Golat between Ain Tellawi and Balad Sinjar

PLANTS FROM IRAQ

Number	Genus and Species	Locality
F & L 461	Achillea oligocephala DC.	Muwasul Tiatan Mukzuk Nuwar
L 275	Achillea Santolina L.	Near Baghdad
F & L 574	Achillea Santolina L.	Tell Es Shur between Tall Afar and Balad Sinjar
F & L 947	Adiantum Capillus-Veneris L.	Rowandiz Gorge
L 38	Adonis aestivalis L.	Rustam Farm near Baghdad
L 95	Adonis aestivalis L.	Near Baghdad
F & L 803	Aegilops Aucheri Boiss.	Jebel Baykhair near Zakho
F & L 450	Aegilops crassa Boiss.	Muwasul Tiatan Mukzuk Nuwar
F & L 686	Aegilops crassa Boiss.	30 km. due west of Balad Sinjar
F & L 552	Aegilops crassa Boiss.	Mir Khasim between Balad Sinjar and Tall Afar
F & L 423	Aegilops squarrosa L.	Jebel Golat between Ain Tellawi and Balad Sinjar
F & L 458	Aeluropus litoralis (Gouin) Parl.	Muwasul Tiatan Mukzuk Nuwar
L 168, 310, 353	Aeluropus litoralis (Gouin) Parl.	Near Baghdad
L 311, 355	Aeluropus repens (Desf.) Parl.	Rustam Farm near Baghdad
L 184	Agropyron squarrosum (Roth) Link.	Rustam Farm near Baghdad
F & L 911	Ajuga Chia Schreb. var. tridactylites Ging.	Jebel Baradost near Diana Rowandiz
L 338	Alhagi maurorum Medic.	Rustam Farm near Baghdad
F & L 783	Alkanna Kotschyana DC.	Jebel Baykhair near Zakho
F & L 540	Alkanna tinctoria (L.) Tausch.	Mir Khasim between Balad Sinjar and Tall Afar
F & L 370	Allium ampeloprasum L.	Haditha (wheatfield)
F & L 304	Allium paniculatum L.	Jebel Golat between Ain Tellawi and Balad Sinjar
F & L 449	Allium paniculatum L.	Muwasul Tiatan Mukzuk Nuwar
F & L 577	Allium paniculatum L.	Tell Es Shur between Tall Afar and Balad Sinjar
F & L 727	Althaea hirsuta L.	Jerwona near Ain Sifni
F & L 841	Althaea Hohenackeri Boiss. & Huet.	Jebel Pikasar near Aqra
F & L 634	Althaea lavateriflora DC.	Jebel Khatchra near Balad Sinjar
F & L 941	Althaea lavateriflora DC.	Jebel Baradost near Diana Rowandiz
L 214	Althaea Ludwigii L.	Rustam Farm near Baghdad
F & L 120	Althaea Ludwigii L.	Rutba
F & L 593	Althaea rosea Cav.	Tell Es Shur between Tall Afar and Balad Sinjar
F & L 668	Alyssum alpestre L. var. obovatum Boiss.	Jebel Khatchra near Balad Sinjar
F & L 605	Alyssum campestre L.	Karya Sheikh Khanis near Balad Sinjar
L 251, 329, 472	Amaranthus graecizans L.	Near Baghdad
L 429	Amaranthus cf. paniculatus L.	Rustam Farm near Baghdad

Number	Genus and Species	Locality
L 143	*Amaranthus retroflexus* L.	Near Baghdad
L 473	*Amaranthus viridis* L.	Near Baghdad
L 138, 296, 413	*Ammi majus* L.	Near Baghdad
L 429	*Ammi majus* L. var. *longiseta* Reichb.	Near Baghdad
F & L 936	*Amygdalus elaeagrifolia* Spach.	Jebel Baradost near Diana Rowandiz
L 358	*Amygdalus spartioides* Spach.	Near Baghdad
L 4	*Anagallis arvensis* L.	Rustam Farm near Baghdad
F & L 813	*Anagyris foetida* L.	Jebel Baykhair near Zakho
F & L 711	*Anagyris foetida* L.	Sheikh Adi near Ain Sifni
F & L 775	*Anagyris foetida* L.	Jebel Baykhair near Zakho
L 151	*Anchusa strigosa* Labill.	Near Baghdad
F & L 399	*Anchusa strigosa* Labill.	Qala Sharqat
F & L 544	*Anchusa strigosa* Labill.	Mir Khasim between Balad Sinjar and Tall Afar
F & L 452	*Anchusa strigosa* Labill.	Muwasul Tiatan Mukzuk Nuwar
L 306, 357	*Andrachne telephioides* L.	Near Baghdad
F & L 114	*Andrachne telephioides* L.	Montafah
F & L 382	*Andrachne telephioides* L.	Telegraph pole M90 between Baiji and Mosul
F & L 632, 661	*Andrachne telephioides* L.	Jebel Khatchra near Balad Sinjar
F & L 401	*Andrachne telephioides* L.	Qala Sharqat
L 274	*Andropogon annulatus* Forsk.	Near Baghdad
F & L 433	*Androsace maxima* L.	Jebel Golat between Ain Tellawi and Balad Sinjar
F & L 952	*Anthemis altissima* L.	Sulaimaniya
L 197	*Anthemis altissima* L.	Near Baghdad
L 129, 239	*Anthemis Cotula* L.	Near Baghdad
F & L 389	*Anthemis Cotula* L.	Telegraph pole M90 between Baiji and Mosul
L 37	*Anthemis hebronica* Boiss. & Kotschy	Near Baghdad
L 36	*Anthemis* cf. *melampodina* Del.	Rustam Farm near Baghdad
F & L 817	*Apocynum venetum* L.	Jebel Baykhair near Zakho
L 66, 439	*Aristida plumosa* L.	Near Baghdad
F & L 786	*Aristolochia maurorum* L.	Jebel Baykhair near Zakho
F & L 75	*Aristolochia maurorum* L.	Mesaida near Amara
F & L 385	*Arnebia decumbens* (Vent.) Kuntze	Telegraph pole M90 between Baiji and Mosul
F & L 127	*Arnebia decumbens* (Vent.) Kuntze	Rutba
L 64	*Arnebia linearifolia* DC.	Near Baghdad
F & L 547	*Artedia squamata* L.	Mir Khasim between Balad Sinjar and Tall Afar
L 438	*Artemisia annua* L.	Rustam Farm near Baghdad

PLANTS FROM IRAQ

Number	Genus and Species	Locality
F & L 794	*Asparagus stipularis* Forsk.	Jebel Baykhair near Zakho
F & L 613	*Asperugo procumbens* L.	Karya Bat Khan near Balad Sinjar
L 134, 522	*Asperugo procumbens* L.	Near Baghdad
L 350	*Asperula arvensis* L.	Near Baghdad
L 56	*Asphodelus tenuifolius* Cav.	Rustam Farm near Baghdad
L 155	*Astragalus alexandrinus* Boiss.	Near Baghdad
F & L 630	*Astragalus chaborasicus* Boiss. & Hausskn.	Jebel Khatchra near Balad Sinjar
L 49, 53	*Astragalus cruciatus* Link.	Near Baghdad
F & L 141	*Astragalus Forskahlei* Boiss.	Rutba
F & L 606	*Astragalus maximus* Willd.	Tell Es Shur between Tall Afar and Balad Sinjar
F & L 91	*Atractylis flava* Desf.	Montafah
F & L 88	*Atriplex leucoclada* Boiss. subsp. *turcomanica* (Moq.) Aellen	Montafah
L 315, 351	*Atriplex leucoclada* Boiss. subsp. *turcomanica* (Moq.) Aellen	Near Baghdad
F & L 59	*Atriplex tatarica* L.	Mesaida near Amara
F & L 7	*Avena fatua* L.	Gatt Al Dwat near Amara
L 199, 279, F & L 488	*Avena fatua* L.	Near Baghdad
F & L 443	*Avena fatua* L.	Jebel Golat between Ain Tellawi and Balad Sinjar
F & L 43	*Bacopa Monniera* (L.) Wettst.	Chahala near Amara
F & L 51	*Bacopa Monniera* (L.) Wettst.	Near Amara
L 383	*Barbarea vulgaris* R. Br.	Near Baghdad
F & L 13	*Beta vulgaris* L. subsp. *lomatogonoides* Aellen	Gatt Al Dwat near Amara
L 564	*Beta vulgaris* L. subsp. *maritima* (L.) Thell. var. *glabra* Aellen	Near Baghdad
L 159	*Beta vulgaris* L. subsp. *vulgaris* (L.) Thell.	Near Baghdad
F & L 440	*Bromus macrostachys* Desf.	Jebel Golat between Ain Tellawi and Balad Sinjar
F & L 897	*Bromus macrostachys* Desf.	Jebel Baradost near Diana Rowandiz
L 187	*Bromus mollis* L.	Near Baghdad
F & L 136	*Bromus tectorum* L.	Rutba
L 179, 256	*Bromus tectorum* var. *grandiflorus* Hook.	Near Baghdad
F & L 773	*Bupleurum aleppicum* Boiss.	Jebel Baykhair near Zakho
F & L 767	*Bupleurum brevicaule* Schlecht.	Baban near Al Qosh
F & L 744	*Bupleurum falcatum* L.	Jerwona near Ain Sifni
F & L 733	*Bupleurum kurdicum* Boiss.	Jerwona near Ain Sifni
F & L 400	*Calendula persica* C. A. Mey.	Qala Sharqat
L 14	*Calligonum polygonoides* L.	Near Baghdad
L 237	*Callipeltis Cucullaria* L.	Near Baghdad

Number	Genus and Species	Locality
F & L 485	*Callipeltis Cucullaria* L.	Jebel Golat between Ain Tellawi and Balad Sinjar
L 101, 477	*Capparis spinosa* L.	Near Baghdad
F & L 653	*Capparis spinosa* L.	Jebel Khatchra near Balad Sinjar
F & L 722	*Carthamus Oxyacantha* M. Bieb.	Jerwona near Ain Sifni
F & L 437	*Carum elegans* Fenzl	Jebel Golat between Ain Tellawi and Balad Sinjar
L 86, 434	*Carum elegans* Fenzl	Near Baghdad
L 166, 465	*Caucalis leptophylla* L.	Near Baghdad
L 193	*Caucalis leptophylla* L.	Rustam Farm near Baghdad
F & L 420	*Caucalis leptophylla* L.	Jebel Golat between Ain Tellawi and Balad Sinjar
F & L 126	*Caylusea canescens* (L.) St. Hil.	Rutba
F & L 837	*Celsia heterophylla* Desf.	Jebel Baykhair near Zakho
F & L 961, 963	*Celsia heterophylla* Desf.	Sulaimaniya
L 120	*Celsia heterophylla* Desf.	Near Baghdad
F & L 376	*Celsia lanceolata* Vent. var. *singarica* Murb. f.	Telegraph pole M90 between Baiji and Mosul
L 227	*Celtis australis* L.	Rustam Farm near Baghdad
F & L 946	*Celtis Tournefortii* Lam.	Jebel Baradost near Diana Rowandiz
F & L 717	*Celtis Tournefortii* Lam.	Sheikh Adi near Ain Sifni
L 406	*Centaurea araneosa* Boiss.	Near Baghdad
F & L 500	*Centaurea Behen* L.	Jebel Golat between Ain Tellawi and Balad Sinjar
F & L 915	*Centaurea depressa* M. Bieb.	Jebel Baradost near Diana Rowandiz
F & L 810	*Centaurea iberica* Trev.	Jebel Baykhair near Zakho
F & L 60	*Centaurea iberica* Trev.	Near Amara
F & L 560	*Centaurea myriocephala* Sch. Bip.	Mir Khasim between Balad Sinjar and Tall Afar
F & L 612	*Centaurea pallescens* Del. var. *hyalolepis* Boiss.	Karya Bat Khan near Balad Sinjar
F & L 387	*Centaurea phyllocephala* Boiss.	Telegraph pole M90 between Baiji and Mosul
F & L 421	*Centaurea phyllocephala* Boiss.	Jebel Golat between Ain Tellawi and Balad Sinjar
F & L 522	*Centaurea phyllocephala* Boiss.	Between Tall Afar and Balad Sinjar
F & L 623	*Centaurea regia* Boiss.	Jebel Khatchra near Balad Sinjar
F & L 723	*Centaurea solstitialis* L.	Jerwona near Ain Sifni
F & L 809	*Centaurea virgata* Lam.	Jebel Baykhair near Zakho
F & L 637	*Centaurea virgata* Lam.	Jebel Khatchra near Balad Sinjar
F & L 445	*Cephalaria syriaca* (L.) Schrad.	Jebel Golat between Ain Tellawi and Balad Sinjar
F & L 524	*Cephalaria syriaca* (L.) Schrad.	Between Tall Afar and Balad Sinjar

Number	Genus and Species	Locality
F & L 835	*Cephalaria syriaca* (L.) Schrad.	Jebel Baykhair near Zakho
F & L 938	*Ceterach officinarum* Willd.	Jebel Baradost near Diana Rowandiz
L 463	*Ceterach officinarum* Willd.	Near Baghdad
F & L 766	*Chamaemelum microcephalum* Boiss.	Baban near Al Qosh
F & L 913	*Chamaemelum microcephalum* Boiss.	Jebel Baradost near Diana Rowandiz
L 434, 451	*Chrozophora tinctoria* (L.) A. Juss.	Near Baghdad
F & L 687	*Chrozophora verbascifolia* (Willd.) A. Juss.	30 km. due west of Balad Sinjar
F & L 825	*Chrozophora verbascifolia* (Willd.) A. Juss.	Jebel Baykhair near Zakho
L 333	*Chrozophora verbascifolia* (Willd.) A. Juss.	Rustam Farm near Baghdad
F & L 848	*Chrysanthemum Parthenium* (L.) Pers.	Rowandiz Gorge
F & L 644, 659	*Chrysophthalmum montanum* (DC.) Boiss.	Jebel Khatchra near Balad Sinjar
F & L 762	*Cicer arietinum* L.	Sheikh Adi near Ain Sifni
F & L 397	*Cichorium divaricatum* Schousb.	Telegraph pole M90 between Baiji and Mosul
L 419	*Cichorium divaricatum* Schousb.	Near Baghdad
F & L 519	*Cichorium Intybus* L. Schrad.	Between Tall Afar and Balad Sinjar
F & L 364	*Citrullus Colocynthis* (L.) Schrad.	Wadi Al Qaim
L 245	*Citrullus Colocynthis* (L.) Schrad.	Near Baghdad
F & L 96	*Cleome Kotschyana* Boiss.	Montafah
F & L 875	*Colladonia crenata* Boiss.	Jebel Baradost near Diana Rowandiz
L 359	*Colladonia crenata* Boiss.	Near Baghdad
F & L 71	*Convolvulus arvensis* L.	Mesaida near Amara
L 316	*Convolvulus Cantabrica* L.	Rustam Farm near Baghdad
L 476	*Convolvulus Cantabrica* L.	Near Baghdad
F & L 745	*Convolvulus Cantabrica* L.	Jerwona near Ain Sifni
F & L 117	*Convolvulus pilosellifolius* Desr.	Montafah
F & L 601	*Convolvulus reticulatus* Choisy	Karya Sheikh Khanis near Balad Sinjar
L 426	*Corchorus olitorius* L.	Near Baghdad
L 437	*Coreopsis tinctoria* Nutt.	Near Baghdad
L 105	*Coronilla varia* L.	Near Baghdad
F & L 712	*Cousinia arbelensis* Winkl. & Bornm.	Sheikh Adi near Ain Sifni
F & L 798	*Cousinia* cf. *Kotschyi* Boiss.	Jebel Baykhair near Zakho
F & L 553	*Cousinia stenocephala* Boiss.	Mir Khasim between Balad Sinjar and Tall Afar
F & L 611	*Cousinia stenocephala* Boiss.	Karya Bat Khan near Balad Sinjar

Number	Genus and Species	Locality
F & L 710	*Crataegus Azarolus* L.	Sheikh Adi near Ain Sifni
F & L 945	*Crataegus Azarolus* L.	Jebel Baradost near Diana Rowandiz
F & L 859	*Crataegus Azarolus* L.	Rowandiz Gorge
F & L 976	*Crataegus Azarolus* L.	Sulaimaniya
F & L 437	*Crepis aspera* L.	Jebel Golat between Ain Tellawi and Balad Sinjar
F & L 645	*Crepis assyriaca* Bornm.	Jebel Khatchra near Balad Sinjar
F & L 870	*Crepis pulchra* L.	Gindian near Diana Rowandiz
F & L 639	*Crepis pulchra* L.	Jebel Khatchra near Balad Sinjar
F & L 912, 933	*Crucianella glauca* A. Rich.	Jebel Baradost near Diana Rowandiz
F & L 772	*Crucianella kurdistanica* Malinowski	Jebel Baykhair near Zakho
F & L 663	*Crupina Crupinastrum* Vis.	Jebel Khatchra near Balad Sinjar
F & L 828	*Crupina Crupinastrum* Vis.	Jebel Baykhair near Zakho
L 326	*Crypsis aculeata* (L.) Ait.	Near Baghdad
L 242	*Cucumis prophetarum* L.	Near Baghdad
F & L 924	*Cuscuta approximata* Bab. var. *urceolata* (Kunze) Yuncker	Jebel Baradost near Diana Rowandiz
F & L 473, 503	*Cuscuta babylonica* Auch.	Jebel Golat between Ain Tellawi and Balad Sinjar
F & L 536, 566	*Cuscuta babylonica* Auch.	Mir Khasim between Balad Sinjar and Tall Afar
F & L 570	*Cuscuta babylonica* Auch.	Tell Es Shur between Tall Afar and Balad Sinjar
F & L 721	*Cuscuta babylonica* Auch.	Sheikh Adi near Ain Sifni
L 301	*Cuscuta Lehmanniana* Bunge	Rustam Farm near Baghdad
L 33	*Cuscuta pedicellata* Ledeb.	Rustam Farm near Baghdad
L 226	*Cydonia oblonga* Mill.	Rustam Farm near Baghdad
L 43	*Cymbopogon Schoenanthus* (L.) Spreng.	Near Baghdad
L 302, 427	*Cynanchum acutum* L.	Rustam Farm near Baghdad
F & L 50	*Cynodon Dactylon* (L.) Pers.	Near Amara
L 280, 322, 444	*Cynodon Dactylon* (L.) Pers.	Near Baghdad
L 319	*Cyperus fuscus* L.	Near Baghdad
L 423	*Cyperus fuscus* L.	Rustam Farm near Baghdad
F & L 512	*Cyperus longus* L.	Between Tall Afar and Balad Sinjar
F & L 308	*Cyperus longus* L.	Jebel Golat between Ain Tellawi and Balad Sinjar
F & L 49	*Cyperus rotundus* L.	Chahala near Amara
L 250, 325	*Cyperus rotundus* L.	Near Baghdad
F & L 899	*Dactylis glomerata* L.	Jebel Baradost near Diana Rowandiz
L 231, 400, 559	*Dalbergia Sissoo* Roxb.	Near Baghdad
L 265	*Daphne acuminata* Boiss.	Near Baghdad

Number	Genus and Species	Locality
F & L 928	*Daphne acuminata* Boiss.	Jebel Baradost near Diana Rowandiz
L 90, 298	*Datura Metel* L.	Near Baghdad
F & L 489	*Daucus aureus* Desf.	Jebel Golat between Ain Tellawi and Balad Sinjar
F & L 734	*Daucus guttatus* Sibth. & Sm.	Jerwona near Ain Sifni
F & L 770	*Delphinium cappadocicum* Boiss.	Jebel Baykhair near Zakho
F & L 741	*Delphinium cappadocicum* Boiss.	Jerwona near Ain Sifni
F & L 830	*Delphinium oliganthum* Boiss.	Jebel Baykhair near Zakho
F & L 691	*Delphinium oliganthum* Boiss.	30 km. due west of Balad Sinjar
F & L 586	*Delphinium oliganthum* Boiss.	Tell Es Shur between Tall Afar and Balad Sinjar
F & L 538	*Delphinium oliganthum* Boiss.	Mir Khasim between Balad Sinjar and Tall Afar
F & L 730	*Delphinium peregrinum* L.	Jerwona near Ain Sifni
F & L 850	*Delphinium peregrinum* L.	Rowandiz Gorge
L 371	*Delphinium rigidum* DC.	Near Baghdad
F & L 52	*Delphinium rigidum* DC.	Near Amara
L 115	*Dianthus anatolicus* Boiss.	Near Baghdad
F & L 422	*Dianthus anatolicus* Boiss.	Jebel Golat between Ain Tellawi and Balad Sinjar
F & L 926	*Dianthus anatolicus* Boiss.	Jebel Baradost near Diana Rowandiz
L 490	*Dianthus Cyri* Fisch. & Mey.	Near Baghdad
L 116, 123	*Dianthus fimbriatus* M. Bieb.	Near Baghdad
F & L 111	*Dianthus pallens* Sibth. & Sm. var. *oxylepis* Boiss.	Montafah
F & L 465	*Dianthus polycladus* Boiss.	Muwasul Tiatan Mukzuk Nuwar
F & L 555	*Dianthus polycladus* Boiss.	Mir Khasim between Balad Sinjar and Tall Afar
L 320	*Digitaria sanguinalis* (L.) Scop.	Near Baghdad
L 309	*Diplotaxis erucoides* (L.) DC.	Rustam Farm near Baghdad
F & L 380	*Diplotaxis Harra* (Forsk.) Boiss.	Telegraph pole M90 between Baiji and Mosul
F & L 90	*Diplotaxis Harra* (Forsk.) Boiss.	Montafah
F & L 430	*Echinaria capitata* (L.) Desf.	Jebel Golat between Ain Tellawi and Balad Sinjar
L 424	*Echinochloa colona* (L.) Link	Near Baghdad
L 290, 428, 508	*Echinochloa Crusgalli* (L.) Beauv.	Near Baghdad
F & L 978	*Echinops Blancheanus* Boiss.	Sulaimaniya
L 407	*Echinops sphaerocephalus* L.	Near Baghdad
F & L 642	*Echinops sphaerocephalus* L.	Jebel Khatchra near Balad Sinjar
F & L 750	*Echium italicum* L.	Jerwona near Ain Sifni
L 314	*Eclipta alba* (L.) Hassk.	Rustam Farm near Baghdad
L 401	*Elaeagnus angustifolia* L. var. *orientalis* (L.) Kuntze	Rustam Farm near Baghdad
F & L 581	*Elymus caput-medusae* L.	Tell Es Shur between Tall Afar and Balad Sinjar

Number	Genus and Species	Locality
F & L 427	*Elymus crinitus* Schreb.	Jebel Golat between Ain Tellawi and Balad Sinjar
F & L 621	*Elymus crinitus* Schreb.	Jebel Khatchra near Balad Sinjar
F & L 428	*Elymus Delileanus* Schult.	Jebel Golat between Ain Tellawi and Balad Sinjar
F & L 97	*Ephedra campylopoda* C. A. Mey.	Montafah
F & L 958	*Epilobium hirsutum* L.	Sulaimaniya
L 387	*Epilobium hirsutum* L.	Near Baghdad
L 141, 247, 278, 470	*Eragrostis cilianensis* (All.) Link	Near Baghdad
L 288	*Eragrostis tenella* (L.) Roem. & Schult.	Near Baghdad
F & L 872	*Eremostachys laciniata* (L.) Bunge	Jebel Baradost near Diana Rowandiz
F & L 801	*Erianthus Hostii* Griseb.	Jebel Baykhair near Zakho
L 418	*Erigeron canadensis* L.	Rustam Farm near Baghdad
F & L 128	*Erodium Ciconium* (L.) Willd.	Rutba
F & L 85	*Erodium cicutarium* (L.) L'Hér.	Montafah
L 132, 206	*Erodium cicutarium* (L.) L'Hér.	Near Baghdad
F & L 394	*Erodium cicutarium* (L.) L'Hér.	Telegraph pole M90 between Baiji and Mosul
F & L 406	*Erodium cicutarium* (L.) L'Hér.	Qala Sharqat
F & L 511	*Erodium cicutarium* (L.) L'Hér.	Between Tall Afar and Balad Sinjar
L 48	*Erodium glaucophyllum* Ait.	Near Baghdad
F & L 125	*Erodium glaucophyllum* Ait.	Rutba
F & L 393	*Erodium glaucophyllum* Ait.	Telegraph pole M90 between Baiji and Mosul
F & L 742	*Erodium gruinum* (L.) Ait.	Jerwona near Ain Sifni
F & L 395	*Erodium laciniatum* (Cav.) Willd.	Telegraph pole M90 between Baiji and Mosul
F & L 558	*Erodium laciniatum* (Cav.) Willd.	Mir Khasim between Balad Sinjar and Tall Afar
L 19, 188	*Eruca sativa* Mill.	Rustam Farm near Baghdad
F & L 94	*Erucaria aleppica* Gaertn.	Montafah
L 20	*Erucaria microcarpa* Boiss.	Rustam Farm near Baghdad
L 405	*Eryngium creticum* Lam.	Near Baghdad
F & L 20	*Erythraea latifolia* Sm.	Gatt Al Dwat near Amara
F & L 822	*Erythraea latifolia* Sm.	Jebel Baykhair near Zakho
L 308, 327, 354	*Euphorbia Chamaesyce* L.	Near Baghdad
F & L 431	*Euphorbia Chamaesyce* L.	Jebel Golat between Ain Tellawi and Balad Sinjar
F & L 118, 139	*Euphorbia Chesneyi* (Kl. & Garcke) Boiss.	Rutba
F & L 666	*Euphorbia craspedia* Boiss.	Jebel Khatchra near Balad Sinjar

Number	Genus and Species	Locality
F & L 896	Euphorbia craspedia Boiss.	Jebel Baradost near Diana Rowandiz
L 372	Euphorbia denticulata Lam.	Near Baghdad
L 61	Euphorbia falcata L.	Near Baghdad
F & L 563	Euphorbia falcata L.	Mir Khasim between Balad Sinjar and Tall Afar
F & L 585	Euphorbia falcata L.	Tell Es Shur between Tall Afar and Balad Sinjar
F & L 853	Euphorbia falcata L. var. rubra Boiss.	Rowandiz Gorge
F & L 966	Euphorbia Gaillardoti Boiss. & Blocki	Sulaimaniya
L 7, 152	Euphorbia Helioscopia L.	Near Baghdad
F & L 743	Euphorbia Helioscopia L.	Jerwona near Ain Sifni
F & L 486	Euphorbia lanata Sieb.	Jebel Golat between Ain Tellawi and Balad Sinjar
F & L 747	Euphorbia macroclada Boiss.	Jerwona near Ain Sifni
L 1	Euphorbia Peplus L.	Near Baghdad
F & L 787	Euphorbia tinctoria Boiss. & Huet.	Jebel Baykhair near Zakho
F & L 934	Euphorbia tinctoria Boiss. & Huet.	Jebel Baradost near Diana Rowandiz
L 307	Euphorbia turcomanica Boiss.	Near Baghdad
F & L 371	Euphorbia cf. oxyodonta Boiss. & Hausskn.	Haditha (wheatfield)
L 15	Fagonia Bruguieri DC.	Rustam Farm near Baghdad
F & L 84, 99	Fagonia Oliverii DC.	Montafah
F & L 886	Fibigia clypeata (L.) Boiss.	Jebel Baradost near Diana Rowandiz
F & L 627	Ficus Carica L. var. rupestris Hausskn.	Jebel Khatchra near Balad Sinjar
F & L 864	Ficus Carica L. var. rupestris Hausskn.	Gindian near Rowandiz
F & L 781	Ficus palmata Forsk.	Jebel Baykhair near Zakho
L 114, 236	Filago spathulatus Presl	Near Baghdad
F & L 419	Filago spathulatus Presl	Jebel Golat between Ain Tellawi and Balad Sinjar
F & L 738	Filago spathulatus Presl	Jerwona near Ain Sifni
L 318, 422	Fimbristylis dichotoma (L.) Vahl	Rustam Farm near Baghdad
F & L 47	Fimbristylis dichotoma (L.) Vahl	Chahala near Amara
L 440	Frankenia Aucheri Jaub. & Spach	Rustam Farm near Baghdad
F & L 44	Frankenia pulverulenta L.	Chahala near Amara
F & L 858, 862	Fraxinus oxyphylla M. Bieb.	Rowandiz Gorge
F & L 792	Fumana arabica (L.) Spach	Jebel Baykhair near Zakho
L 180	Fumaria parviflora Lam.	Rustam Farm near Baghdad
L 32	Gagea reticulata (Pall.) R. & S.	Rustam Farm near Baghdad
F & L 917	Galium adhaerens Boiss. & Bal.	Jebel Baradost near Diana Rowandiz

Number	Genus and Species	Locality
L 349	*Galium coronatum* Sibth. & Sm.	Near Baghdad
F & L 636	*Galium coronatum* Sibth. & Sm. var. *stenophyllum* Boiss.	Jebel Khatchra near Balad Sinjar
L 370	*Galium mite* Boiss. & Hohen.	Near Baghdad
F & L 887	*Galium mite* Boiss. & Hohen.	Jebel Baradost near Diana Rowandiz
F & L 726	*Galium nigricans* Boiss.	Jerwona near Ain Sifni
F & L 754	*Galium tricorne* With.	Sheikh Adi near Ain Sifni
F & L 799	*Galium verum* L.	Jebel Baykhair near Zakho
F & L 418	*Garhadiolus Hedypnois* (Fisch. & Mey.) Jaub. & Spach	Jebel Golat between Ain Tellawi and Balad Sinjar
F & L 849	*Gastrocotyle hispida* (Forsk.) Bunge	Rowandiz Gorge
F & L 925	*Gentiana Olivieri* Griseb.	Jebel Baradost near Diana Rowandiz
F & L 68	*Geranium dissectum* L.	Mesaida near Amara
F & L 973	*Geranium dissectum* L.	Sulaimaniya
F & L 425	*Geranium rotundifolium* L.	Jebel Golat between Ain Tellawi and Balad Sinjar
L 344	*Gladiolus atroviolaceus* Boiss.	Near Baghdad
F & L 576	*Gladiolus atroviolaceus* Boiss.	Tell Es Shur between Tall Afar and Balad Sinjar
L 154	*Glaucium corniculatum* (L.) Curt.	Near Baghdad
F & L 466	*Glaucium corniculatum* (L.) Curt.	Muwasul Tiatan Mukzuk Nuwar
F & L 481	*Glaucium corniculatum* (L.) Curt.	Jebel Golat between Ain Tellawi and Balad Sinjar
F & L 138	*Glaucium grandiflorum* Boiss. & Huet.	Rutba
L 304, 411	*Glinus lotoides* L.	Rustam Farm near Baghdad
F & L 752	*Glycyrrhiza glabra* L.	Jerwona near Ain Sifni
F & L 962	*Glycyrrhiza glabra* L.	Sulaimaniya
F & L 833	*Gnaphalium luteo-album* L.	Jebel Baykhair near Zakho
F & L 969	*Gypsophila platyphylla* Boiss.	Sulaimaniya
F & L 689	*Gypsophila porrigens* (L.) Boiss.	30 km. due west of Balad Sinjar
L 44	*Gypsophila Rokejeka* Del.	Near Baghdad
F & L 793	*Gypsophila ruscifolia* Boiss.	Jebel Baykhair near Zakho
F & L 546	*Haplophyllum Buxbaumii* (Poir.) Boiss.	Mir Khasim between Balad Sinjar and Tall Afar
F & L 618	*Haplophyllum Buxbaumii* (Poir.) Boiss.	Jebel Khatchra near Balad Sinjar
F & L 413	*Haplophyllum propinquum* Spach	Jebel Golat between Ain Tellawi and Balad Sinjar
F & L 121, 129	*Haplophyllum propinquum* Spach	Rutba

Number	Genus and Species	Locality

F & L 101, 115..*Haplophyllum tuberculatum*
　　　　　　　Forsk....................Montafah
L 409...........*Haplophyllum tuberculatum*
　　　　　　　Forsk....................Near Baghdad
F & L 796.......*Hedysarum pannosum* Boiss....Jebel Baykhair near Zakho
L 299..........*Heleochloa alopecuroides*
　　　　　　　(Schrad.) Host.............Near Baghdad
F & L 960.......*Heleochloa schoenoides* (L.)
　　　　　　　Host......................Sulaimaniya
L 16...........*Helianthemum salicifolium* (L.)
　　　　　　　Mill......................Rustam Farm near Baghdad
F & L 561.......*Helianthemum salicifolium* (L.)
　　　　　　　Mill......................Mir Khasim between Balad Sinjar and Tall Afar
L 135..........*Helianthemum salicifolium* (L.)
　　　　　　　Mill......................Near Baghdad
F & L 453.......*Helichrysum graveolens* Boiss....Muwasul Tiatan Mukzuk Nuwar
F & L 554.......*Helichrysum graveolens* Boiss....Mir Khasim between Balad Sinjar and Tall Afar
F & L 614.......*Helichrysum graveolens* Boiss....Karya Bat Khan near Balad Sinjar
F & L 660.......*Helichrysum graveolens* Boiss....Jebel Khatchra near Balad Sinjar
F & L 776.......*Helichrysum graveolens* Boiss....Jebel Baykhair near Zakho
F & L 595.......*Helicophyllum crassipes* (Boiss.)
　　　　　　　Schott....................Tell Es Shur between Tall Afar and Balad Sinjar
L 408..........*Heliotropium Eichwaldi* Steud...Near Baghdad
L 305, 415......*Heliotropium supinum* L.......Near Baghdad
F & L 402.......*Heliotropium supinum* L.......Qala Sharqat
F & L 868.......*Heliotropium supinum* L.......Gindian near Diana Rowandiz
L 75...........*Heliotropium undulatum* Vahl..Near Baghdad
F & L 113.......*Heliotropium undulatum* Vahl..Montafah
L 190..........*Herniaria cinerea* DC.........Rustam Farm near Baghdad
L 39...........*Herniaria hemistemon* Gay.....Near Baghdad
F & L 103.......*Herniaria incana* Lam.........Montafah
F & L 119.......*Herniaria incana* Lam.........Rutba
F & L 426.......*Heteranthelium piliferum* (Russ.)
　　　　　　　Hochst....................Jebel Golat between Ain Tellawi and Balad Sinjar
L 243..........*Hibiscus Trionum* L...........Near Baghdad
F & L 948.......*Hibiscus Trionum* L...........Sulaimaniya
L 51...........*Hippocrepis cornigera* Boiss...Near Baghdad
F & L 667.......*Hippomarathrum scabrum*
　　　　　　　(Fenzl) Boiss.............Jebel Khatchra near Balad Sinjar
F & L 439.......*Hordeum bulbosum* L...........Jebel Golat between Ain Tellawi and Balad Sinjar
F & L 900.......*Hordeum bulbosum* L...........Jebel Baradost near Diana Rowandiz
F & L 58........*Hordeum maritimum* With......Mesaida near Amara
L 131..........*Hordeum murinum* L...........Near Baghdad

Number	Genus and Species	Locality
F & L 377	*Hordeum murinum* L.	Telegraph pole M90 between Baiji and Mosul
F & L 641	*Hordeum spontaneum* Koch	Jebel Khatchra near Balad Sinjar
F & L 365	*Hyoscyamus albus* L.	Mile 170 west of H-3 Pipe-line Station
F & L 647	*Hyoscyamus albus* L.	Jebel Khatchra near Balad Sinjar
F & L 309, 484	*Hyoscyamus pusillus* L.	Jebel Golat between Ain Tellawi and Balad Sinjar
F & L 434	*Hypecoum procumbens* L.	Jebel Golat between Ain Tellawi and Balad Sinjar
L 133	*Hypericum crispum* L.	Near Baghdad
F & L 412	*Hypericum crispum* L.	Jebel Golat between Ain Tellawi and Balad Sinjar
F & L 456	*Hypericum crispum* L.	Muwasul Tiatan Mukzuk Nuwar
F & L 685	*Hypericum crispum* L.	30 km. due west of Balad Sinjar
F & L 548	*Hypericum helianthemoides* (Spach) Boiss.	Mir Khasim between Balad Sinjar and Tall Afar
F & L 455	*Hypericum helianthemoides* (Spach) Boiss.	Muwasul Tiatan Mukzuk Nuwar
L 121	*Hypericum scabrum* L.	Near Baghdad
F & L 884	*Hypericum scabrum* L.	Jebel Baradost near Diana Rowandiz
L 98	*Iberis odorata* L.	Near Baghdad
F & L 478	*Inula divaricata* (Cass.) Boiss.	Jebel Golat between Ain Tellawi and Balad Sinjar
F & L 821	*Inula squarrosa* L.	Jebel Baykhair near Zakho
F & L 836	*Ipomoea purpurea* (L.) Lam.	Jebel Baykhair near Zakho
L 235	*Isatis aleppica* Scop.	Near Baghdad
F & L 716	*Juglans regia* L.	Sheikh Adi near Ain Sifni
L 456	*Juncus acutus* L.	Near Baghdad
F & L 807	*Juncus effusus* L.	Jebel Baykhair near Zakho
L 269	*Juncus effusus* L.	Near Baghdad
F & L 521	*Juncus pyramidatus* Laharpe	Between Tall Afar and Balad Sinjar
F & L 46	*Jussiaea repens* L.	Chahala near Amara
F & L 57	*Koeleria phleoides* (Vill.) Pers.	Mesaida near Amara
L 50, 192	*Koelpinia linearis* Pall.	Near Baghdad
F & L 416	*Koelpinia linearis* Pall.	Jebel Golat between Ain Tellawi and Balad Sinjar
F & L 619	*Lactuca cretica* Desf.	Jebel Khatchra near Balad Sinjar
L 102	*Lactuca saligna* L.	Near Baghdad
F & L 805	*Lactuca sativa* L.	Jebel Baykhair near Zakho
F & L 827	*Lactuca tuberosa* Jacq.	Jebel Baykhair near Zakho
F & L 768	*Lallemantia iberica* (M. Bieb.) Fisch. & Mey.	Baban near Al Qosh

Number	Genus and Species	Locality
F & L 905	*Lallemantia peltata* (L.) Fisch. & Mey.	Jebel Baradost near Diana Rowandiz
L 127	*Lamium amplexicaule* L.	Near Baghdad
L 382	*Lamium maculatum* L.	Near Baghdad
L 62	*Lappula spinocarpa* (Forsk.) Aschers.	Near Baghdad
F & L 124	*Lappula spinocarpa* (Forsk.) Aschers.	Rutba
F & L 780	*Lathyrus annuus* L.	Jebel Baykhair near Zakho
L 541	*Lathyrus Aphaca* L.	Near Baghdad
L 178, 203	*Lathyrus Cicera* L.	Near Baghdad
L 516	*Lepidium Draba* L.	Near Baghdad
F & L 34	*Lepidium Draba* L.	Chahala near Amara
F & L 66	*Lepidium Draba* L.	Mesaida near Amara
F & L 834	*Lepidium latifolium* L.	Jebel Baykhair near Zakho
F & L 471	*Lepidium perfoliatum* L.	Muwasul Tiatan Mukzuk Nuwar
L 3	*Lepidium sativum* L.	Near Baghdad
F & L 11	*Lepidium sativum* L.	Gatt Al Dwat near Amara
F & L 36, 38	*Limnanthemum nymphoides* (L.) Link	Chahala near Amara
F & L 30	*Limnanthemum nymphoides* (L.) Link	Gatt Al Dwat near Amara
L 241, 389, 412	*Linaria Elatine* (L.) Mill.	Near Baghdad
F & L 824	*Linum angustifolium* Huds.	Jebel Baykhair near Zakho
L 148, 346	*Linum flavum* L.	Near Baghdad
L 219, 220	*Linum grandiflorum* Desf.	Rustam Farm near Baghdad
F & L 143	*Linum mucronatum* Bertol.	Between H-4 and H-5 Pipe-line Stations
L 336	*Lippia nodiflora* (L.) Michx.	Rustam Farm near Baghdad
F & L 18	*Lippia nodiflora* (L.) Michx.	Gatt Al Dwat near Amara
L 207	*Lippia nodiflora* (L.) Michx.	Near Baghdad
F & L 6	*Lolium temulentum* L.	Gatt Al Dwat near Amara
L 85, 139, 281	*Lolium temulentum* L.	Near Baghdad
F & L 922	*Lotus Gebetia* Vent.	Jebel Baradost near Diana Rowandiz
F & L 92	*Lotus lanuginosus* Vent.	Montafah
F & L 306	*Lotus tenuifolius* Reichb.	Jebel Golat between Ain Tellawi and Balad Sinjar
L 425	*Luffa cylindrica* (L.) Roem.	Rustam Farm near Baghdad
L 27	*Lycium barbarum* L.	Rustam Farm near Baghdad
F & L 27	*Lycium barbarum* L.	Chahala near Amara
F & L 746	*Lythrum Hyssopifolia* L.	Jerwona near Ain Sifni
L 128, 386	*Lythrum Salicaria* L. var. *tomentosum* DC.	Near Baghdad
L 2, 176	*Malcolmia africana* (L.) R. Br.	Near Baghdad
L 24	*Malcolmia Bungei* Boiss.	Near Baghdad

Number	Genus and Species	Locality
L 108	*Malcolmia crenulata* (DC.) Boiss.	Near Baghdad
L 25, 52, 84, 189	*Malcolmia torulosa* (Desf.) Boiss.	Near Baghdad
L 8	*Malva parviflora* L.	Near Baghdad
F & L 64	*Malva parviflora* L.	Mesaida near Amara
F & L 110	*Malva parviflora* L.	Montafah
L 521	*Malva rotundifolia* L.	Near Baghdad
F & L 464	*Marrubium radiatum* Del.	Muwasul Tiatan Mukzuk Nuwar
F & L 388	*Mathiola oxyceras* DC.	Telegraph pole M90 between Baiji and Mosul
L 21	*Mathiola oxyceras* DC.	Near Baghdad
L 144, 196	*Matricaria aurea* (L.) Boiss.	Near Baghdad
F & L 523	*Medicago denticulata* Willd.	Between Tall Afar and Balad Sinjar
F & L 567	*Medicago denticulata* Willd.	Mir Khasim between Balad Sinjar and Tall Afar
F & L 971	*Medicago Gerardi* Waldst. & Kit.	Sulaimaniya
F & L 756	*Medicago orbicularis* All.	Sheikh Adi near Ain Sifni
L 268, 332	*Medicago sativa* L.	Near Baghdad
F & L 927	*Melandrium eriocalycinum* Boiss.	Jebel Baradost near Diana Rowandiz
F & L 898	*Melica Cupani* Guss.	Jebel Baradost near Diana Rowandiz
F & L 840	*Micromeria Juliana* (L.) Benth. var. *myrtifolia* Boiss.	Jebel Pikasar near Aqra
F & L 483	*Micropus erectus* L.	Jebel Golat between Ain Tellawi and Balad Sinjar
L 72, 185	*Micropus supinus* L.	Near Baghdad
F & L 384	*Moltkia coerulea* (Willd.) Lehm.	Telegraph pole M90 between Baiji and Mosul
F & L 411	*Moltkia coerulea* (Willd.) Lehm.	Jebel Golat between Ain Tellawi and Balad Sinjar
F & L 603	*Moltkia coerulea* (Willd.) Lehm.	Karya Sheikh Khanis near Balad Sinjar
L 70, 74	*Moltkia collosa* (Vahl) Wettst.	Near Baghdad
L 147	*Moluccella laevis* L.	Near Baghdad
F & L 502	*Moluccella laevis* L.	Jebel Golat between Ain Tellawi and Balad Sinjar
L 228	*Morus alba* L.	Rustam Farm near Baghdad
F & L 906	*Muscari comosum* (L.) Mill.	Jebel Baradost near Diana Rowandiz
L 87	*Myrtus communis* L.	Rustam Farm near Baghdad
F & L 559	*Nigella arvensis* L.	Mir Khasim between Balad Sinjar and Tall Afar
L 174	*Nigella sativa* L.	Near Baghdad
L 156	*Obione flabellum* (Bunge) Ulbr.	Near Baghdad
F & L 707	*Olea europaea* L.	Sheikh Adi near Ain Sifni
F & L 527	*Oliveria orientalis* DC.	Ain Tellawi near Tall Afar
F & L 795	*Onobrychis caput-galli* (L.) Lam.	Jebel Baykhair near Zakho

PLANTS FROM IRAQ 181

Number	Genus and Species	Locality
F & L 790	*Onobrychis galegifolia* Boiss.	Jebel Baykhair near Zakho
F & L 98	*Onobrychis lanata* Boiss.	Montafah
F & L 964	*Ononis antiquorum* L.	Sulaimaniya
F & L 748	*Ononis mitissima* L.	Jerwona near Ain Sifni
F & L 749	*Ononis sicula* Guss.	Jerwona near Ain Sifni
F & L 429	*Ononis sicula* Guss.	Jebel Golat between Ain Tellawi and Balad Sinjar
F & L 133	*Onopordon heteracanthum* C. A. Mey.	Rutba
F & L 643	*Onopordon illyricum* L.	Jebel Khatchra near Balad Sinjar
F & L 977	*Onopordon illyricum* L.	Sulaimaniya
F & L 444	*Onosma aleppicum* Boiss.	Jebel Golat between Ain Tellawi and Balad Sinjar
F & L 597	*Onosma aleppicum* Boiss.	Karya Sheikh Khanis near Balad Sinjar
F & L 662	*Onosma flavum* (Lehm.) Vatke	Jebel Khatchra near Balad Sinjar
F & L 600	*Onosma sericeum* Willd.	Karya Sheikh Khanis near Balad Sinjar
F & L 719	*Onosma sericeum* Willd.	Sheikh Adi near Ain Sifni
L 110	*Ornithogalum narbonense* L.	Near Baghdad
F & L 812	*Paliurus aculeatus* Lam.	Jebel Baykhair near Zakho
F & L 705	*Paliurus aculeatus* Lam.	Sheikh Adi near Ain Sifni
F & L 940	*Paliurus aculeatus* Lam.	Jebel Baradost near Diana Rowandiz
F & L 728	*Pallenis spinosa* (L.) Cass.	Jerwona near Ain Sifni
L 297	*Panicum miliaceum* L.	Near Baghdad
F & L 396	*Papaver Rhoeas* L.	Telegraph pole M90 between Baiji and Mosul
F & L 890	*Paracaryum cristatum* (Lam.) Boiss.	Jebel Baradost near Diana Rowandiz
F & L 367	*Parietaria alsinefolia* Del.	Wadi Al Hajal near Haditha
F & L 565	*Parietaria alsinefolia* Del.	Mir Khasim between Balad Sinjar and Tall Afar
F & L 564	*Parietaria debilis* Forst.	Mir Khasim between Balad Sinjar and Tall Afar
L 254, 345	*Parietaria judaica* L.	Near Baghdad
F & L 703	*Parietaria judaica* L.	Sheikh Adi near Ain Sifni
L 76	*Paronychia argentea* Lam.	Near Baghdad
F & L 102	*Paronychia argentea* Lam.	Montafah
L 270	*Paronychia argentea* Lam.	Near Baghdad
F & L 602	*Paronychia capitata* (L.) Lam.	Karya Sheikh Khanis near Balad Sinjar
L 287	*Paspalum distichum* L.	Near Baghdad
F & L 368	*Peganum Harmala* L.	Wadi Al Hajal near Haditha
F & L 26	*Peganum Harmala* L.	Chahala near Amara
L 63, 259	*Peganum Harmala* L.	Near Baghdad
F & L 130	*Peganum Harmala* L.	Rutba

Number	Genus and Species	Locality
F & L 137	*Phagnalon rupestre* (L.) DC.	Rutba
L 233	*Phalaris brachystachys* Link	Near Baghdad
F & L 441	*Phalaris brachystachys* Link	Jebel Golat between Ain Tellawi and Balad Sinjar
L 554	*Phalaris minor* Retz.	Near Baghdad
L 282, 284	*Phalaris paradoxa* L.	Near Baghdad
F & L 442	*Phalaris paradoxa* L.	Jebel Golat between Ain Tellawi and Balad Sinjar
L 375	*Phlomis Bruguieri* Desf.	Near Baghdad
F & L 506, 508	*Phlomis Bruguieri* Desf.	Jebel Golat between Ain Tellawi and Balad Sinjar
F & L 610	*Phlomis Bruguieri* Desf.	Karya Bat Khan near Balad Sinjar
F & L 697	*Phlomis linearis* Boiss. & Bal.	30 km. due west of Balad Sinjar
L 376	*Phlomis orientalis* Mill.	Near Baghdad
F & L 507	*Phlomis orientalis* Mill.	Jebel Golat between Ain Tellawi and Balad Sinjar
F & L 846	*Phlomis orientalis* Mill.	Rowandiz Gorge
L 339	*Phragmites communis* (L.) Trin.	Near Baghdad
F & L 700	*Physocaulos nodosus* (L.) Tausch	Sheikh Adi near Ain Sifni
F & L 724	*Pimpinella Kotschyana* Boiss.	Jerwona near Ain Sifni
F & L 839	*Pimpinella Kotschyana* Boiss.	Jebel Pikasar near Aqra
F & L 379	*Pimpinella peregrina* L.	Telegraph pole M90 between Baiji and Mosul
F & L 753	*Pinus halepensis* Mill.	Baban near Al Qosh
F & L 714	*Pistacia mutica* Fisch. & Mey.	Sheikh Adi near Ain Sifni
F & L 652, 654	*Pistacia Terebinthus* L.	Jebel Khatchra near Balad Sinjar
F & L 708	*Pistacia Terebinthus* L.	Sheikh Adi near Ain Sifni
F & L 942, 944	*Pistacia Terebinthus* L.	Jebel Baradost near Diana Rowandiz
L 221	*Pisum sativum* L.	Near Baghdad
L 67, 216	*Plantago Coronopus* L.	Near Baghdad
L 485	*Plantago lanceolata* L.	Near Baghdad
F & L 54	*Plantago lanceolata* L.	Mesaida near Amara
L 205, 217, 317, 430	*Plantago lanceolata* L.	Near Baghdad
F & L 820	*Plantago lanceolata* L.	Jebel Baykhair near Zakho
L 35	*Plantago Loeflingii* L.	Rustam Farm near Baghdad
F & L 29	*Plantago Loeflingii* L.	Chahala near Amara
L 34	*Plantago ovata* Forsk.	Near Baghdad
F & L 383	*Plantago ovata* Forsk.	Telegraph pole M90 between Baiji and Mosul
F & L 424, 487	*Plantago Psyllium* L.	Jebel Golat between Ain Tellawi and Balad Sinjar
F & L 860	*Platanus orientalis* L.	Rowandiz Gorge
L 126	*Poa bulbosa* L.	Near Baghdad
L 69	*Poa persica* Trin.	Near Baghdad

Number	Genus and Species	Locality
F & L 901	*Poa persica* Trin.	Jebel Baradost near Diana Rowandiz
F & L 303	*Poa tatarica* Fisch.	Jebel Golat between Ain Tellawi and Balad Sinjar
L 112, 391	*Polygonum aviculare* L.	Near Baghdad
L 303, 328	*Polygonum Bellardi* All.	Near Baghdad
F & L 72	*Polygonum Bellardi* All.	Mesaida near Amara
F & L 755	*Polygonum cognatum* Meisn.	Sheikh Adi near Ain Sifni
F & L 954	*Polygonum nodosum* Pers.	Sulaimaniya
L 289	*Polygonum Persicaria* L.	Near Baghdad
F & L 2	*Polygonum serrulatum* Lag.	Gatt Al Dwat near Amara
F & L 305	*Polypogon monspeliensis* (L.) Desf.	Jebel Golat between Ain Tellawi and Balad Sinjar
L 283	*Polypogon monspeliensis* (L.) Desf.	Near Baghdad
F & L 48	*Polypogon monspeliensis* (L.) Desf.	Chahala near Amara
F & L 979	*Populus deltoides* Marsh.	Sulaimaniya
L 223	*Populus euphratica* Oliv.	Rustam Farm near Baghdad
F & L 77	*Populus euphratica* Oliv.	Mesaida near Amara
F & L 4	*Potamogeton lucens* L.	Gatt Al Dwat near Amara
F & L 819	*Potentilla fallacina* Blocki	Jebel Baykhair near Zakho
L 107, 238	*Poterium verrucosum* Ehrenb.	Near Baghdad
F & L 970	*Poterium verrucosum* Ehrenb.	Sulaimaniya
F & L 604	*Poterium verrucosum* Ehrenb.	Karya Sheikh Khanis near Balad Sinjar
F & L 633	*Poterium verrucosum* Ehrenb.	Jebel Khatchra near Balad Sinjar
F & L 541	*Poterium verrucosum* Ehrenb.	Mir Khasim between Balad Sinjar and Tall Afar
F & L 725	*Poterium villosum* Sibth. & Sm.	Jerwona near Ain Sifni
L 384	*Prangos ferulacea* Lindl.	Near Baghdad
L 230	*Prosopis juliflora* DC.	Rustam Farm near Baghdad
F & L 80	*Prosopis Stephaniana* (Willd.) Kunth	Near Amara
F & L 63	*Prosopis Stephaniana* (Willd.) Kunth	Mesaida near Amara
L 170	*Prunus Amygdalus* Stokes	Near Baghdad
F & L 865	*Prunus cerasifera* Ehrh. var. *divaricata* (Ledeb.) Bailey	Gindian near Rowandiz
F & L 974	*Prunus instititia* L.	Sulaimaniya
L 153	*Prunus microcarpa* C. A. Mey.	Near Baghdad
F & L 709	*Prunus microcarpa* C. A. Mey.	Sheikh Adi near Ain Sifni
F & L 769	*Prunus microcarpa* C. A. Mey.	Jebel Baykhair near Zakho
F & L 878	*Prunus microcarpa* C. A. Mey.	Jebel Baradost near Diana Rowandiz
F & L 153	*Pterocephalus involucratus* Spreng.	Between H-4 and H-5 Pipe-line Stations

Number	Genus and Species	Locality
F & L 657	*Pterocephalus Putkianus* Boiss. & Kotschy	Jebel Khatchra near Balad Sinjar
F & L 843	*Pterocephalus strictus* Boiss. & Hohen	Jebel Pikasar near Aqra
L 26, 136, 294, 417	*Pulicaria crispa* (Forsk.) Sch. Bip	Near Baghdad
F & L 89	*Pulicaria crispa* (Forsk.) Sch. Bip	Montafah
F & L 41	*Pulicaria dysenterica* (L.) Gaertn	Chahala near Amara
F & L 863	*Pyrus syriaca* Boiss	Gindian near Rowandiz
F & L 616	*Quercus Aegilops* L	Jebel Khatchra near Balad Sinjar
F & L 713	*Quercus Aegilops* L	Sheikh Adi near Ain Sifni
F & L 778	*Quercus Aegilops* L	Jebel Baykhair near Zakho
F & L 880	*Quercus Aegilops* L	Jebel Baradost near Diana Rowandiz
F & L 881, 935	*Quercus dschorochensis* K. Koch	Jebel Baradost near Diana Rowandiz
F & L 943	*Quercus persica* Jaub. & Spach	Jebel Baradost near Diana Rowandiz
L 58	*Ranunculus aquatilis* L	Rustam Farm near Baghdad
L 97	*Ranunculus arvensis* L	Near Baghdad
F & L 869	*Ranunculus cassius* Boiss	Gindian near Diana Rowandiz
F & L 736	*Ranunculus cassius* Boiss	Jerwona near Ain Sifni
L 182	*Ranunculus lomatocarpus* Fisch. & Mey	Near Baghdad
F & L 831	*Ranunculus lomatocarpus* Fisch. & Mey	Jebel Baykhair near Zakho
L 264	*Ranunculus myriophyllus* DC	Near Baghdad
F & L 1	*Ranunculus pantothrix* Brot	Gatt Al Dwat near Amara
F & L 40	*Ranunculus pantothrix* Brot	Chahala near Amara
F & L 61	*Raphanus sativus* L	Near Amara
F & L 640	*Reseda alba* L	Jebel Khatchra near Balad Sinjar
L 40	*Reseda lutea* L	Near Baghdad
F & L 438	*Reseda muricata* Presl	Jebel Golat between Ain Tellawi and Balad Sinjar
F & L 655	*Rhamnus punctata* Boiss	Jebel Khatchra near Balad Sinjar
F & L 802	*Rhaphis gryllus* (L.) Desv	Jebel Baykhair near Zakho
F & L 631	*Rhus Coriaria* L	Jebel Khatchra near Balad Sinjar
F & L 706, 718	*Rhus Coriaria* L	Sheikh Adi near Ain Sifni
F & L 446	*Roripa Nasturtium-aquaticum* (L.) Schinz & Thell	Jebel Golat between Ain Tellawi and Balad Sinjar
L 276	*Rubus discolor* Weihe & Nees	Near Baghdad
F & L 751	*Rubus discolor* Weihe & Nees	Jerwona near Ain Sifni
F & L 814	*Rubus discolor* Weihe & Nees	Jebel Baykhair near Zakho

PLANTS FROM IRAQ

Number	Genus and Species	Locality
F & L 56	*Rumex dentatus* L. var. *pleiodon* Boiss.	Mesaida near Amara
L 292	*Rumex dentatus* L. var. *pleiodon* Boiss.	Rustam Farm near Baghdad
L 210	*Rumex obtusifolius* L.	Near Baghdad
F & L 14	*Rumex obtusifolius* L.	Gatt Al Dwat near Amara
L 519	*Rumex pulcher* L.	Near Baghdad
F & L 520	*Rumex pulcher* L.	Between Tall Afar and Balad Sinjar
F & L 729	*Rumex pulcher* L.	Jerwona near Ain Sifni
L 125, 446	*Rumex roseus* L.	Near Baghdad
F & L 895	*Rumex tuberosus* L.	Jebel Baradost near Diana Rowandiz
L 11	*Salix acmophylla* Boiss.	Rustam Farm near Baghdad
L 59, 99	*Salix acmophylla* Boiss.	Near Baghdad
L 557	*Salix amygdalina* L.	Near Baghdad
F & L 975	*Salix Safsaf* Forsk.	Sulaimaniya
F & L 861	*Salix Safsaf* Forsk.	Rowandiz Gorge
F & L 33	*Salix Safsaf* Forsk.	Chahala near Amara
F & L 782	*Salvia acetabulosa* L. var. *simplicifolia* Boiss.	Jebel Baykhair near Zakho
F & L 665	*Salvia acetabulosa* L. var. *simplicifolia* Boiss.	Jebel Khatchra near Balad Sinjar
F & L 100	*Salvia controversa* Ten.	Montafah
F & L 788	*Salvia* cf. *kurdica* Boiss. & Hohen.	Jebel Baykhair near Zakho
F & L 664	*Salvia palaestina* Benth.	Jebel Khatchra near Balad Sinjar
F & L 607	*Salvia palaestina* Benth.	Karya Bat Khan near Balad Sinjar
F & L 468	*Salvia palaestina* Benth.	Muwasul Tiatan Mukzuk Nuwar
F & L 542	*Salvia palaestina* Benth.	Mir Khasim between Balad Sinjar and Tall Afar
F & L 582, 594	*Salvia syriaca* L.	Tell Es Shur between Tall Afar and Balad Sinjar
L 146	*Salvia Szovitsiana* Bunge	Near Baghdad
F & L 39	*Salvinia natans* (L.) All.	Chahala near Amara
L 201, 277, 343	*Saponaria Vaccaria* L.	Near Baghdad
F & L 765	*Saponaria Vaccaria* L.	Baban near Al Qosh
F & L 392	*Scabiosa Olivieri* Coult.	Telegraph pole M90 between Baiji and Mosul
F & L 432	*Scabiosa Olivieri* Coult.	Jebel Golat between Ain Tellawi and Balad Sinjar
L 104, 167	*Scabiosa palaestina* L.	Near Baghdad
F & L 459	*Scabiosa palaestina* L.	Muwasul Tiatan Mukzuk Nuwar
F & L 537	*Scabiosa palaestina* L.	Mir Khasim between Balad Sinjar and Tall Afar
L 362	*Scandix iberica* M. Bieb.	Near Baghdad
L 258	*Scandix Pecten-Veneris* L.	Near Baghdad

Number	Genus and Species	Locality
F & L 800	*Schoenus nigricans* L.	Jebel Baykhair near Zakho
F & L 480	*Scirpus Holoschoenus* L.	Jebel Golat between Ain Tellawi and Balad Sinjar
L 161	*Scirpus littoralis* Schrad.	Near Baghdad
F & L 5	*Scirpus littoralis* Schrad.	Gatt Al Dwat near Amara
L 31	*Scirpus maritimus* L.	Rustam Farm near Baghdad
F & L 955	*Scirpus maritimus* L.	Sulaimaniya
F & L 968	*Scolymus maculatus* L.	Sulaimaniya
L 94, 183	*Scorpiurus sulcata* L.	Near Baghdad
F & L 972	*Scorpiurus sulcata* L.	Sulaimaniya
F & L 608	*Scorzonera papposa* DC.	Karya Bat Khan near Balad Sinjar
L 447	*Scorzonera papposa* DC.	Near Baghdad
F & L 578	*Scorzonera papposa* DC.	Tell Es Shur between Tall Afar and Balad Sinjar
F & L 599	*Scrophularia xanthoglossa* Boiss.	Karya Sheikh Khanis near Balad Sinjar
F & L 720	*Scrophularia xanthoglossa* Boiss.	Sheikh Adi near Ain Sifni
F & L 669	*Scutellaria cretacea* Boiss. & Hausskn.	Jebel Khatchra near Balad Sinjar
F & L 914	*Scutellaria peregrina* L. var. *Sibthorpii* Boiss. & Reut.	Jebel Baradost near Diana Rowandiz
F & L 105	*Senecio coronopifolius* Desf.	Montafah
F & L 123	*Senecio coronopifolius* Desf.	Rutba
F & L 390	*Senecio coronopifolius* Desf.	Telegraph pole M90 between Baiji and Mosul
F & L 408	*Senecio coronopifolius* Desf.	Jebel Golat between Ain Tellawi and Balad Sinjar
L 459	*Senecio coronopifolius* Desf.	Near Baghdad
F & L 832	*Senecio coronopifolius* Desf.	Jebel Baykhair near Zakho
F & L 447	*Serratula cerinthefolia* Sibth. & Sm.	Muwasul Tiatan Mukzuk Nuwar
F & L 658	*Serratula cerinthefolia* Sibth. & Sm.	Jebel Khatchra near Balad Sinjar
L 396	*Sesbania aegyptiaca* Pers.	Near Baghdad
L 271, 321, 474	*Setaria lutescens* (Weig.) Hubb.	Near Baghdad
F & L 903	*Sideritis libanotica* Lab. var. *incana* Boiss.	Jebel Baradost near Diana Rowandiz
F & L 373	*Silene Behen* L.	Haditha (wheatfield)
F & L 904	*Silene chloraefolia* Sm.	Jebel Baradost near Diana Rowandiz
L 96, 460	*Silene conoidea* L.	Near Baghdad
F & L 950	*Silene conoidea* L.	Sulaimaniya
F & L 699	*Silene dichotoma* Ehrh.	Sheikh Adi near Ain Sifni

PLANTS FROM IRAQ

Number	Genus and Species	Locality
F & L 590	*Silene longipetala* Vent.	Tell Es Shur between Tall Afar and Balad Sinjar
L 5	*Silene rubella* L.	Rustam Farm near Baghdad
F & L 407	*Silene setacea* Viv.	Qala Sharqat
F & L 777	*Silene stenobotrys* Boiss. & Hausskn.	Jebel Baykhair near Zakho
L 109	*Sisymbrium damascenum* Boiss. & Gaill.	Near Baghdad
F & L 414	*Sisymbrium septulatum* DC.	Jebel Golat between Ain Tellawi and Balad Sinjar
F & L 24	*Solanum nigrum* L.	Chahala near Amara
F & L 956	*Solanum villosum* Mill.	Sulaimaniya
L 402	*Solanum villosum* Mill.	Rustam Farm near Baghdad
L 100, 293	*Sonchus asper* (L.) Vill.	Near Baghdad
L 295, 323	*Sorghum halepense* (L.) Pers.	Near Baghdad
F & L 808	*Sorghum halepense* (L.) Pers.	Jebel Baykhair near Zakho
L 45	*Spergularia rubra* (L.) Presl	Near Baghdad
F & L 104	*Spergularia rubra* (L.) Presl	Montafah
L 10	*Spergularia salina* Presl	Near Baghdad
F & L 62	*Spergularia salina* Presl	Near Amara
F & L 69	*Spergularia salina* Presl	Mesaida near Amara
F & L 957	*Stachys pubescens* Ten.	Sulaimaniya
L 200, 218	*Statice spicata* Willd.	Rustam Farm near Baghdad
F & L 310	*Statice spicata* Willd.	Jebel Golat between Ain Tellawi and Balad Sinjar
F & L 475	*Sterigmostemum sulphureum* (Russ.) Bornm.	Jebel Golat between Ain Tellawi and Balad Sinjar
L 60	*Stipa tortilis* Desf.	Near Baghdad
F & L 702	*Symphytum* cf. *kurdicum* Boiss. & Hausskn.	Sheikh Adi near Ain Sifni
F & L 79	*Tamarix florida* Bunge	Near Amara
L 478	*Tamarix laxa* Willd.	Near Baghdad
L 399	*Tamarix leptostachya* Bunge	Rustam Farm near Baghdad
L 42	*Tamarix macrocarpa* Bunge	Rustam Farm near Baghdad
L 404	*Tamarix macrocarpa* Bunge	Near Baghdad
F & L 17	*Tamarix pentandra* Pall.	Gatt Al Dwat near Amara
F & L 816	*Tamus communis* L.	Jebel Baykhair near Zakho
L 249	*Tecoma radicans* (L.) DC.	Near Baghdad
F & L 847	*Teucrium parviflorum* Schreb.	Rowandiz Gorge
F & L 457	*Teucrium Polium* L.	Muwasul Tiatan Mukzuk Nuwar
L 267	*Teucrium Polium* L.	Near Baghdad
F & L 398	*Teucrium Polium* L.	Qala Sharqat
F & L 482	*Teucrium Polium* L.	Jebel Golat between Ain Tellawi and Balad Sinjar
F & L 572	*Teucrium Polium* L.	Tell Es Shur between Tall Afar and Balad Sinjar

Number	Genus and Species	Locality
F & L 893, 929	*Teucrium Polium* L.	Jebel Baradost near Diana Rowandiz
F & L 505	*Teucrium pruinosum* Boiss.	Jebel Golat between Ain Tellawi and Balad Sinjar
L 163	*Texiera glastifolia* (DC.) Jaub. & Spach	Near Baghdad
F & L 791	*Thymbra spicata* L.	Jebel Baykhair near Zakho
F & L 132	*Thymus Kotschyanus* Boiss. & Hohen.	Jebel Baykhair near Zakho
F & L 132	*Thymus Kotschyanus* Boiss. & Hohen.	Rutba
F & L 629	*Thymus syriacus* Boiss.	Jebel Khatchra near Balad Sinjar
L 181	*Tragopogon majus* Jacq.	Near Baghdad
L 442	*Tribulus macropterus* Boiss.	Near Baghdad
F & L 920	*Trifolium formosum* Urv.	Jebel Baradost near Diana Rowandiz
L 324	*Trifolium formosum* Urv.	Near Baghdad
L 432	*Trifolium galilaeum* Boiss.	Near Baghdad
F & L 650	*Trifolium pilulare* Boiss.	Jebel Khatchra near Balad Sinjar
F & L 732	*Trifolium purpureum* Loisel.	Jerwona near Ain Sifni
F & L 759	*Trifolium purpureum* Loisel.	Sheikh Adi near Ain Sifni
F & L 949	*Trifolium resupinatum* L.	Sulaimaniya
L 255	*Trifolium stellatum* L.	Near Baghdad
L 77, 164	*Trigonella caelesyriaca* Boiss.	Near Baghdad
F & L 45	*Trigonella Foenumgraecum* L.	Chahala near Amara
L 140	*Trigonella stellata* Forsk.	Near Baghdad
L 204	*Trigonella uncata* Boiss. & Noë.	Rustam Farm near Baghdad
F & L 19	*Trigonella uncata* Boiss. & Noë.	Gatt Al Dwat near Amara
F & L 28	*Triticum aestivum* L.	Chahala near Amara
L 29	*Triticum aestivum* L.	Near Baghdad
F & L 87	*Triticum aestivum* L.	Montafah
F & L 460	*Triticum aestivum* L.	Muwasul Tiatan Mukzuk Nuwar
F & L 575, 580	*Triticum aestivum* L.	Tell Es Shur between Tall Afar and Balad Sinjar
F & L 550	*Turgenia latifolia* (L.) Hoffm.	Mir Khasim between Balad Sinjar and Tall Afar
F & L 649	*Umbilicus intermedius* Boiss.	Jebel Khatchra near Balad Sinjar
F & L 760	*Umbilicus intermedius* Boiss.	Sheikh Adi near Ain Sifni
L 122	*Urtica dioica* L.	Near Baghdad
F & L 625	*Verbascum Andrusi* Post	Jebel Khatchra near Balad Sinjar
F & L 789	*Verbascum laetum* Boiss. & Hausskn.	Jebel Baykhair near Zakho
F & L 477	*Verbascum tripolitanum* Boiss.	Jebel Golat between Ain Tellawi and Balad Sinjar
F & L 515	*Verbascum tripolitanum* Boiss.	Between Tall Afar and Balad Sinjar
L 106, 377	*Verbena officinalis* L.	Near Baghdad
F & L 516	*Verbena officinalis* L.	Between Tall Afar and Balad Sinjar

Plants from Iraq

Number	Genus and Species	Locality
F & L 806	*Verbena officinalis* L.	Jebel Baykhair near Zakho
F & L 967	*Verbena officinalis* L.	Sulaimaniya
F & L 921	*Veronica aleppica* Boiss.	Jebel Baradost near Diana Rowandiz
L 165	*Veronica Anagallis* L.	Near Baghdad
F & L 372	*Veronica Anagallis* L.	Haditha (wheatfield)
F & L 513	*Veronica Anagallis* L.	Between Tall Afar and Balad Sinjar
F & L 693	*Veronica Anagallis* L.	30 km. due west of Balad Sinjar
F & L 10	*Veronica Anagallis* L.	Gatt Al Dwat near Amara
L 257, 464	*Veronica hederaefolia* L.	Near Baghdad
F & L 374	*Veronica hederaefolia* L.	Haditha (wheatfield)
F & L 583, 589	*Vicia angustifolia* Roth	Tell Es Shur between Tall Afar and Balad Sinjar
L 511	*Vicia angustifolia* Roth	Near Baghdad
F & L 764	*Vicia angustifolia* Roth	Sheikh Adi near Ain Sifni
L 229	*Vicia Faba* L.	Near Baghdad
F & L 3	*Vicia peregrina* L.	Gatt Al Dwat near Amara
L 6	*Vicia peregrina* L.	Near Baghdad
F & L 923	*Vicia tenuifolia* Roth	Jebel Baradost near Diana Rowandiz
F & L 761	*Vitis vinifera* L.	Sheikh Adi near Ain Sifni
F & L 857	*Vitis vinifera* L.	Rowandiz Gorge
F & L 842	*Wendlandia Kotschyi* Boiss. & Hohen.	Jebel Pikasar near Aqra
L 334	*Xanthium Strumarium* L.	Near Baghdad
F & L 739	*Ziziphora capitata* L.	Jerwona near Ain Sifni
F & L 757	*Ziziphora capitata* L.	Sheikh Adi near Ain Sifni
L 215	*Ziziphora taurica* M. Bieb.	Near Baghdad
F & L 551	*Ziziphora tenuior* L.	Mir Khasim between Balad Sinjar and Tall Afar
L 240	*Ziziphora tenuior* L.	Rustam Farm near Baghdad
L 546	*Zizyphus Spina-Christi* Willd. var. *inermis* Boiss.	Near Baghdad
F & L 454	*Zoegea Leptaurea* L.	Muwasul Tiatan Mukzuk Nuwar
F & L 549	*Zoegea Leptaurea* L.	Mir Khasim between Balad Sinjar and Tall Afar
F & L 609	*Zoegea Leptaurea* L.	Karya Bat Khan near Balad Sinjar
F & L 638	*Zoegea Leptaurea* L.	Jebel Khatchra near Balad Sinjar
F & L 695	*Zoegea Leptaurea* L.	30 km. due west of Balad Sinjar
F & L 391	*Zollikoferia mucronata* (Forsk.) Boiss.	Telegraph pole M90 between Baiji and Mosul
L 342	*Zozimia absinthifolia* (Vent.) DC.	Near Baghdad
F & L 95	*Zozimia absinthifolia* (Vent.) DC.	Montafah
L 103	*Zygophyllum Fabago* L.	Near Baghdad

GEOGRAPHICAL LIST OF PLANTS

In the preceding table the plants have been arranged in alphabetical sequence. Since it is important to determine the range and distribution of genera and species the collection has been rearranged according to the following localities.

Area	Localities
Northwest	Balad Sinjar and Tall Afar
North	Zakho, Al Qosh, Sheikh Adi, Jerwona, Baiji, Haditha, and Qala Sharqat
Northeast	Aqra, Rowandiz, and Sulaimaniya
Central	Baghdad
Southeast	Amara and Hor al Hawiza
West	Rutba

JEBEL KHATCHRA NEAR BALAD SINJAR

Achillea aleppica DC.
Althaea lavateriflora DC.
Alyssum alpestre L. var. *obovatum* Boiss.
Andrachne telephioides L.
Astragalus chaborasicus (Boiss. & Hausskn.)
Capparis spinosa L.
Centaurea regia Boiss.
Centaurea virgata Lam.
Chrysophthalmum montanum (DC.) Boiss.
Crepis assyriaca Bornm.
Crepis pulchra L.
Crupina Crupinastrum Vis.
Echinops sphaerocephalus L.
Elymus crinitus Schreb.
Euphorbia craspedia Boiss.
Ficus Carica L. var. *rupestris* Hausskn.
Galium coronatum Sibth. & Sm. var. *stenophyllum* Boiss.
Haplophyllum Buxbaumii (Poir.) Boiss.
Helichrysum graveolens Boiss.
Hippomarathrum scabrum (Fenzl) Boiss.
Hordeum spontaneum Koch
Hyoscyamus albus L.
Lactuca cretica Desf.
Onopordon illyricum L.
Onosma flavum (Lehm.) Vatke
Pistacia Terebinthus L.
Poterium verrucosum Ehrenb.
Pterocephalus Putkianus Boiss. & Kotschy
Quercus Aegilops L.
Reseda alba L.
Rhamnus punctata Boiss.
Rhus Coriaria L.
Salvia acetabulosa L. var. *simplicifolia* Boiss.
Salvia palaestina Benth.
Scutellaria cretacea Boiss. & Hausskn.
Serratula cerinthefolia Sibth. & Sm.
Thymus syriacus Boiss.
Trifolium pilulare Boiss.
Umbilicus intermedius Boiss.
Verbascum Andrusi Post
Zoegea Leptaurea L.

JEBEL GOLAT NEAR BALAD SINJAR

Achillea conferta DC.
Achillea micrantha M. Bieb.
Achillea oligocephala DC.
Aegilops squarrosa L.
Allium paniculatum L.
Androsace maxima L.
Avena fatua L.
Bromus macrostachys Desf.
Callipeltis Cucullaria L.
Carum elegans Fenzl
Caucalis leptophylla L.
Centaurea Behen L.
Centaurea phyllocephala Boiss.
Cephalaria syriaca (L.) Schrad.
Crepis aspera L.
Cuscuta babylonica Auch.
Cyperus longus L.
Daucus aureus Desf.
Dianthus anatolicus Boiss.
Echinaria capitata (L.) Desf.
Elymus crinitus Schreb.
Elymus Delileanus Schult.
Euphorbia Chamaesyce L.
Euphorbia lanata Sieb.
Filago spathulatus Presl
Garhadiolus Hedypnois (Fisch. & Mey.) Jaub. & Spach
Geranium rotundifolium L.
Glaucium corniculatum (L.) Curt.
Haplophyllum propinquum Spach

JEBEL GOLAT NEAR BALAD SINJAR—continued

Heteranthelium piliferum (Russ.) Hochst.
Hordeum bulbosum L.
Hyoscyamus pusillus L.
Hypecoum procumbens L.
Hypericum crispum L.
Inula divaricata (Cass.) Boiss.
Koelpinia linearis Pall.
Lotus tenuifolius Reichb.
Micropus erectus L.
Moltkia coerulea (Willd.) Lehm.
Moluccella laevis L.
Ononis sicula Guss.
Onosma aleppicum Boiss.
Phalaris brachystachys Link.
Phalaris paradoxa L.
Phlomis Bruguieri Desf.
Phlomis orientalis Mill.
Plantago Psyllium L.
Poa tatarica Fisch.
Polypogon monspeliensis (L.) Desf.
Reseda muricata Presl
Roripa Nasturtium-aquaticum (L.) Schinz & Thell.
Scabiosa Olivieri Coult.
Scirpus Holoschoenus L.
Senecio coronopifolius Desf.
Sisymbrium septulatum DC.
Statice spicata Willd.
Sterigmostemum sulphureum (Russ.) Bornm.
Teucrium Polium L.
Teucrium pruinosum Boiss.
Verbascum tripolitanum Boiss.

BETWEEN TALL AFAR AND BALAD SINJAR

Achillea conferta DC.
Achillea micrantha M. Bieb.
Achillea oligocephala DC.
Aegilops crassa Boiss.
Alkanna tinctoria (L.) Tausch
Alyssum campestre L.
Anchusa strigosa Labill.
Artedia squamata L.
Asperugo procumbens L.
Centaurea myriocephala Sch. Bip.
Centaurea pallescens Del. var. *hyalolepis* Boiss.
Centaurea phyllocephala Boiss.
Cephalaria syriaca (L.) Schrad.
Chrozophora verbascifolia (Willd.) A. Juss.
Cichorium Intybus L.
Convolvulus reticulatus Choisy
Cousinia stenocephala Boiss.
Cuscuta babylonica Auch.
Cyperus longus L.
Delphinium oliganthum Boiss.
Dianthus polycladus Boiss.
Erodium cicutarium (L.) L'Hér.
Erodium laciniatum (Cav.) Willd.
Euphorbia falcata L.
Gypsophila porrigens (L.) Boiss.
Haplophyllum Buxbaumii (Poir.) Boiss.
Helianthemum salicifolium (L.) Mill.
Helichrysum graveolens Boiss.
Hypericum crispum L.
Hypericum helianthemoides (Spach) Boiss.
Juncus pyramidatus Laharpe
Medicago denticulata Willd.
Moltkia coerulea (Willd.) Lehm.
Nigella arvensis L.
Oliveria orientalis DC.
Onosma aleppicum Boiss.
Onosma sericeum Willd.
Parietaria alsinefolia Del.
Parietaria debilis Forst.
Paronychia capitata (L.) Lam.
Phlomis Bruguieri Desf.
Phlomis linearis Boiss. & Bal.
Poterium verrucosum Ehrenb.
Rumex pulcher L.
Salvia palaestina Benth.
Scorzonera papposa DC.
Scrophularia xanthoglossa Boiss.
Turgenia latifolia (L.) Hoffm.
Verbascum tripolitanum Boiss.
Verbena officinalis L.
Veronica Anagallis L.
Zizyphora tenuior L.
Zoegea Leptaurea L.

TELL ES SHUR BETWEEN TALL AFAR AND BALAD SINJAR

Achillea micrantha M. Bieb.
Achillea Santolina L.
Allium paniculatum L.
Althaea rosea Cav.
Astragalus maximus Willd.
Cuscuta babylonica Auch.
Delphinium oliganthum Boiss.
Elymus caput-medusae L.
Gladiolus atroviolaceus Boiss.
Helicophyllum crassipes (Boiss.) Schott
Salvia syriaca L.
Scorzonera papposa DC.
Silene longipetala Vent.
Teucrium Polium L.
Triticum aestivum L.
Vicia angustifolia Roth

Jebel Baykhair near Zakho

Aegilops Aucheri Boiss.
Alkanna Kotschyana DC.
Anagyris foetida L.
Apocynum venetum L.
Aristolochia maurorum L.
Asparagus stipularis Forsk.
Bupleurum aleppicum Boiss.
Celsia heterophylla Desf.
Centaurea iberica Trev.
Centaurea virgata Lam.
Cephalaria syriaca (L.) Schrad.
Chrozophora verbascifolia (Willd.) A. Juss.
Cousinia cf. Kotschyi Boiss.
Crucianella kurdistanica Malinowski
Crupina Crupinastrum Vis.
Delphinium cappadocicum Boiss.
Delphinium oliganthum Boiss.
Erianthus Hostii Griseb.
Erythraea latifolia Sm.
Euphorbia tinctoria Boiss. & Huet.
Ficus palmata Forsk.
Fumana arabica (L.) Spach
Galium verum L.
Gnaphalium luteo-album L.
Gypsophila ruscifolia Boiss.
Hedysarum pannosum Boiss.
Helichrysum graveolens Boiss.
Inula squarrosa L.
Ipomoea purpurea (L.) Lam.
Juncus effusus L.
Lactuca sativa L.
Lathyrus annuus L.
Lepidium latifolium L.
Linum angustifolium Huds.
Onobrychis caput-galli (L.) Lam.
Onobrychis galegifolia Boiss.
Paliurus aculeatus Lam.
Plantago lanceolata L.
Potentilla fallacina Blocki
Prunus microcarpa C. A. Mey.
Quercus Aegilops L.
Ranunculus lomatocarpus Fisch. & Mey.
Rhaphis Gryllus (L.) Desv.
Rubus discolor Weihe & Nees
Salvia acetabulosa L. var. simplicifolia Boiss.
Salvia cf. kurdica Boiss. & Hohen.
Schoenus nigricans L.
Senecio coronopifolius Desf.
Silene stenobotrys Boiss. & Hausskn.
Sorghum halepense (L.) Pers.
Tamus communis L.
Thymbra spicata L.
Verbascum laetum Boiss. & Hausskn.
Verbena officinalis L.

Baban near Al Qosh

Bupleurum brevicaule Schlecht.
Chamaemelum microcephalum Boiss.
Lallemantia iberica (M. Bieb.) Fisch. & Mey.
Pinus halepensis Mill.
Saponaria Vaccaria L.

Sheikh Adi near Ain Sifni

Achillea micrantha M. Bieb.
Anagyris foetida L.
Celtis Tournefortii Lam.
Cicer arietinum L.
Cousinia arbelensis Winkl. & Bornm.
Crataegus Azarolus L.
Cuscuta babylonica Auch.
Galium tricorne With.
Juglans regia L.
Medicago orbicularis All.
Olea europaea L.
Onosma sericeum Willd.
Paliurus aculeatus Lam.
Parietaria judaica L.
Physocaulos nodosus (L.) Tausch
Pistacia mutica Fisch. & Mey.
Pistacia Terebinthus L.
Polygonum cognatum Meisn.
Prunus microcarpa C. A. Mey.
Quercus Aegilops L.
Rhus Coriaria L.
Scrophularia xanthoglossa Boiss.
Silene dichotoma Ehrh.
Symphytum cf. kurdicum Boiss. & Hausskn.
Trifolium purpureum Loisel.
Umbilicus intermedius Boiss.
Vicia angustifolia Roth
Vitis vinifera L.
Ziziphora capitata L.

Jerwona near Ain Sifni

Althaea hirsuta L.
Bupleurum falcatum L.
Bupleurum kurdicum Boiss.
Carthamus Oxyacantha M. Bieb.
Centaurea solstitialis L.
Convolvulus Cantabrica L.

Jerwona near Ain Sifni—continued

Daucus guttatus Sibth. & Sm.
Delphinium cappadocicum Boiss.
Delphinium peregrinum L.
Echium italicum L.
Erodium gruinum (L.) Ait.
Euphorbia Helioscopia L.
Euphorbia macroclada Boiss.
Filago spathulatus Presl
Galium nigricans Boiss.
Glycyrrhiza glabra L.
Lythrum hyssopifolia L.

Ononis mitissima L.
Ononis sicula Guss.
Pallenis spinosa (L.) Cass.
Pimpinella Kotschyana Boiss.
Poterium villosum Sibth. & Sm.
Ranunculus cassius Boiss.
Rubus discolor Weihe & Nees
Rumex pulcher L.
Trifolium purpureum Loisel.
Ziziphora capitata L.

Between Baiji and Mosul

Achillea conferta DC.
Andrachne telephioides L.
Anthemis Cotula L.
Arnebia decumbens (Vent.) Kuntze
Celsia lanceolata Vent. var. singarica Murb.
Centaurea phyllocephala Boiss.
Cichorium divaricatum Schousb.
Diplotaxis Harra (Forsk.) Boiss.
Erodium cicutarium (L.) L'Hér.
Erodium glaucophyllum Ait.

Erodium laciniatum (Cav.) Willd.
Hordeum murinum L.
Mathiola oxyceras DC.
Moltkia coerulea (Willd.) Lehm.
Papaver Rhoeas L.
Pimpinella peregrina L.
Plantago ovata Forsk.
Scabiosa Olivieri Coult.
Senecio coronopifolius Desf.
Zollikoferia mucronata (Forsk.) Boiss.

Haditha

Allium ampeloprasum L.
Euphorbia cf. oxyodonta Boiss. & Hausskn.
Parietaria alsinefolia Del.

Peganum Harmala L.
Silene Behen L.
Veronica Anagallis L.
Veronica hederaefolia L.

Qala Sharqat

Anchusa strigosa Labill.
Andrachne telephioides L.
Calendula persica C. A. Mey.
Erodium cicutarium (L.) L'Hér.

Heliotropium supinum L.
Silene setacea Viv.
Teucrium Polium L.

Jebel Pikasar near Aqra

Althaea Hohenackeri Boiss. & Huet.
Micromeria Juliana (L.) Benth. var. myrtifolia Boiss.

Pimpinella Kotschyana Boiss.
Pterocephalus strictus Boiss. & Hohen.
Wendlandia Kotschyi Boiss. & Hohen.

Rowandiz Area

Adiantum Capillus-Veneris L.
Chrysanthemum Parthenium (L.) Pers.
Crataegus Azarolus L.
Crepis pulchra L.
Delphinium peregrinum L.
Euphorbia falcata L. var. rubra Boiss.
Ficus Carica L. var. rupestris Hausskn.
Fraxinus oxyphylla M. Bieb.
Gastrocotyle hispida (Forsk.) Bunge
Heliotropium supinum L.

Phlomis orientalis Mill.
Platanus orientalis L.
Prunus cerasifera Ehrh. var. divaricata (Ledeb.) Bailey
Pyrus syriaca Boiss.
Ranunculus cassius Boiss.
Salix Safsaf Forsk.
Teucrium parviflorum Schreb.
Vitis vinifera L.

JEBEL BARADOST NEAR ROWANDIZ

Acanthus longistylis Freyn.
Acer monspessulanum L.
Ajuga Chia Schreb. var. tridactylites Ging.
Althaea lavateriflora DC.
Amygdalus elaeagrifolia Spach
Bromus macrostachys Desf.
Celtis Tournefortii Lam.
Centaurea depressa M. Bieb.
Ceterach officinarum Willd.
Chamaemelum microcephalum Boiss.
Colladonia crenata Boiss.
Crataegus Azarolus L.
Crucianella glauca A. Rich.
Cuscuta approximata Bab. var. urceolata (Kunze) Yuncker
Dactylis glomerata L.
Daphne acuminata Boiss.
Dianthus anatolicus Boiss.
Eremostachys laciniata (L.) Bunge
Euphorbia craspedia Boiss.
Euphorbia tinctoria Boiss. & Huet.
Fibigia clypeata (L.) Boiss.
Galium adhaerens Boiss. & Bal.
Galium mite Boiss. & Hohen.
Gentiana Olivieri Griseb.
Hordeum bulbosum L.
Hypericum scabrum L.
Lallemantia peltata (L.) Fisch. & Mey.
Lotus Gebelia Vent.
Melandrium eriocalycinum Boiss.
Melica Cupani Guss.
Muscari comosum (L.) Mill.
Paliurus aculeatus Lam.
Paracaryum cristatum (Lam.) Boiss.
Pistacia Terebinthus L.
Poa persica Trin.
Prunus microcarpa C. A. Mey.
Quercus Aegilops L.
Quercus dischorochensis K. Koch
Quercus persica Jaub. & Spach
Rumex tuberosus L.
Scutellaria peregrina L. var. Sibthorpii Boiss. & Reut.
Sideritis libanotica Lab. var. incana Boiss.
Silene chloraefolia Sm.
Teucrium Polium L.
Trifolium formosum Urv.
Veronica aleppica Boiss.
Vicia tenuifolia Roth

SULAIMANIYA

Anthemis altissima L.
Celsia heterophylla Desf.
Crataegus Azarolus L.
Echinops Blancheanus Boiss.
Epilobium hirsutum L.
Euphorbia Gaillardoti Boiss. & Blocki
Geranium dissectum L.
Glycyrrhiza glabra L.
Gypsophila platyphylla Boiss.
Heleochloa schoenoides (L.) Host
Hibiscus Trionum L.
Medicago Gerardi Waldst. & Kit.
Ononis antiquorum L.
Onopordon illyricum L.
Polygonum nodosum Pers.
Populus deltoides Marsh.
Poterium verrucosum Ehrenb.
Prunus instititia L.
Salix Safsaf Forsk.
Scirpus maritimus L.
Scolymus maculatus L.
Scorpiurus sulcata L.
Silene conoidea L.
Solanum villosum Mill.
Stachys pubescens Ten.
Trifolium resupinatum L.
Verbena officinalis L.

BAGHDAD

Acanthophyllum microcephalum Boiss.
Achillea falcata L.
Achillea micrantha M. Bieb.
Achillea Santolina L.
Adonis aestivalis L.
Aeluropus litoralis (Gouin) Parl.
Aeluropus repens (Desf.) Parl.
Agropyron squarrosum (Roth) Link.
Alhagi maurorum Medic.
Althaea Ludwigii L.
Amaranthus graecizans L.
Amaranthus cf. paniculatus L.
Amaranthus retroflexus L.
Amaranthus viridis L.
Ammi majus L.
Ammi majus L. var. longiseta Reichb.
Amygdalus spartioides Spach
Anagallis arvensis L.
Anchusa strigosa Labill.
Andrachne telephioides L.
Andropogon annulatus Forsk.
Anthemis altissima L.
Anthemis Cotula L.
Anthemis hebronica Boiss. & Kotschy
Anthemis cf. melampodina Del.
Aristida plumosa L.

BAGHDAD—*continued*

Arnebia linearifolia DC.
Artemisia annua L.
Asperugo procumbens L.
Asperula arvensis L.
Asphodelus tenuifolius Cav.
Astragalus alexandrinus Boiss.
Astragalus cruciatus Link.
Atriplex leucoclada Boiss. subsp. *turcomanica* (Moq.) Aellen
Avena fatua L.
Barbarea vulgaris R. Br.
Beta vulgaris L. subsp. *maritima* (L.) Thell. var. *glabra* Aellen
Beta vulgaris L. subsp. *vulgaris* (L.) Thell.
Bromus mollis L.
Bromus tectorum var. *grandiflorus* Hook.
Calligonum polygonoides L.
Callipeltis Cucullaria L.
Capparis spinosa L.
Carum elegans Fenzl
Caucalis leptophylla L.
Celsia heterophylla Desf.
Celtis australis L.
Centaurea araneosa Boiss.
Ceterach officinarum Willd.
Chrozophora tinctoria (L.) A. Juss.
Chrozophora verbascifolia (Willd.) A. Juss.
Cichorium divaricatum Schousb.
Citrullus Colocynthis (L.) Schrad.
Colladonia crenata Boiss.
Convolvulus Cantabrica L.
Corchorus olitorius L.
Coreopsis tinctoria Nutt.
Coronilla varia L.
Crypsis aculeata (L.) Ait.
Cucumis prophetarum L.
Cuscuta Lehmanniana Bunge
Cuscuta pedicellata Ledeb.
Cydonia oblonga Mill.
Cymbopogon Schoenanthus (L.) Spreng.
Cynanchum acutum L.
Cynodon Dactylon (L.) Pers.
Cyperus fuscus L.
Cyperus rotundus L.
Dalbergia Sissoo Roxb.
Daphne acuminata Boiss.
Datura Metel L.
Delphinium rigidum DC.
Dianthus anatolicus Boiss.
Dianthus Cyri Fisch. & Mey.
Dianthus fimbriatus M. Bieb.
Digitaria sanguinalis (L.) Scop.
Diplotaxis erucoides (L.) DC.
Echinochloa colona (L.) Link
Echinochloa Crusgalli (L.) Beauv.
Echinops sphaerocephalus L.
Eclipta alba (L.) Hassk.
Elaeagnus angustifolia L. var. *orientalis* (L.) Kuntze

Epilobium hirsutum L.
Eragrostis cilianensis (All.) Link
Eragrostis tenella (L.) Roem. & Schult.
Erigeron canadensis L.
Erodium cicutarium (L.) L'Hér.
Erodium glaucophyllum Ait.
Eruca sativa Mill.
Erucaria microcarpa Boiss.
Eryngium creticum Lam.
Euphorbia Chamaesyce L.
Euphorbia denticulata Lam.
Euphorbia falcata L.
Euphorbia Helioscopia L.
Euphorbia Peplus L.
Euphorbia turcomanica Boiss.
Fagonia Bruguieri DC.
Filago spathulatus Presl
Fimbristylis dichotoma (L.) Vahl
Frankenia Aucheri Jaub. & Spach
Fumaria parviflora Lam.
Gagea reticulata (Pall.) R. & S.
Galium coronatum Sibth. & Sm.
Galium mite Boiss. & Hohen.
Gladiolus atroviolaceus Boiss.
Glaucium corniculatum (L.) Curt.
Glinus lotoides L.
Gypsophila Rokejeka Del.
Haplophyllum tuberculatum Forsk.
Heleochloa alopecuroides (Schrad.) Host
Helianthemum salicifolium (L.) Mill.
Heliotropium Eichwaldi Steud.
Heliotropium supinum L.
Heliotropium undulatum Vahl
Herniaria cinerea DC.
Herniaria hemistemon Gay
Hibiscus Trionum L.
Hippocrepis cornigera Boiss.
Hordeum murinum L.
Hypericum crispum L.
Hypericum scabrum L.
Iberis odorata L.
Isatis aleppica Scop.
Juncus acutus L.
Juncus effusus L.
Koelpinia linearis Pall.
Lactuca saligna L.
Lamium amplexicaule L.
Lamium maculatum L.
Lappula spinocarpa (Forsk.) Aschers.
Lathyrus Aphaca L.
Lathyrus Cicera L.
Lepidium Draba L.
Lepidium sativum L.
Linaria Elatine (L.) Mill.
Linum flavum L.
Linum grandiflorum Desf.
Lippia nodiflora (L.) Michx.
Lolium temulentum L.
Luffa cylindrica (L.) Roem.
Lycium barbarum L.

BAGHDAD—*continued*

Lythrum Salicaria L. var. *tomentosum* DC.
Malcolmia africana (L.) R. Br.
Malcolmia Bungei Boiss.
Malcolmia crenulata (DC.) Boiss.
Malcolmia torulosa (Desf.) Boiss.
Malva parviflora L.
Malva rotundifolia L.
Mathiola oxyceras DC.
Matricaria aurea (L.) Boiss.
Medicago sativa L.
Micropus supinus L.
Moltkia collosa (Vahl) Wettst.
Moluccella laevis L.
Morus alba L.
Myrtus communis L.
Nigella sativa L.
Obione flabellum (Bunge) Ulbr.
Ornithogalum narbonnense L.
Panicum miliaceum L.
Parietaria judaica L.
Paronychia argentea Lam.
Paspalum distichum L.
Peganum Harmala L.
Phalaris brachystachys Link
Phalaris minor Retz.
Phalaris paradoxa L.
Phlomis Bruguieri Desf.
Phlomis orientalis Mill.
Phragmites communis (L.) Trin.
Pisum sativum L.
Plantago Coronopus L.
Plantago lanceolata L.
Plantago Loeflingii L.
Plantago ovata Forsk.
Poa bulbosa L.
Poa persica Trin.
Polygonum aviculare L.
Polygonum Bellardi All.
Polygonum Persicaria L.
Polypogon monspeliensis (L.) Desf.
Populus euphratica Oliv.
Poterium verrucosum Ehrenb.
Prangos ferulacea Lindl.
Prosopis juliflora DC.
Prunus Amygdalus Stokes
Prunus microcarpa C. A. Mey.
Pulicaria crispa (Forsk.) Sch. Bip.
Ranunculus aquatilis L.
Ranunculus arvensis L.
Ranunculus lomatocarpus Fisch. & Mey.
Ranunculus myriophyllus DC.
Reseda lutea L.
Rubus discolor Weihe & Nees
Rumex dentatus L. var. *pleiodon* Boiss.
Rumex obtusifolius L.

Rumex pulcher L.
Rumex roseus L.
Salix acmophylla Boiss.
Salix amygdalina L.
Salvia Szovitsiana Bunge
Saponaria Vaccaria L.
Scabiosa palaestina L.
Scandix iberica M. Bieb.
Scandix Pecten-Veneris L.
Scirpus littoralis Schrad.
Scirpus maritimus L.
Scorpiurus sulcata L.
Scorzonera papposa DC.
Senecio coronopifolius Desf.
Sesbania aegyptiaca Pers.
Setaria lutescens (Weig.) Hubb.
Silene conoidea L.
Silene rubella L.
Sisymbrium damascenum Boiss. & Gaill.
Solanum villosum Mill.
Sonchus asper (L.) Vill.
Sorghum halepense (L.) Pers.
Spergularia rubra (L.) Presl
Spergularia salina Presl
Statice spicata Willd.
Stipa tortilis Desf.
Tamarix laxa Willd.
Tamarix leptostachya Bunge
Tamarix macrocarpa Bunge
Tecoma radicans (L.) DC.
Teucrium Polium L.
Texiera glastifolia (DC.) Jaub. & Spach
Tragopogon majus Jacq.
Tribulus macropterus Boiss.
Trifolium formosum Urv.
Trifolium galilaeum Boiss.
Trifolium stellatum L.
Trigonella caelesyriaca Boiss.
Trigonella stellata Forsk.
Trigonella uncata Boiss. & Noë
Triticum aestivum L.
Urtica dioica L.
Verbena officinalis L.
Veronica Anagallis L.
Veronica hederaefolia L.
Vicia angustifolia Roth
Vicia Faba L.
Vicia peregrina L.
Xanthium Strumarium L.
Ziziphora taurica M. Bieb.
Ziziphora tenuior L.
Zizyphus Spina-Christi Willd. var. *inermis* Boiss.
Zozimia absinthifolia (Vent.) DC.
Zygophyllum Fabago L.

AMARA

Aristolochia maurorum L.
Atriplex tatarica L.

Avena fatua L.
Bacopa Monniera (L.) Wettst.

AMARA—*continued*

Beta vulgaris L. subsp. *lomatogonoides* Aellen
Centaurea iberica Trev.
Convolvulus arvensis L.
Cynodon Dactylon (L.) Pers.
Cyperus rotundus L.
Delphinium rigidum DC.
Erythraea latifolia Sm.
Fimbristylis dichotoma (L.) Vahl
Frankenia pulverulenta L.
Geranium dissectum L.
Hordeum maritimum With.
Jussiaea repens L.
Koeleria phleoides (Vill.) Pers.
Lepidium Draba L.
Lepidium sativum L.
Limnanthemum nymphoides (L.) Link
Lippia nodiflora (L.) Michx.
Lolium temulentum L.
Lycium barbarum L.
Malva parviflora L.
Peganum Harmala L.
Plantago lanceolata L.
Plantago Loeflingii L.
Polygonum Bellardi All.
Polygonum serrulatum Lag.
Polypogon monspeliensis (L.) Desf.
Populus ephtratica Oliv.
Potamogeton lucens L.
Prosopis Stephaniana (Willd.) Kunth
Pulicaria dysenterica (L.) Gaertn.
Ranunculus pantothrix Brot.
Raphanus sativus L.
Rumex dentatus L. var. *pleiodon* Boiss.
Rumex obtusifolius L.
Salix Safsaf Forsk.
Salvinia natans (L.) All.
Scirpus littoralis Schrad.
Solanum nigrum L.
Spergularia salina Presl
Tamarix florida Bunge
Tamarix pentandra Pall.
Trigonella Foenumgraecum L.
Trigonella uncata Boiss. & Noë
Triticum aestivum L.
Veronica Anagallis L.
Vicia peregrina L.

RUTBA

Althaea Ludwigii L.
Arnebia decumbens (Vent.) Kuntze
Astragalus Forskahlei Boiss.
Bromus tectorum L.
Caylusea canescens (L.) St. Hil.
Erodium Ciconium (L.) Willd.
Erodium glaucophyllum Ait.
Euphorbia Chesneyi (Kl. & Garcke) Boiss.
Glaucium grandiflorum Boiss. & Huet.
Haplophyllum propinquum Spach
Herniaria incana Lam.
Lappula spinocarpa (Forsk.) Aschers.
Onopordon heteracanthum C. A. Mey.
Peganum Harmala L.
Phagnalon rupestre (L.) DC.
Senecio coronopifolius Desf.
Thymus Kotschyanus Boiss. & Hohen.

GLOSSARY

The colloquial words as used in Iraq have been listed with the classical forms in parentheses. In Iraq the letter *k* is usually pronounced *ch* and the letter *q* as *g*. In the glossary the diacritical marks have been checked by Mr. Abdul-Majid Abbass and Mr. Jassim Khalaf, Iraq Government students at the University of Chicago.

Badinjun (Bādinjān), 22. Brinjals.
Bagulla (Baqal, pl. *Buqūl)*, 22. Beans.
Baslah (pl. *Bassal)*, 22. Onion.
Battikha (pl. *Battikh)*, 22. Melon.
Charid (Kurud), 22. Water lift.
Chawi (Kawi), 39, 66, 135, 139. Branding scar.
Chāi (Shāi), 115. Tea.
Chupattis (Hindi), 115. Unleavened cakes.
Dukhn, 34. Millet.
Fallahin, 25. Cultivators.
Fijla (pl. *Fijil)*, 22. Radish.
Gahwah (Qahwah), see *Kahwa.*
Haj, 31. Pilgrimage.
Hennā (Hinnā), 39. Henna.
Huntah (Hintah), 22. Wheat.
Ithra (Thira), 22. Maize.
Jidri, 112. Smallpox.
Kahwa, 115. Coffee.
Kanaqina (local Arabic), 110. Quinine.
Kawi, see *Chawi.*
Kessereh (Kasrah), 111, 112. Catchment basin.
Khiara (pl. *Khiar)*, 22. Cucumber.
Kibrit, 28. Sulphur.

Kubeli, 37. Eye lotion.
Kuhl, 39. Kohl.
Liwa, 32. Administrative district.
Māsh, 22. Mash.
Mazūt, 24. Oil used on animals.
Mutasarrif, 32. Governor of a district (*liwa*).
Na'ura (pl. *Newa'ir)*, 22, 23. Noria or Persian water wheel.
Qadha, 26. Political division.
Qīr, 28. Bitumen.
Quffah, 24. Gufa.
Qura (pl. *Quwar)*, 24. Kiln.
Qutn, 22. Cotton.
Sha'ir, 22. Barley.
Shajarat armut, 22. Pear tree.
Shajarat rumman, 22. Pomegranate tree.
Shajarat tiffah, 22. Apple tree.
Shajarat tukki, 22. Mulberry tree.
Shakhtur (pl. *Shakhatīr)*, 24. Barge.
Simsim, 22. Sesame.
Sukham, 151. Soot.
Sūq, 110. Market; bazaar.
Tamr, 115. Dates.
Tabuqa (pl. *Tabūq)*, 24. Brick.
Timmin, 22. Rice.
Tīna (pl. *Tīn)*, 34. Fig.

BIBLIOGRAPHY

The following bibliographical references have been used in the preparation of this Report. No attempt has been made to compile all the references to this area but rather those selected writings which bear strictly on the land and the people of the Upper Euphrates region. The reader is referred to the selected bibliography and notes on sources in Grant (1937).

Assistance rendered by libraries both at home and abroad has been acknowledged in the Preface.

Abbreviations

- AA American Anthropologist
- AJA American Journal of Archaeology
- AJPA American Journal of Physical Anthropology
- AJSL American Journal of Semitic Languages and Literature
- BRSGI Bollettino della Reale Società Geografica Italiana
- FMNH Field Museum of Natural History
- GJ Geographical Journal. *See also* JRGS
- GR Geographical Review
- JBNHS Journal of the Bombay Natural History Society
- JRAI Journal of the Royal Anthropological Institute of Great Britain and Ireland
- JRAS Journal of the Royal Asiatic Society
- JRCAS Journal of the Royal Central Asian Society
- JRGS Journal of the Royal Geographical Society
- NH Natural History
- OES Oriental Explorations and Studies, American Geographical Society. New York
- RSTMH Transactions of the Royal Society of Tropical Medicine and Hygiene. London

AITCHISON, J. E. T.
  1890. Notes on the products of western Afghanistan and N. E. Persia. Edinburgh.

ANDREW, SIR WILLIAM
  1882. Euphrates Valley route to India, in connection with the Central Asian and Egyptian questions. London.

ASHKENAZI, TOVIA
  1938. Tribus semi-nomades de la Palestine du nord. Paris.

AYROUT, HENRY HABIB
  1938. Moeurs et coutumes des fellahs. Paris.

BLANCHARD, RAOUL
  1925. La route du désert de Syrie. Annales de Géographie, vol. 34, pp. 235–243. Paris.
  1929. La Mésopotamie. Géographie Universelle, vol. 8, pp. 215–232. Paris.

BLUNT, LADY ANNE
  1879. Bedouin tribes of the Euphrates. 2 vols. London.

BOESCH, HANS H.
  1939. El-Iraq. Economic Geography, vol. 15, No. 4, pp. 325–361.

BOISSIER, EDMOND
  1867–84. Flora orientalis. Geneva.

BORNMÜLLER, J.
  1917. Zur Flora des nördlichen Syriens. Notizbl. Bot. Gart. Berlin, vol. 7, No. 63, pp. 1–44. Berlin-Dahlem.
BOUCHEMAN, ALBERT DE
  1934. Matériel de la vie bédouine. Documents d'Etudes Orientales, vol. 3. Institut Français de Damas, Damascus.
BURKILL, I. H.
  1909. A working list of the flowering plants of Baluchistan. Calcutta.
BUXTON, L. H. DUDLEY and RICE, DAVID TALBOT
  1931. Report on the human remains found at Kish. JRAI, vol. 61, pp. 57–119.
CARRUTHERS, DOUGLAS
  1918. The great desert caravan route, Aleppo to Basra. GJ, vol. 52, pp. 157–184.
  1938. Introduction and notes in Northern Najd. A journey from Jerusalem to Anaiza in Qasim. London.
CHARLES, H.
  1939. Tribus moutonnières du Moyen-Euphrate. Documents d'Etudes Orientales, vol. 8. Institut Français de Damas, Beirut.
CHINA, W. E.
  1938. Hemiptera from Iraq, Iran and Arabia. FMNH, Zool. Ser., vol. 20, No. 32, pp. 427–437.
CLAWSON, M. DON
  1936. The Shammar Bedouin dental survey. The Dental Magazine and Oral Topics, vol. 53, Nos. 2, 3, February, March. London.
CLEMOW, F. G.
  1916. The Shiah pilgrimage and the sanitary defences of Mesopotamia and the Turco-Persian frontier. The Lancet, August 12, 19, and September 2. London.
COLES, F. E.
  1938. Dust storms in Iraq. Professional Notes No. 84, vol. 6, No. 4. Meteorological Office, Air Ministry. London.
COON, CARLETON STEVENS
  1939. The races of Europe. New York.
DOUGHTY, CHARLES M.
  1926. Travels in Arabia Deserta. London.
DOWSON, V. H. W.
  1921–23. Dates and date cultivation of the Iraq. Pts. 1–3. Printed for the Agricultural Directorate of Iraq. Cambridge, England.
  1939. Provisional list of the date palms of the Iraq. Tropical Agriculture, vol. 16, No. 7, pp. 164–168. Trinidad.
DYMOCK, WILLIAM
  1885. The vegetable materia medica of western India. Ed. 2. Bombay.
——, WARDEN, CHARLES JAMES HISLOP, and HOOPER, DAVID
  1889–93. Pharmacographia indica. 3 vols. Bombay.
EPSTEIN, ELIHU
  1940. Al Jezireh. JRCAS, vol. 27, Pt. 1, pp. 68–82.
FIELD, HENRY
  1926. New discoveries at Kish: A great temple; 5000-years old pottery. Illustrated London News, vol. 79, No. 2054, p. 395, September 4.

1929a. Early man in North Arabia. Amer. Mus. Nat. Hist., NH, vol. 29, pp. 33–44.
1929b. The Field Museum–Oxford University Joint Expedition to Kish, Mesopotamia, 1923–29. FMNH, Anthr. Leaflet No. 28.
1931a. Among the Beduins of North Arabia. Open Court, vol. 45, pp. 577–595. Chicago.
1931b. The Field Museum–Oxford University Joint Expedition to Kish. Art and Archaeology, No. 5, pp. 243–252, and No. 6, pp. 323–334. Washington.
1932a. The ancient and modern inhabitants of Arabia. Open Court, vol. 46, pp. 847–871. Chicago.
1932b. The cradle of *Homo sapiens*. AJA, vol. 36, pp. 426–430.
1932c. Human remains from Jemdet Nasr, Mesopotamia. JRAS, Pt. 4, pp. 967–970.
1932d. Ancient wheat and barley from Kish, Mesopotamia. AA, new ser., vol. 34, pp. 303–309.
1933. The antiquity of man in Southwestern Asia. AA, new ser., vol. 35, pp. 51–62.
1934. Sulle caratteristiche geografiche dell' Arabia settentrionale. BRSGI, vol. 11, pp. 3–13.
1935a. Arabs of central Iraq, their history, ethnology and physical characters. Introduction by Sir Arthur Keith. FMNH, Anthr. Mem., vol. 4.
1935b. The Field Museum Anthropological Expedition to the Near East, 1934. Science, vol. 81, No. 2093, p. 146.
1935c. *Ibid.* The Oriental Institute Archaeological Report on the Near East. AJSL, vol. 51, pp. 207–209.
1936. The Arabs of Iraq. AJPA, vol. 21, pp. 49–56.
1937a. Oryx and ibex as cult animals in Arabia. Man, vol. 37, No. 69. London.
1937b. Jews of Sandur, Iraq. Asia, vol. 37, pp. 708–710.
1937c. *See* Hooper, David.
1939a. The physical characters of the modern inhabitants of Iran. The Asiatic Review, vol. 35, No. 123, pp. 572–576. London.
1939b. Contributions to the anthropology of Iran. FMNH, Anthr. Ser., vol. 29.

FRAZER, SIR JAMES GEORGE
1924. The golden bough. London.

GILLIAT-SMITH, B., and TURRILL, W. B.
1930. On the flora of the Nearer East. Kew Bull., Nos. 7–10. London.

GOVERNMENT OF IRAQ PUBLICATIONS
1929. Maps of Iraq with notes for visitors. Baghdad.

GRANT, CHRISTIANA PHELPS
1937. The Syrian Desert. London.

GUARMANI, CARLO
1938. Northern Najd. A journey from Jerusalem to Anaiza in Qasim. Trans. by Lady Capel-Cure. London.

GUEST, EVAN
1933. Notes on plants and plant products with their colloquial names in Iraq. Bull. No. 27, Department of Agriculture, Iraq. Baghdad.

HANDBOOK OF ARABIA
1920. General. Vol. 1. London.

HARRISON, PAUL W.
1924. The Arab at home. New York.

HOOPER, DAVID, and FIELD, HENRY
  1937. Useful plants and drugs of Iran and Iraq. FMNH, Bot. Ser., vol. 9, No. 3, pp. 71–241.

HUDSON, ELLIS HERNDON
  1928. Trypanosomiasis among the Bedouin Arabs of the Syrian Desert. U. S. Naval Med. Bull., vol. 26, No. 4. Washington, D.C.
  1938. The significance of bejel. Reprinted from Publication No. 6 of the American Association for the Advancement of Science, pp. 35–39.
  1939. Can syphilis exist apart from sex? N.Y. State Jour. of Med., vol. 39, No. 19, pp. 1840–45.

IONIDES, M. G.
  1937. The régime of the rivers Euphrates and Tigris. New York.

JAMALI, M. F.
  1934. The new Iraq. Problems of Bedouin education. New York.

KEITH, SIR ARTHUR
  1935. Introduction *in* Arabs of central Iraq, their history, ethnology and physical characters. FMNH, Anthr. Mem., vol. 4, pp. 11–76.

——, and KROGMAN, W. M.
  1932. The racial characteristics of the southern Arabs (pp. 301–333) *in* "Arabia Felix" by Bertram Thomas. New York.

KENNEDY, WALTER P.
  1935. The polynuclear count in an Iraq population. RSTMH, vol. 28, No. 5, pp. 475–480.
  1937a. Some additions to the fauna of Iraq. JBNHS, vol. 39, pp. 745–749. Bombay.
  1937b. The macropolycyte in health and disease in Iraq. Journal of Pathology and Bacteriology, vol. 44, No. 3, pp. 701–704. Edinburgh.
  1937c. The leucocyte picture in Iraq. RSTMH, vol. 31, No. 3, pp. 309–332.

—— and MACKAY, IAN
  1935. Further studies on the polynuclear count in Iraq. RSTMH, vol. 29, No. 3, pp. 291–298.
  1936. The normal leucocyte picture in a hot climate. Journal of Physiology, vol. 87, No. 4, pp. 336–344. London.
  See also MACKAY, IAN

KROGMAN, W. M., *see* KEITH, SIR ARTHUR

LAUFER, BERTHOLD
  1919. Sino-Iranica. FMNH, Anthr. Ser., vol. 15, No. 3, pp. 185–630.
  1934. The Noria or Persian wheel. Oriental studies in honour of Dasturji Saheb Cursetji Erachji Pavry, pp. 238–250. Oxford.

LAWRENCE, T. E.
  1926. Seven pillars of wisdom. London.

LYDE, LIONEL W.
  1933. The continent of Asia. London.

MACKAY, IAN, and KENNEDY, WALTER P.
  1936. Some cases of non-gonococcal urethritis in the Near East. Journal of the Royal Army Medical Corps, pp. 194–197. London.
  See also KENNEDY, WALTER P.

MUSIL, ALOIS
  1927a. Arabia Deserta. OES, No. 2. American Geographical Society. New York.
  1927b. The Middle Euphrates. OES, No. 3. American Geographical Society. New York.

1928. The manners and customs of the Rwala Bedouins. OES, No. 6. American Geographical Society. New York.

OPPENHEIM, MAX FREIHERR VON
1939. Die Beduinen, vol. 1. Leipzig.

POST, G. E.
1896. Flora of Syria, Palestine, and Sinai. Beirut.

RASWAN, CARL R.
1930. Tribal areas and migration lines of the North Arabian Bedouins. GR, vol. 20, pp. 494-502.
1935. Black tents of Arabia. Boston.
1936. Moeurs et coutumes des Bédouins. Paris.

RICE, D. TALBOT, see BUXTON, L. H. DUDLEY

SAMUELSSON, GUNNAR
1933a. Lycochloa, eine neue Gramineen-Gattung aus Syrien. Ark. Bot., vol. 25A, No. 8, pp. 1-6. Stockholm.
1933b. Rumex pictus Forsk. und einige verwandte Arten. Ber. Schwei. Bot. Gesell., vol. 42, Pt. 2, pp. 770-779. Bern.
1935. Notes on two collections of plants from Syria, Palestine, Transjordan and Iraq. Särtryck ur Svensk Botanisk Tidskrift, vol. 29, Pt. 3. Uppsala.
1938. Cives novae florae syricacae. Repert. Spec. Nov. Beihefte, vol. 100, pp. 38-49. Berlin-Dahlem.

SCHLIMMER, J. L.
1874. Terminologie médico-pharmaceutique et anthropologique française-persane. Teheran.

SCHMIDT, KARL P.
1930. Reptiles of Marshall Field North Arabian Desert Expedition, 1927-28. FMNH, Zool. Ser., vol. 17, pp. 223-230.
1939. Reptiles and amphibians from Southwestern Asia. FMNH, Zool. Ser., vol. 24, pp. 49-92.

STAMP, L. DUDLEY
1929. Asia. London.

SUMMERSCALE, J. P.
1938. Report on economic and commercial conditions in Iraq. Department of Overseas Trade, No. 699. London.

SYDOW, H.
1935. Ein Beitrag zur Kenntnis der parasitischen Pilze des Mittelmeergebiets. Svensk Botanisk Tidskrift, vol. 29, Pt. 1, pp. 65-78.

SYKES, MARK
1907. Journeys in North Mesopotamia. GJ, vol. 30, pp. 237-254, 284-398.

THIÉBAUT, J.
1936. Flore Libano-Syrienne. Mém. Inst. d'Egypte, vol. 11. Cairo.

TROTTER, MILDRED
1936. The hair of the Arabs of central Iraq. AJPA, vol. 21, pp. 423-428.

UVAROV, B. P.
1938. Orthoptera from Iraq and Iran. FMNH, Zool. Ser., vol. 20, pp. 439-451.

VAVILOV, N. I.
1934. Agricultural Afghanistan [In Russian]. Leningrad.

WILLCOCKS, SIR WILLIAM
1911. The irrigation of Mesopotamia. London.

TRIBES REFERRED TO IN CHAPTER V

In the following table each tribe is listed in alphabetical order. The prefixes Al, Al bu, and Bani follow the tribal names.

Minor tribe, section, or sub-section	Main tribe or confederation	Minor tribe, section, or sub-section	Main tribe or confederation
Abaidat, Al	Baqqarah	Faiyadah, Al bu	Dulaim
Abd, Al bu	Dulaim	Faiyadah	Haiwat
Abdullah	Anaiza	Falahat, Al bu	Dulaim
Aithah, Al bu	Dulaim	Fallujiyin	Haiwat
Ajaj, Al bu	Dulaim	Farraj, Al bu	Dulaim
Ajarjah, Al	Aqaidat	Farraj Allah	Kubais, Bani
Ajrah, Al	Anaiza	Fuqarah	Anaiza
Akash, Al bu	Dulaim	Furjah	Anaiza
Ali, Al	Aqaidat		
Ali, Al	Baqqarah	Ghadir, Al bu	Dulaim
Ali, Al bu	Dulaim	Ghanim, Al bu	Baqqarah
Ali al Jasim, Al bu	Dulaim	Ghazail, Al bu	Dulaim
Aliyat, Al bu	Aqaidat	Ghurrah, Al bu	Dulaim
Alwan, Al bu	Dulaim	Guraibawiyin, Al	Haiwat
Amarah, Al	Anaiza		
Amarat	Anaiza	Haddad, Al bu	Dulaim
Annas, Al	Haiwat	Haidah, Al bu	Kubais, Bani
Aql, Al bu	Dulaim	Haiwat	Zoba
Arab, Al bu	Dulaim	Hajjaj	Anaiza
Araf, Al bu	Dulaim	Hajji Isa, *Bait*	Kubais, Bani
Ashahin, Al	Baqqarah	Halabsah, Al bu	Dulaim
Ashja	Anaiza	Hamad, Al bu	Kubais, Bani
Ashshihah, Al bu	Dulaim	Hamad al Dhiyab, Al bu	Dulaim
Assaf, Al bu	Dulaim	Hamad al Hussain, Al bu	Dulaim
Ataifat	Anaiza	Hamdan, Al bu	Baqqarah
Ausaj, Al bu	Dulaim	Hammamid	Anaiza
Azzah, Al	Chitadah	Hamudi, Al	Aqaidat
Azzam, Al bu	Dulaim	Hantush, Al bu	Dulaim
		Hardan, Al bu	Aqaidat
Badran, Al bu	Baqqarah	Hardha, Al	Anaiza
Baiqat	Ar Rahhaliya	Harub, Al	Ar Rahhaliya
Bajaidah, Al	Anaiza	Hasanah	Anaiza
Baqqarah, Al	Dulaim	Hassan, Al bu	Dulaim
Barghuth, Al	Chitadah	Hassan, Al bu	Aqaidat
Budur	Anaiza	Hassun, Al	Aqaidat
Butainat, Al	Anaiza	Hawa, Al bu	Dulaim
		Hatim, Al bu	Dulaim
Chitadah	Zoba	Hazalat, Al	Anaiza
		Hazim, Al bu	Dulaim
Dahaman, Al	Anaiza	Hiblan, Al	Anaiza
Dahamshar, Al	Anaiza	Hilal, Al bu	Dulaim
Dariah, *Bait*	Kubais, Bani	Hitawiyin	Zoba
Dhanna Majid	Anaiza	Hulaiyil, Al	Haiwat
Dhiyab, Al bu	Dulaim	Humaid, Al	Chitadah
Dimim, Al	Aqaidat	Huntush, Al bu	Dulaim
Dilamah, Al	Anaiza	Huraiwat, Al bu	Dulaim
Dughaiyim, Al	Faddaghah	Hussain al Ali, Al bu	Dulaim
Duhail, Al bu	Dulaim	Hussani, Al	Anaiza
Dukhaiyil, Al	Dulaim		
Dulaim Qartan	Zoba	Idhar, Al	Aqaidat
Duran	Anaiza	Isa, Al	Aqaidat
Fadan	Anaiza	Isa, Al bu	Dulaim
Fahad, Al bu	Dulaim		

INDEX OF TRIBES: CHAPTER V

Minor tribe, section, or sub-section	Main tribe or confederation	Minor tribe, section, or sub-section	Main tribe or confederation
Jabal, Al	Anaiza	Muridh	Anaiza
Jabar, Al bu	Dulaim	Musa, Al bu	Baqqarah
Jadan, Al bu	Dulaim	Musa, Al bu	Dulaim
Jadu, Al	Aqaidat	Musaib, Al	Anaiza
Jaghaifah, Al bu	Dulaim	Musalikh	Anaiza
Jalaid, Al	Anaiza	Musalihah	Dulaim
Jalal, Al	Anaiza	Mushahidah, Al	Aqaidat
Jasim, Al bu	Dulaim	Mutarafah, Al	Anaiza
Jifal, Al	Anaiza		
Juhaish, Al bu	Dulaim	Nabbizah, Al	Baqqarah
Jumailah	Dulaim	Nabit, Al	Faddaghah
		Nasrah, Al	Anaiza
Kaka	Anaiza	Nassar, Al	Faddaghah
Kawakibah	Anaiza	Nimr, Al bu	Dulaim
Khalaf, Al bu	Dulaim	Nusair	Anaiza
Khalifah, Al bu	Dulaim		
Khalil, Al bu	Haiwat	Qaan, Al bu	Aqaidat
Khamis, Al bu	Dulaim	Qadrau, Al	Aqaidat
Khamishat, Al	Anaiza	Qara-Ghul	Dulaim
Khammas, Al	Chitadah	Qartan, Al	Dulaim
Khanfar, Al	Aqaidat	Qumzan, Al	Chitadah
Khanjar, Al	Baqqarah	Quraifa, Al bu	Dulaim
Khashtah, Al	Anaiza	Quraiti, Al bu	Dulaim
Khurushiyin	Dulaim	Quran, Al	Aqaidat
Kulaib, Al bu	Dulaim		
		Rad, Al	Dulaim
Luhaib	Dulaim	Radhi, Al	Chitadah
		Rahamah, Al bu	Aqaidat
Madlij, Al bu	Dulaim	Raihan, Al bu	Dulaim
Majawadah, Al	Aqaidat	Ramlah, Al bu	Dulaim
Mahal, Al bu	Dulaim	Rudaini, Al bu	Dulaim
Maish, Al bu	Baqqarah	Rus, Al	Anaiza
Malahimah, Al	Dulaim	Ruwalla	Anaiza
Malhud, Al	Anaiza		
Manayi	Anaiza	Saadan, Al	Dulaim
Mani, Al bu	Dulaim	Sbaa, Al	Anaiza
Marasimah, Al	Aqaidat	Salatin, Al	Anaiza
Mashadiqah	Anaiza	Salih, Al bu	Dulaim
Mashittah	Anaiza	Salih al Ali, Al bu	Dulaim
Mathluthah, *Bait*	Kubais, Bani	Salman, Al bu	Ar Rahhaliya
Matrad, Al bu	Dulaim	Salqah, Al	Anaiza
Miri, Al bu	Aqaidat	Samalah, Al bu	Dulaim
Miri, Al bu	Dulaim	Sanid, Al	Anaiza
Mish, Al bu	Baqqarah	Saqr, Al bu	Dulaim
Mudhaiyan, Al	Anaiza	Saqra	Anaiza
Mufarraj, Al bu	Faddaghah	Sarai, Al bu	Aqaidat
Muhaid, Al	Anaiza	Sari, Al	Anaiza
Muhallaf, Al	Anaiza	Saudah, Al bu	Dulaim
Muhamdah, Al bu	Dulaim	Shaar	Zoba
Muhammad, Al	Aqaidat	Shaban, Al bu	Dulaim
Muhammad al Dhiyab, Al bu	Dulaim	Shaddid	Kubais, Bani
		Shahab, Al bu	Dulaim
Muhammad al Jasim, Al bu	Dulaim	Shaitat, Al	Aqaidat
		Shimlan, Al	Anaiza
Muhanna, Al bu	Dulaim	Shiti	Dulaim
Mujbil, Al bu	Dulaim	Shuait, Al	Aqaidat
Mukatharah, Al	Anaiza	Shumailat, Al	Anaiza
Mukhaiyat, Al	Anaiza	Shuwartan	Dulaim
Mulahimah, Al	Dulaim	Subaihat	Dulaim
Muqallad, Al bu	Dulaim	Subaikhan, Al	Aqaidat

Minor tribe, section, or sub-section	Main tribe or confederation	Minor tribe, section, or sub-section	Main tribe or confederation
Sumaidi, Al bu	Dulaim	Tuluh	Anaiza
Sumail, Al	Chitadah	Tuwaisat, Al bu	Dulaim
Sumailat	Dulaim		
Suqur, Al	Anaiza	Ubaid, Al bu	Dulaim
Suwailmat, Al	Anaiza	Ubidah, Al	Anaiza
Suwalma	Anaiza	Ujur, Al bu	Dulaim
		Watbah, Al	Anaiza
Taha, Al bu	Dulaim	Wulud Ali	Anaiza
Taiyib, Al bu	Dulaim		
Tamah, Al bu	Dulaim	Zabanah, Al	Anaiza
Taumah, Al	Aqaidat	Zaid, Bani	Dulaim
Thulth, Al	Aqaidat	Zubar, Al	Chitadah

DULAIMIS ILLUSTRATED IN PLATES

1007: Plate 29
1009: Plate 25
1010: Plate 8
1011: Plate 6
1012: Plate 20
1013: Plates 2, 3
1016: Plate 33
1017: Plate 15
1018: Plate 9
1019: Plate 13
1020: Plate 32
1021: Plate 11
1022: Plate 31
1023: Plate 11
1024: Plate 30
1025: Plate 22
1026: Plate 30
1027: Plate 24
1028: Plate 19
1030: Plate 27
1033: Plate 26
1034: Plate 13
1035: Plate 26
1036: Plate 27
1037: Plate 5
1039: Plate 5
1040: Plate 21
1041: Plate 15
1042: Plate 33
1044: Plate 7
1045: Plate 14
1046: Plate 12

1047: Plate 22
1048: Plate 8
1049: Plate 9
1050: Plate 10
1051: Plate 34
1052: Plate 4
1053: Plate 6
1054: Plate 7
1055: Plates 16, 17
1057: Plate 31
1058: Plate 32
1059: Plate 12
1060: Plate 35
1061: Plate 28
1063: Plate 29
1064: Plate 34
1065: Plate 10
1066: Plate 19
1067: Plate 24
1080: Plate 4
1081: Plate 23
1082: Plate 28
1083: Plate 35
1084: Plate 18
1085: Plate 21
1086: Plate 23
1087: Plate 20
1088: Plate 25
1092: Plate 18
1093: Plate 14
1124: Plate 36

ANAIZA TRIBESMEN ILLUSTRATED IN PLATES

1571: Plates 40, 41
1572: Plate 38
1573: Plate 39
1575: Plate 39
1576: Plate 45
1577: Plate 43
1578: Plate 47
1579: Plate 47
1580: Plate 44
1581: Plate 44

1582: Plate 46
1583: Plate 43
1584: Plate 38
1585: Plate 45
1586: Plate 42
1587: Plate 42
1588: Plate 37
1589: Plate 37
1592: Plate 46

TRIBAL NAMES APPEARING ON MAP OF IRAQ (A)

Abbas: o, 20
Abuda: o, 21
'Afaj: n, 20
Afshār: j, 21–22
Ahl Al Kut: p, 21
Ahmadawand: l, 21–22
'Ajib: o, 20
Ako: j, 19
Al Ajarja: l, 15
Alattab: o, 21
Al bu Abbas: l, 18
Al bu 'Ajil: l, 18–19
Al bu 'Amir: m, 19; n, 19
Al bu Atalla: o, 20–21
Al bu Badran: j–k, 17
Al bu Darraj: o, 21
Al bu Dhiyab: m, 18
Al bu Fahad: m, 18
Al bu Faraj: m, 20–21
Al bu Ghuwainim: o, 20–21
Al bu Hamad: j, 18
Al bu Hamdan: k, 19
Al bu Hassan: o, 20
Al bu Husain: j, 18
Al Buisa: m, 18
Al bu Jaiyash: o, 20
Al bu Mahal: l, 16–17
Al bu Muhammad: o, 22
Al bu Nail: n–o, 19
Al bu Nashi: o, 20
Al bu Nimir: l–m, 17–18
Al bu Nisan: l, 18–19
Al bu Rudaini: l–m, 16–17; m, 17–18
Al bu Sa'ad: o, 21
Al bu Sali: o, 21
Al bu Sarai: k, 15
Al bu Sultan: n, 19
Al Hasan: p, 21
Al Hatim: o, 20–21
Al Humaid: n–o, 21
Al Ibrahim: p–o, 21
Al Idhar: l, 15–16
Aliqan: i, 16
Al Ismail: p, 21
Al Jabar: o, 20–21
Al Jumai'an: p, 21
Al Maiya: p, 22
Al Majawada: l, 15
Al Manashra: o, 20
Al Munaisin: p, 22
Al Muslib: o, 21
Al Sa'ad: p–o, 22
Al Saba': n, 15
Al Sali: o, 20
Al Shatat: l, 15
Al Sudan: o, 22
Al Suwa'id: o, 22
Al Tulph: l, 15

'Amarat: n, 16; n, 18
Ambuqiya: m, 19
Aqaidat, j, 17; k–l, 15
Aqail, o, 21
Aq'ra: o–n, 19
Artushi: i–j, 17–18
Asachrat: p, 21
Ashair al Saba: j, 18
Auramani: k, 21
'Awasid: n, 19
Ayyash: o, 19
Azairij: o, 21
Aznaur: j–i, 16
'Azza: l, 19
Azzubaid: n, 19

Babajani: k, 21; l, 20–21
Bahahitha: n–o, 20
Baiyat: l, 19
Bajlan: l, 20
Balik: j, 19
Balikian: j, 19
Bani Ard: o, 19–20
Bani Hasan: n, 18–19
Bani Huchaim: o, 19–20
Bani Khaiqan: p, 21
Bani Kubais: m, 17
Bani Lam: n, 21; n, 21–22
Bani Rabia: n, 20–21
Bani Rabi'a: m, 20
Bani Rikab: n–o, 20–21
Bani Said: o, 21
Bani Salama: o, 19
Bani Sali: o, 22
Bani Tamim: m, 19; m, 20; m–l, 19
Bani Turuf: n, 19; o, 22
Bani Uqba: m, 20
Bani Wais: m, 20
Bani Zaid: o, 20; o, 21
Bani Zuraj: o, 20
Baqqara: j, 15; k, 15
Baradost: j, 19
Barkat: o, 20
Barush: j, 18–19
Barwari Bala: i, 18
Barwari Jir: i–j, 18
Barwariya: i, 17
Barzan: i, 19
Baz: i, 18
Begzadeh: i, 19
Belavar: l, 21–22
Besheri: i, 16
Bilbas: j, 19–20
Budair: o, 20
Budur: o, 20
Buhtui: l, 21–22
Buzzun:[1] o, 21

[1] Buzzun, Isa, Muraiyan listed as one tribe on the map.

INDEX OF TRIBAL NAMES: IRAQ

Chabsha: o, 19
Chahardauli: k, 22
Chal: i, 18
Chaldaean: j, 18
Challabiyin: n, 20
Chechen: j, 15
Chichan: m, 19–20
Chingini: k, 20
Chitada: m, 18–19
Chunan: i, 15

Daaja Saʻadan: n, 20
Dachcha: o, 20–21
Dainiya: m, 20
Dakhori: i, 15–16
Dakshuri: i, 16
Dalabha: n, 20
Dargala: j, 19
Dashi: i, 15
Daudi: k–l, 19
Dawar: n, 20
Derevri: i, 16
Dershau: i, 16–17
Dhafir: p–q, 19–20–21
Dhawālim: o, 20
Dilfiya: m, 20
Dilo: k, 20; l, 20
Dinavar: l, 22
Dizai: k, 18–19
Dola Bila: j, 19
Dola Goran: j, 19
Dola Mairi: j, 19
Dola Majal: j, 19
Dolka: j, 19
Doski: i, 19
Dulaim: l–m, 16–18
Duski: i, 17–18

Eiru: i, 17

Fadʻan: n, 15
Faddagha: m, 19
Fartus: o, 20
Fatla: n, 19; o, 19

Galbaghi: k, 21
Garsan: i, 16–17; i, 17
Gaurak: j, 20
Gavadan: i–j, 17
Geravi: i, 18
Geshki: l, 21–22
Gezh: l, 19–20; l, 20
Ghazalat: o, 19
Ghazzi: o, 20–21
Ghurair: m, 19
Girdi: i, 19; j, 18–19; j, 19
Goyan: i, 17–18
Guli: i, 17–18
Gurān: l, 20–21; l, 21

Hachcham: o–n, 21
Hairuni: i, 16–17
Haiwat: m, 18–19

Hajjan: j, 17
Hamad: m, 20; n, 19–20
Hamawand: k, 19–20
Hamza: n, 20
Haruti: j, 19
Hassanan: j, 17
Haverki: i, 16
Hawāzin: q, 22
Herki: i, 19–20; j, 18
Humaidat: o, 19
Husainat: p, 21
Hwatim: n, 19

Ibrahim: o, 19
Isa: o, 21
Ismail Uzairi: k, 20

Jabbari: k, 19–20
Jaf: j, 21; k, 20; l, 20
Jaghaifa: l, 16–17
Jalālawand: l, 21; m, 22
Jaliha: n, 19; o, 20
Jannabiyin: m–n, 18–19
Jelian: i, 17
Jilu: i, 18–19
Jomani: i, 16
Jubur: j, 17; k, 18; l–m, 19–20; m, 19; n, 19
Jubur (Khabur): k–j, 15–16
Juhaish: j, 17; n, 19–20
Jumaila: m, 18–19
Jumur: l, 22
Juwaibir: o, 20
Juwarin: p, 21

Kafrushi Shinki: k, 20
Kakai: k, 19
Kakawand: l, 22
Kalawand: l, 22
Kalawi: j, 19
Kalendalan: i, 15
Kalhūr: l, 20–21; m, 20
Kamangar: l, 21–22
Karkhiya Bawiya: m, 19
Khafaja: n, 19; o, 19; o, 21
Khala Jan: i, 15
Khamisya: p, 21
Khazail: n–o, 19; o, 20
Khazraj: m, 18–19
Khizil: l, 22
Khudabandalu: k, 22; l, 22
Khurkhura: k, 21
Kichan: i, 17
Kolmetchma: i, 16
Kopa: j, 19
Kuliai: l, 22
Kushnao: j, 19

Lak: k, 19
Lakk: k, 22
Lughawiyin: o–n, 21

Ma'dan: m, 20
Mahalami: i, 16
Mahmedan: i, 18
Majawir: o, 19–20
Malawaha: j, 17
Mamkhoran: i, 18
Mamush: j, 20
Manda: j, 20
Mandumi: k, 22
Mangur Zudi Manda: j, 20
Mansur: o, 19
Mantik: k, 19
Marra Pizdher: j, 20
Masūd: n, 19
Mazi: i, 15
Merivani: k, 21
Metini: i, 15
Milli: i, 15; j, 15
Miran Begi: j, 18
Mirsinan: i, 15–16
Mizuri: i, 18–19; j, 18
Mu'alla: m, 20
Mu'amara: n, 19
Muamara: j, 17
Muhamda: m, 18
Muhsin: o, 20
Mujamma: m, 18; m, 19
Mujarra: p, 21
Mukhadhara: o, 20
Mukri: j, 20–21
Muraiyan: o, 21–22
Mushahida: m, 19
Mutair: q, 21–22
Mutaiwid: j, 16
Muzaira: o–p, 22

Naida: m, 20
Najdat Dafafa: m, 19
Naodasht: j, 19
Nashwa or Khulut: p, 22
Nassun: o, 21
Nerva: i, 18
Non-tribal Kurd: j, 19
Non-tribal Kurd and Arab: j, 18–19
Nuchiyan: i, 19

Ojagh: j, 20
Omarmi: l, 20
Oramar: i, 18–19
Osmānawand: l, 21; m, 22

Paīrawand: l, 22
Palani: l, 20
Penjinara: i, 16
Pinianish: i, 18
Pirahasani: j, 19
Piran: j, 19
Pizdher: j, 20

Qarahalus: m, 20
Qarakhul: o, 21
Qara Papāq: j, 20

Qarqariya: j, 17
Qubadi: l, 21
Qulu: j, 18–19
Qurait: n, 19

Raikan: i, 18
Reshkotanli: i, 16
Rowandok: j, 19
Rudaini: m, 20
Rumm: j, 19
Rustambegi: l, 21

Sadā: m, 20; o, 20
Sadiq: o, 19
Sa'id: n, 20
Sakhwar: l, 19–20
Sarchef: j, 21
Sargalu Sheikhs: k, 20
Shabbana: n, 19–20
Shaikhan: k, 20
Shammar Jarba: k–l, 17–18
Shammar Toqa: m–n, 19–20
Shaqarqi: j–i, 21–22
Sharabiyin: j, 15
Sharaf Biyani: l–k, 20
Shasavan: j–i, 21–22
Shebek Christian: j, 18
Sheikh Bizaini: j, 18; k, 19
Sheikh Ismail: k, 22
Sheikhs of Quala' Sedka: k, 19–20
Shekak: i, 19
Sherikan: i, 15
Shernakh: i, 17
Shibil: o, 19
Shillana: j–k, 19–20
Shirwan: j, 19
Shovan: i, 17
Shu'aiba: o, 20
Shuan: k, 19
Shuraifat: p, 21
Sihoi: i, 17
Silivani: i, 17
Sindi: i, 17
Sinjabi: l, 20; l, 21
Sinn: j, 19
Sirokhli: i, 16–17
Slopi: i, 17
Sor: i, 15
Sturki: i, 16
Sufran: o, 20
Sukuk: m, 19
Sulduz: i, 20
Surchi: j, 18–19
Surgichi: i, 15–16
Sursur: l–k, 21–22

Tai: j, 16
Taiyan: i, 17
Talabani: k, 19; l, 20
Tall 'Afaris: j, 17
Tanzi: i, 16–17
Tiari: i, 18

Index of Tribal Names: Iraq

Tilehkuh: j, 21
Tkhuma: i, 18
Toba: o, 20
Toqiya: o, 21
Tufail: n, 19
Turcoman Arab: j, 18

'Ubaid: l, 19
'Umairīyāt: m, 20

Waladbegi: l, 20–21; l, 21

Yasar: n, 18–19
Yassar: n, 19
Yezidi: j, 16; j, 17–18

Zaiyad: o, 19; o, 20; o–n, 20
Zangana: l, 20; l–k, 20
Zarari: j, 18–19
Zedik: i–j, 18–19
Zend: l, 20
Zibari: j, 18–19
Zudi: j, 20

TRIBAL NAMES APPEARING ON MAP OF IRAN (B)

Abad: p, 24
Abdul Khān: o, 23
Abdul Rezai: p, 27-28
Abulvardi: p, 27
Afshār: j, 23
Agha Jari: p, 24; p, 25
Airizaumari: o, 24
Aiyasham: o, 23
Alamdar: n, 24
Alaswand: o, 24
Al bu Hamdan: n, 23
Al Duhaim: o, 23
Ali Muradi: p, 27-28
Al Kathir: n, 23; o, 23
Al Khamīs: o, 24
Al Ruwaīyan: o, 23
Alwanīeh: o, 24
Amarlū: j, 24
Amla (Lur): n, 23
Anafijah: o, 23
Andakah: n, 24
Arab: n, 23
Aushar: p, 24

Baghdādī: k, 24; k, 25
Bahārwand: n, 23
Bairanawand: m, 23
Bait Saad: o, 23
Bakhtīāri: m, 24; n, 23; n, 24-25; o, 25
Bakīsh: p, 26
Bāla Girīeh: m, 23; n, 23
Bandari: p, 24
Bani Abdullahi: q, 28
Bani Khālid: o, 24
Bani Tamim: o, 23
Bani Turuf: o, 23
Barangird: o, 24
Baseri: p, 27-28; q, 27; q, 28
Bāvi: p, 26
Bawasat: n, 24
Bawieh (Bāvīeh): p-o, 23; o, 24
Boir Ahmadi: p, 26; o, 26
Boiramides: n, 24
Bulāwāso: o, 24
Burujird: n, 23

Chaab i Dubais: n, 23
Chāb: p, 22-23
Chaman-i-Urga: n, 24
Charasi: p, 24
Chavari: l, 22-23
Cherūm: p, 24-25
Chigini: m, 23; j, 24

Dailam: o, 23
Dalwand: m, 23
Darashur: q, 26
Darazi: p-q, 27

Dinarūni: n, 24
Dindārlū: q, 27-28
Dīrakwand: n, 23

'Emadi: p, 28

Farsi: p, 28

Gandali: o, 24
Garrai: p, 27
Gashtil: p, 24
Gazistun: n, 24
Ghiāsvand: j, 24
Ghuri: p, 27
Gūklān Turkomāns: i, 30
Gundalis: n, 24
Gundalzu: o, 23-24
Gurgha: o, 24
Gurgi: p, 24

Haft Lang: n, 23
Haidari: p, 24
Hajjilu: k, 23
Hamaid: o, 23-24
Hannai: q, 28
Hardan: o, 23
Hawāshim: o, 23

Inānlū: k, 24; k, 25; k, 26

Jāāfarbai ak Atehbai: i, 29
Jabbareh Arab: p, 27
Ja'fari: p, 24
Jalīlavand: j, 24
Jāneki Sardsīr: o, 25
Jani Khan Arab: p, 28
Jumur: k, 23

Kāid Rahmat: m, 23
Kākāvand: j, 24
Karohi: o, 24
Khalkhal: i, 23
Khamseh: p, 27; p, 27-28; q, 27; q, 28; q, 29
Khazraj: o, 23
Khidr-i-Surkh: o, 24
Khudabandalu: k, 23; l, 23
Khusrui: q, 28
Khwājahvand: j, 25-26
Kurdbaiglū: i, 22-23
Kurd-u-Turk: j, 28
Kuruni: p, 27

Labu Haji: q, 27
Labu Muhammadi: p, 28
Laki: p, 25
Lakk (Lek): k, 22-23
Lashani: q, 28
Lur: n, 23

Index of Tribal Names: Iran

Ma'afī: j, 25
Makawandi: o, 24
Mamassani: p, 26; q, 26; q, 27
Mir: n, 23
Mishwand: m, 23
Mizdaj: n, 25
Muhaisin: p-o, 23; p, 22-23
Mujazi: n, 24
Mūmianwand: m, 23
Murad ali Wand: n, 23
Muris: n, 24
Mutur: p, 24

Naqd'Ali: p, 28
Nargasin: n, 24
Nasīr: o, 24
Nidharat: p, 24
Nūyi Silai: o, 25-26

Papi: n, 23
Pir Islami: p, 28

Qajār: j, 29
Qalawand: n, 23
Qanawati: p, 24
Qaraguzlu: k-l, 23
Qāshqāī: o, 26; p, 26; p-o, 27; q, 25; q, 26; q, 27

Rashvand: j, 25
Rustam: p, 26

Sagwand: m, 23; n, 23
Saiyidali: o, 23
Saiyidān: o, 24
Sakhtsar: j, 25
Salāmāt: o, 23-24
Sha'abuni: p, 24
Shāhsavan: k, 26
Shaikh Mamu: p, 24-25
Shaiwand: n, 24
Shatrānlū: i, 23
Sheni: o, 24
Sherafah: o, 22-23
Shir Ali: p, 24
Shiri: p, 28
Shishbulūki: p, 27
Shuraifat: p, 24
Silsileh: m, 23
Suluklu: p, 27
Surkha: n, 23

Tafarakha: o, 24
Talish: i, 23-24
Turkashawand: l, 23
Tushmals: n, 24

Yamūt Turkomāns: i, 29; i, 30

Zangīna: o, 24
Zeloi: n, 24
Zirgan: o, 23

INDEX

Abbass, Abdul-Majid, 12, 198
Abu Ghuraib Canal, description of, 18
Abu Kemal, 17; northern limit of cultivation of date palm at, 21; population of, 28; Sunnis in, 28
Agricultural products, 22–23
Akeydat, *see* Aqaidat
Al Abaidat, 95
Al Ajarjah, 93
Al Ali, 94
Al Annas, 102
Al Azzah, 101
Al Barghuth, 101
Al bu Aliyat, 95
Al bu Alwan, 96
Al bu Badran, 95
Al bu Dhiyab, 96
Al bu Fahad, 97
Al bu Ghanim, 95
Al bu Haidah, 101
Al bu Hamad, 101
Al bu Hamdan, 95
Al bu Hardan, 93
Al bu Hardan (Section), 94
Al bu Hassan, 95
Al bu Isa, 97
Al bu Khalifah, 98
Al bu Khalil, 102
Al bu Maish, 96
Al bu Miri, 94
Al bu Mish, 96
Al bu Mufarraj, 102
Al bu Muhamdah, 98
Al bu Muhammad, venereal disease among, 116
Al bu Musa, 96
Al bu Qaan, 95
Al bu Rahamah, 95
Al bu Rudaini, 99
Al bu Salman, 101
Al bu Sarai, 95
Al Dimim, 93
Al Dughaiyim, 102
Al Guraibawiyin, 102
Al Hamudi, 94
Al Harub, 101
Al Hassun, 94
Al Hulaiyil, 102
Al Humaid, 101
Ali Jaudat, 9
Al Isa, 94
Al Jadu, 95
Al Khammas, 101
Al Khanfar, 95
Al Khanjar, 95
Al Majawadah, 94
Al Marasimah, 94
Al Muhammad, 94
Al Mushahidah, 94
Al Nabbizah, 96
Al Nabit, 102
Al Nassar, 102
Al Qadrau, 94
Al Qumzan, 101
Al Quran, 95
Al Radhi, 101
Al Saadan, 34, 100
Al Sbaa, 93
Al Shaitat, 95
Al Shuait, 95
Al Subaikhan, 94
Al Sumail, 101
Al Taumah, 94
Al Thulth, 95
Al Zubar, 101
Amara, classification of land surface of, 106–107; flora of, 196; population of, in 1930, 108, in 1935, registered, 105, unregistered, 104
Amarat, habitat of, 27; relations with other tribes, 27, 34; tribal list of, 91
Amphibians, 24
Ana, 17; Jews in, 28; population of, 28, in 1882, 28; Sunnis in, 28
Anaiza tribesmen (nineteen males measured), 11, 12, 13, 26, 27, 54–74
age of, 63, 70; groupings, 63
bigonial breadth of, 70
bizygomatic breadth of, 70; groupings, 70
blondism among, 64
body hair of, 64; compared to Arabs of central Iraq, 64
camels of, 55; exports of, 55
cauterization among, 66
cephalic index of, 68, 70; groupings, 68; compared to Proto-Mediterranean mean, 68
demography of, 63
disease among, 66. *See also* Pathology
ears of, measurements and indices of, 70
eyes of, 64; groupings, 64
eye slits of, 64
facial measurements and indices of, 68–69, 70; groupings, 68–69, 70, 74
facial types of, 73; ram-faced among, 73–74
fronto-parietal index of, 70
hair of, 64; groupings, 64
head breadth of, 67, 70; groupings, 67
head length of, 70
health of, 65
horses of, breeds of, 55
kohl used by, 66
lips of, 65

214

INDEX 215

minimum frontal diameter of, 67, 70;
  groupings, 67, 69
morphological characters of, groupings, 63–66
musculature of, 65
nasal breadth and height of, 69, 70;
  groupings, 69, 70
nasal index of, 69, 70; groupings, 69
nasal profile, 65; groupings, 65
nasal tip and wings of, 65; groupings,
  65
Negroid element among, in nose of,
  65, 69; in skin color of, 63
nomadism among, 54–55
origin of, 54
photographic analyses of, 70–71
provenance of, 62
racial position of, 71
raw data: measurements, indices, and
  morphological characters of, 72–
  73
sitting height of, 67, 70; groupings,
  67, 69
skin color of, 63; compared to the
  Arab, 63; to the European, 63
statistical analyses of, groupings, 66–
  70
stature of, 66, 70; groupings, 67
stock, see camels, horses
tattooing among, 66
teeth of, 65; groupings, 65
trade of, geographical facilities for, 55
tribal feuds of, 55
tribal list of, 91–93
tribes and sub-tribes of, 56–61
vital statistics of, 62
zygo-frontal index of, 70
zygo-gonial index of, 70
Anthropometric data, abbreviations,
  list of, used for, 33; selection of,
  32–33, 75, 122, 131
Apple trees, 22
Aqaidat, tribal list of, 93–95
Arabs, attitude toward disease, 110,
  toward pain, 119, toward medical
  treatment, 117, 118–119; four types
  of, 26–27; in Raqqa, 28; racial
  position of, 89–90; use of herbs by,
  22
  Anthropometric data: age, cephalic
    indices and head measurements
    on, from Baghdad, children, 126,
    female, 125, male, 123–124, from
    nineteen towns, 124–125, from
    six towns, female, 125–126, from
    three tribes, 126, from various
    tribes of Iraq, children, 126.
    See also Baghdad, individuals
    measured in Royal Hospital of
Armenians, 13
Artificial cranial deformation, absence
  of, 115
Asellia murraiana, 157

Assyrians, 11, 13
Aziziya Canal, description of, 18

Baban, flora of, 192
Badgers, 23, 160
Baghdad, Central School for Girls of,
  151; classification of land surface
  of, 106–107; flora of, 194; health
  inspection at, 120; Iraq Museum
  in, 11; population of, in 1930, 108,
  in 1935, registered, 105, unregistered, 104; Royal College of Medicine in, 8, 9, 15, 118, 121
  Anthropometric data: individuals
    measured in Royal Hospital of,
    13, 131
Arabs, twenty-three male, 131
  age of, 132, 139
  bigonial breadth of, 139
  bizygomatic breadth of, 139
  blondism among, 133
  brow-ridges of, 137
  cauterization among, 135
  cephalic index of, 137, 139; group-
    ings, 137
  demography of, 132
  diseases of, 135. *See also* Pathology
  ears of, measurements and indices
    of, 139
  eyes of, 133; groupings, 133
  facial measurements and indices of,
    137, 139; groupings, 137–138
  fronto-parietal index of, 139
  hair of, 132–133; groupings, 132
  head breadth of, 136, 139; group-
    ings, 136
  head form of, 136
  head length of, 139
  lips of, 135
  minimum frontal diameter of, 136,
    139; groupings, 137
  morphological characters of, 132–
    135
  nasal breadth and height of, 138,
    139; groupings, 138
  nasal index of, 138, 139; groupings,
    138
  nasal profile of, 133; groupings, 134
  nasal septum of, 133; groupings,
    134
  nasal tip and wings of, 133; group-
    ings, 134
  Negroid element among, in eyes of,
    133, in lips of, 135, in nose of,
    133, 138, in skin color of, 132
  physical appearance of, 135
  prognathism, alveolar, among, 134
  provenance of, 131
  raw data: measurements, indices
    and morphological characters
    of, 141–142
  sitting height of, 136, 139; group-
    ings, 136

skin color of, 132
smallpox among, 135
statistical analyses of, 135-142
stature of, 135-136, 139; groupings, 136
tattooing among, 135
teeth of, 134; groupings, 134
zygo-frontal index of, 139
zygo-gonial index of, 139
males omitted from the statistical analyses, 138-140
bigonial breadth of, 140
bizygomatic breadth of, 140
cauterization among, 139-140
cephalic index of, 140
diseases of, 139-140. *See also* Pathology
ears of, measurements and indices of, 140
eyes of, 139-140
facial form of, 139
facial measurements and indices of, 140
fronto-parietal index of, 140
head breadth and length of, 140
head form of, 139-140
minimum frontal diameter of, 140
Mongoloid type among, 139
nasal breadth and height of, 140
nasal form of, 139-140
nasal index of, 140
provenance of, 138-139
raw data: measurements, indices and morphological characters of, 141-142
sitting height of, 140
stature of, 140
teeth of, 139-140
zygo-frontal index of, 140
zygo-gonial index of, 140
Arabs, twenty female, 143
age of, 143, 150; groupings, 143
bigonial breadth of, 150
bilharziasis among, 147
bizygomatic breadth of, 150
blondism among, 144
cauterization among, 147
cephalic index of, 148, 150; groupings, 148
demography of, 143
diseases among, 146-147. *See also* Pathology
ears of, measurements and indices of, 150
eyes of, 144; groupings, 144
facial measurements and indices of, 148-149, 150; groupings, 149
fronto-parietal index of, 150
gonorrhea among, 147
hair of, 144; groupings, 144
head breadth of, 148, 150; groupings, 148
head length of, 150

malars of, 146
minimum frontal diameter of, 148, 150; groupings, 148
morphological characters of twenty Arab women, 144-147
nasal breadth and height of, 149, 150; groupings, 149
nasal index of, 149, 150; groupings, 149
nasal profile of, 145; groupings, 145
nasal septum of, 145; groupings, 145
nasal tip and wings of, 145; groupings, 145
Negroid blood among, 147
physical appearance of, 146-147
prognathism, alveolar, among, 146
provenance of, 143
raw data: measurements, indices and morphological characters of, 153-155
sitting height of, 147, 150; groupings, 147
skin color of, 144
smallpox among, 147
statistical analyses of, groupings, 147-150
stature of, 147, 150; groupings, 147
tattooing of, 146, 147
teeth of, 145-146; groupings, 146; notes on, 146
zygo-frontal index of, 150
zygo-gonial index of, 150
females omitted from the statistical analyses, 150-151
blondism among, 151
cauterization among, 151
diseases of, 150-151
eyes of, 150-151
head form of, 150-151
nasal septum, inclination of, 151
Negroid blood among, 150, 151
nose of, 150, 151
physical appearance and type of, 150-151
prognathism, alveolar, among, 150, 151
provenance of, 150
raw data: measurements, indices and morphological characters of, 153-155
tattooing among, 151
teeth of, 150-151
females, including statistical and omitted series, 152
bigonial breadth of, 152
bizygomatic breadth of, 152
cephalic index of, 152
ears of, measurements and indices of, 152
facial measurements and indices of, 152
fronto-parietal index of, 152

INDEX

head breadth and length of, 152
minimum frontal diameter of, 152
nasal breadth and height of, 152
nasal index of, 152
raw data: measurements, indices and morphological characters of, 153–155
sitting height of, 152
stature of, 152
zygo-frontal index of, 152
zygo-gonial index of, 152
girls of, eleven, 151–152
  blondism among, 151
  body hair of, 152
  diseases of, 152. *See also* Pathology
  eyes of, 151
  hair of, 152
  lips of, 152
  Negroid blood among, 152
  nose of, 152
  physiognomy of, 152
  provenance of, 151
  raw data: measurements, indices and morphological characters of, 153–155
  skin color of, 152
  teeth of, 152
Ba'ij Beduins (35 individuals), 13, 86
  age of, 86; compared to Iraq Soldiers, 76; to Kish Arabs, 76
  beards among, 87
  bigonial breadth of, 86; compared to Iraq Soldiers, 76; to Kish Arabs, 76
  bizygomatic breadth of, 86; compared to Iraq Soldiers, 76; to Kish Arabs, 76
  body hair of, 87
  cephalic index of, 86; compared to Iraq Soldiers, 76; to Kish Arabs, 76; groupings, 86
  chest development of, 89
  ears of, measurements and indices of, 86; compared to Iraq Soldiers, 76; to Kish Arabs, 76
  eyes of, groupings, 88
  facial index of, groupings, 86
  facial measurements and indices of, 86; compared to Iraq Soldiers, 76; to Kish Arabs, 76
  fronto-parietal index of, 86; compared to Iraq Soldiers, 76; to Kish Arabs, 76
  hair of, groupings, 87
  head breadth and length of, 86; compared to Iraq Soldiers, 76; to Kish Arabs, 76
  head hair, 87
  health of, 89
  leg length of, 86; compared to Iraq Soldiers, 76; to Kish Arabs, 76
  minimum frontal diameter of, 86; compared to Iraq Soldiers, 76; to Kish Arabs, 76
  morphological characters of, groupings, 87–89
  musculature of, 89
  nasal breadth and height of, 86; compared to Iraq Soldiers, 76; to Kish Arabs, 76
  nasal index of, 86; compared to Iraq Soldiers, 76; to Kish Arabs, 76; groupings, 86
  nasal profile of, groupings, 88
  nasal tip and wings of, groupings, 88
  sitting height of, 86; compared to Iraq Soldiers, 76; to Kish Arabs, 76
  statistical analyses of, groupings, 86
  stature of, 86; compared to Iraq Soldiers, 76; to Kish Arabs, 76; groupings, 86
  tattooing among, 89
  teeth of, groupings, 88
  vital statistics of, 87
  zygo-frontal index of, 86; compared to Iraq Soldiers, 76; to Kish Arabs, 76
  zygo-gonial index of, 86; compared to Iraq Soldiers, 76; to Kish Arabs, 76
Baiji, flora from north of, 193
Baiqat, 101
Bait Dariah, 101
Bait Hajji Isa, 101
Bait Mathluthah, 101
Bani Kubais, 101
Bani Zaid, 34; tribal list of, 100
Baqqarah, 95–96
Barley, 22
Basra, classification of land surface of, 106–107; population of, in 1930, 108, in 1935, registered, 105, unregistered, 104
Beans, 22
Beduins, 23, 26, 31, 55; age, cephalic indices and head measurements of, from Mosul *Liwa*, 127, *see also* Ba'ij Beduins; use of herbs by, 22, 25
Belikh River, 18
Birds, 23–24
Bitumen, 28; cholera epidemic averted by, 30; uses of, 24
Bitumen wells, mention by Herodotus of, 30
Boars, 23, 161
Bornmüller, Joseph, 165
Boundaries, 17
Brady, Ethel, 10
Breasted, James H., 9
Brinjals, 22
British Museum, 15
British Oil Development Company, 156
British Royal Air Force Headquarters, 30
Browne, W. E., 27

Burnett, John, 9
Buxton, L. H. Dudley, 7, 8, 81, 82

Camels, 23; oil used as remedy for, 24; use of, for irrigation, 22, 23
Canis aureus, 159
Canis pallipes, 159
Capra blythi, 162
Chaldeans, 13
Cheetahs, 23
Chitadah, 34; tribal list of, 101
Christians, 26; age, cephalic indices and head measurements of, from Baghdad, males, 128, females, 129, from Mosul, males, 128, females, 129, from Tabriz, 130, from Tell Kaif, 128, from Urmia, 130, *see also* Turks; in Deir-ez-Zor, 28
Circassians in Raqqa, 28
Clawson, M. Don, 9
Clemow, F. G., 120
Climate, 20-22
Coon, Carleton S., 10
Cornwallis, Kinahan, 9
Cotton, 22
Cranial deformation, *see* Artificial cranial deformation
Cucumbers, 22

Date palm, 22, 28, 30; limit of cultivation of, 21
Deir-ez-Zor, Arabs in, 28; Jews in, 28; population of, in 1882, 28; Syrian Catholics in, 28
Dekker, J. H., 156
Dinka, Philippus, 156
Diwaniya, Ad, classification of land surface of, 106-107; population of, in 1930, 108, in 1935, registered, 105, unregistered, 104
Diyala, classification of land surface of, 106-107; population of, in 1930, 108, in 1935, registered, 105, unregistered, 104
Dowson, Ernest, 15, 103, 106
Dowson, V. H. W., 22
Drower, E. S., 156
Dulaim, classification of land surface of, 106-107; population of, in 1930, 108, in 1935, registered, 105, unregistered, 104
  Anthropometric data (137 males measured): 13, 26, 27, 33-54
  age of, 34-35, 43; groupings, 35
  baldness among, 35
  bigonial breadth of, 43
  bizygomatic breadth of, 43; groupings, 43
  blindness among, 37
  blondism among, 36, 37
  blood samples of, 38
  body hair of, 35
  cauterization among, 39
  cephalic index of, 40-41, 43; groupings, 41
  disease among, 38. *See also* Pathology
  ears of, measurements and indices of, 43
  eyes of, 36-37; groupings, 36
  eye slits of, 36
  facial measurements and indices of, 41, 43; groupings, 41, 43
  facial types of, 73; ram-faced among, 73
  fronto-parietal index of, 43
  hair of, 35; groupings, 36
  head breadth of, 40, 43; groupings, 40
  head length of, 43
  health of, 38
  henna used by, 39
  kohl used by, 39
  lips of, 37
  minimum frontal diameter of, 40, 43; groupings, 40, 43
  morphological characters of, groupings, 35-39
  musculature of, 38
  nasal breadth and height of, 42, 43; groupings, 42, 43
  nasal index of, 42, 43; groupings, 42
  nasal profile of, 37; groupings, 37
  nasal tip and wings of, 37
  Negroid element among, in hair of, 35; in nose of, 37, 42; in skin color of, 35
  nomadism among, 33-34; in eastern Shamiya, 33, 34; in Jazira, 33, 34
  origin of, 33
  photographic analyses of, 44
  racial position of, 44-45; compared to Beduin, 45; to settled Arab, 45
  raw data: measurements, indices and morphological characters of, 46-54
  religious affiliations of, 33
  sitting height of, 40, 43; groupings, 40, 42
  skin color of, 35; compared to the Kish Arab, 35; to the southern European, 35; to the Arab in the area from the "Fertile Crescent" to Morocco, 35
  statistical analysis of, groupings, 39-54
  stature of, 39, 43; groupings, 39
  tattooing among, 39
  teeth of, 37-38; groupings, 37
  tribal list of, 96-100
  tribal relations of, 34
  zygo-frontal index of, 43
  zygo-gonial index of, 43

Dulaim Qartan, 34; tribal list of, 102
Dunkley, G. W., 9

Eastwood, Austin, 156
Edmonds, C. J., 15, 103
Education, increasing facilities for, 31
Epidemics, danger of, 31
Eptesicus hingstoni, 158
Eptesicus walli, 159
Erbil, classification of land surface of, 106–107; population of, in 1930, 108, in 1935, registered, 105, unregistered, 104
Euphrates River, canals adjoining, 18, changing channels of, 20; course of, 17; flood seasons of, 18, 20; tributaries of, 18, 22

Fadan, 27, 54; tribal list of, 92
Faddaghah, 34; tribal list of, 102
Fahad Beg, 27
Faiyadah, tribal list of, 102
Fallahin, 25
Falluja, Al, 17, buildings of, 30; land of, under cultivation, 30; location of, 30; population of, 30
Fallujiyin, 102
Farraj Allah, *see* Shaddid
Fauna, 23–24, 156–162
Felis chaus, 159
Field, Marshall, 8
Field Museum Anthropological Expedition to the Near East, 8, 9, 156, 163, 165
Field Museum–Oxford University Joint Expedition to Kish, Iraq, 7, 15, 81, 110, 111
Flint implements, 28
Foxes, 23, 159
Frankfort, Henri, 9
Frayha, Anis, 12
Frazer, James, 113
Fruit trees, 22, 28

Gazella, 23, 162
Gerhard, Peter, 11, 12
Ghazi ibn Faisal, 8, 9
Gossypium, 166
Grazing, 23
Grice, C. R., 9
Guest, Evan, 165
Gufas, manufacture of, 24
Gypsies, 13

Habbaniya Lake, 11, 30; environs of, 29
Haditha, flora of, 193
Hail, 20
Haiwat, 34; tribal list of, 102
Hamad, 54–55
Harrison, Paul W., 116, 117
Harvard University, Institute of Geographical Exploration of, 11; Laboratory of Anthropology of, 75; Peabody Museum of, *see* Peabody Museum; Widener Library of, 11
Hasanah, 27, 54
Health, 31
Hemiechinus auritus, 157
Hemiptera, 24
Herodotus, 30
Herpestes persicus, 159
Hilla, classification of land surface of, 106–107; population of, in 1930, 108, in 1935, registered, 105, unregistered, 104
Hill, Arthur, 165
History, 24–26
Hitawiyin, 34; tribal list of, 102
Hit, bitumen wells at, 24, 28, 30; historical references to, 30; date palms at, 28; fruit trees at, 28; gufas manufactured at, 24; Jews in, 30; lime manufactured at, 24; location of, 28; population of, 30, in 1882, 28; salt pans at, 24; sulphur at, 28; uses of bitumen at, 24
Holt, A. L., 24
Hooper, David, 118
Hooton, E. A., 8, 10, 32, 75
Hordeum, 166
Horwood, A. R., 165
Hudson, E. H., 116
Humidity, relative, 20
Hyaena hyaena, 23, 160
Hydar, Rustam, 166
Hyena, 23

Ibn Rashid, 28
Idhar, Al, 93
Insects, 24, 163–164
Iraq, area of, 103, in 1920, 103; census, agricultural, need for, in, 107–108; communications with, 19; cultivable land of, 103, in Irrigation Zone, 103, 106, in Rainfall Zone, 103, 106–107; density of population of, 103; development of Public Welfare of, 120–121; economic and commercial conditions in, 24; geographical position of, 14; hospitals in, 121; nomadism restricted in, 11; population of, in 1919, 103
Iraq Petroleum Company, 9, 24, 27, 34, 55, 121, 139, 156; health conditions improved by, 121
Iraq Soldiers (222 individuals measured at Hilla Army Camp), 13, 83
age of, 83; compared to Ba'ij Beduins, 76; to Kish Arabs, 76
bigonial breadth of, 83; compared to Ba'ij Beduins, 76; to Kish Arabs, 76
bizygomatic breadth of, 83; compared to Ba'ij Beduins, 76; to Kish Arabs, 76

cephalic index of, 83; compared to Ba'ij Beduins, 76; to Kish Arabs, 76; groupings, 83
chest development of, 85
diseases among, 85. See also Pathology
ear measurements and indices of, 83; compared to Ba'ij Beduins, 76; to Kish Arabs, 76
eyes of, groupings, 85
facial index of, 83; compared to Ba'ij Beduins, 76; to Kish Arabs, 76; groupings, 83
facial measurements and indices of, 83; compared to Ba'ij Beduins, 76; to Kish Arabs, 76
fronto-parietal index of, 83; compared to Ba'ij Beduins, 76; to Kish Arabs, 76
hair of, groupings, 84
head breadth of, 83; compared to Ba'ij Beduins, 76; to Kish Arabs, 76
head length of, 83; compared to Ba'ij Beduins, 76; to Kish Arabs, 76
health of, 85
leg length of, 83; compared to Ba'ij Beduins, 76; to Kish Arabs, 76
minimum frontal diameter of, 83; compared to Ba'ij Beduins, 76; to Kish Arabs, 76
musculature of, 85
nasal alae of, groupings, 84
nasal breadth and height of, 83; compared to Ba'ij Beduins, 76; to Kish Arabs, 76
nasal index of, 83; compared to Ba'ij Beduins, 76; to Kish Arabs, 76; groupings, 84
nasal profile of, groupings, 84
sitting height of, 83; compared to Ba'ij Beduins, 76; to Kish Arabs, 76
stature of, 83; compared to Ba'ij Beduins, 76; to Kish Arabs, 76; groupings, 83
tattooing among, 86
teeth of, groupings, 85
vital statistics of, 84
zygo-frontal index of, 83; compared to Ba'ij Beduins, 76; to Kish Arabs, 76
zygo-gonial index of, 83; compared to Ba'ij Beduins, 76; to Kish Arabs, 76
Irrigation, methods of, 22–23

Jackal, 23
Jaculus Loftusi, 161
Jaladiya, limestone quarry at, 24
Jamali, M. F., 31
Jazira, Al, 17

Jebel Baradost, flora of, 194
Jebel Baykhair, flora of, 192
Jebel Enaze, Paleolithic implements found on, 28; source of Wadi Hauran on, 28
Jebel Golat, flora of, 190
Jebel Khatchra, flora of, 190
Jebel Pikasar, flora of, 193
Jemdet Nasr, 7, 111, 116; excavations at, 7; location of, 111; painted pottery at, 7; water supply at, 111–112
Jerwona, flora of, 192
Jews, 13, 26; age, cephalic indices and head measurements of, from Baghdad, female, 129, male, 129, from Erbil, 129, from Kirkuk, 129; at Ana, 28; at Deir-ez-Zor, 28; at Hit, 30
Jumailah, 97

Karbala, classification of land surface of, 106–107; population of, in 1930, 108, in 1935, registered, 105, unregistered, 104
Keith, Arthur, 8, 11, 32, 89–90
Kennedy, Walter P., 8, 9, 156
Kew Herbarium, 165
Khabur River, 11, 18
Khalaf, Jassim, 12, 198
Khurushiyin, 34; tribal list of, 100
Kirkuk, classification of land surface of, 106–107; population of, in 1930, 108, in 1935, registered, 105, unregistered, 104
Kirkuk, Iraq Petroleum Company Hospital at, 121; population of, 108
Kish, 7; first crossing by automobile to Tigris from, 116
Kish Arabs (359 individuals measured), 13, 76
age of, 76; compared to Ba'ij Beduins, 76; to Iraq Soldiers, 76
animals, domesticated, affection for, 119; wild, cruelty to, 119
attitude toward medical treatment of, 118–119
beards among, 78
bigonial breadth of, 76; compared to Ba'ij Beduins, 76; to Iraq Soldiers, 76
bizygomatic breadth of, 76; compared to Ba'ij Beduins, 76; to Iraq Soldiers, 76
blindness among, 81
body hair among, 78
brow-ridges of, 78
cephalic index of, 76; compared to Ba'ij Beduins, 76; to Iraq Soldiers, 76; groupings, 77
chest development of, 80
constitution of, 119, 120
Darwin's point among, 81

INDEX

dental condition of, 113–114
diet of, 115
diseases among, 81. *See also* Pathology
ears of, helix of, 81; lobe of, 81; measurements and indices of, 76; compared to Ba'ij Beduins, 76, to Iraq Soldiers, 76
eyebrows of, 79
eyes of, groupings, 79
facial hair of, 78
facial measurements and indices of, 76; compared to Ba'ij Beduins, 76; to Iraq Soldiers, 76; groupings, 77
fatalism of, 110
fronto-parietal index of, 76; compared to Ba'ij Beduins, 76; to Iraq Soldiers, 76
glabella of, 78
hair of, groupings, 78
head breadth and length of, 76; compared to Ba'ij Beduins, 76; to Iraq Soldiers, 76
health of, 81, 110–121
henna, use of, 81
insensitivity to pain of, 119
leg length of, 76; compared to Ba'ij Beduins, 76; to Iraq Soldiers, 76
lips of, 79
malars of, 78
minimum frontal diameter of, 76; compared to Ba'ij Beduins, 76; to Iraq Soldiers, 76
morphological characters of, 77–80
musculature of, 80
nasal breadth and height of, 76; compared to Ba'ij Beduins, 76; to Iraq Soldiers, 76
nasal bridge of, 79
nasal index of, 76; compared to Ba'ij Beduins, 76; to Iraq Soldiers, 76; groupings, 77
nasal profile of, 80
nasal septum of, 79
nasal tip and wings of, 80
prognathism, alveolar, among, 78; facial, 78
remedies used by, 118
scapulae of, 80
sitting height of, 76; compared to Ba'ij Beduins, 76; to Iraq Soldiers, 76
skin color of, 77
stature of, 76; compared to Ba'ij Beduins, 76; to Iraq Soldiers, 76; groupings, 77
tattooing among, 81
teeth of, groupings, 80
ventral disorders of, cause of, 115
vital statistics of, 77
zygo-frontal index of, 76; compared to Ba'ij Beduins, 76; to Iraq Soldiers, 76

zygo-gonial index of, 76; compared to Ba'ij Beduins, 76; to Iraq Soldiers, 76
Kish Workmen (100 individuals measured), 13, 81–82
bigonial breadth of, 82
bizygomatic breadth of, 82
cephalic index of, 82
eyes of, groupings, 82
facial measurements and indices of, 82
fronto-parietal index of, 82
hair of, groupings, 82
head breadth and length of, 82
minimum frontal diameter of, 82
nasal breadth and height of, 82
nasal index of, 82
stature of, 82
zygo-frontal index of, 82
zygo-gonial index of, 82
Knabenshue, Paul S., 8
Kubaisa, population of, 28; Sunnis in, 28
Kurds, 13; age, cephalic indices and head measurements of, from Ali Sharwan, 130, from Erbil, female, 127, male, 127, from Hussain Kuli Khan, 130, from Kermanshah, 130, from Khanaqin, 127, from Kirkuk, female, 127, male, 127, from Mosul, 127, from Pestako, 130, from Sulaimaniya, 127, from Tabriz, female, 130, male, 130, from Tehran, 130, from Waly, 130
Kut, classification of land surface of, 106–107; population of, in 1930, 108, in 1935, registered, 105, unregistered, 104

Langdon, Stephen, 7
Lathrop, Barbour, 7
Laufer, Berthold, 8
Lazar, Yusuf, 8, 9, 15, 16, 156, 163, 164, 165
Lepidoptera, 164
Lepus connori, 161
Lime, 24
Limestone quarries, 24
Liponycteris magnus, 157
Londonderry, Lord, 9
Luhaib, 34; tribal list of, 100
Lutra lutra, 160

Maize, 22
Majawadah, Al, 94
Malak, Gabriel, 9
Malaria, prophylaxis against, 111
Manufacturing, 24
Marsh Arabs, 13
Martes foiana, 160
Martin, Paul S., 10
Martin, Richard A., 8, 9, 11, 12, 156
Mash, 22
McLeod, T. H., 9
Mediterranean Race, 16

222   ANTHROPOLOGY OF IRAQ

Meles meles, 160
Melons, 22
Mellivora wilsoni, 160
Midhat Pasha, 25
Mihran, H., 9
Mineral resources, 24
Mosul, classification of land surface of, 106–107; population of, in 1930, 108, in 1935, registered, 105, unregistered, 104
Muhallaf, 27; tribal list of, 92
Mulberry trees, 22
Muntafiq, classification of land surface of, 106–107; population of, in 1930, 108, in 1935, registered, 105, unregistered, 104
Murray, Wallace, 8
Myotis omari, 157

Nafatha, oil from, 24
Nasiriya, An, 13
Natural History Museum, Stockholm, herbarium specimens in, 165
Negroids, 30
Nesokia buxtoni, 161
New York Public Library, 11
Nimr, Al bu, 98
Noria, 22–23
Nuri ibn Shalan, 27, 28, 54
Nychiodes(?) *divergaria*, 163

Oil, 24
Omar Pasha, 25
Onions, 22
Oppenheim, Max Freiherr von, 11
Oriental Institute, *see* University of Chicago
Orthoptera, 24

Pahlavi, Riza Shah, *see* Riza Shah Pahlavi
Paleolithic flint implements, 28
Pathology, attitudes toward, *see* Arabs, attitude toward disease, etc., treatment of disease
  abdomen, distention of, 147
  acromegaly, 119
  amputation, 117
  arthritis, rheumatoid, 117
  ataxia, locomotor, 116
  "Baghdad boil," 81, 85, 114, 146, 151, 152
  bejel, 116
  bilharziasis, 147
  blood-letting, 118
  chicken pox, 37, 38, 81, 85
  cholera, 30, 31, 115, 120–121
  constitution, 38, 119–120
  deformation, of the arm, 115; of the ear, 113; of the hands, 115; of the lips, 115
  dental condition, 37, 38, 113–114; groupings, 37, 65, 80
    broken teeth, 38, 114, 134, 146
    caries, groupings of, 80, 85, 88, 113
    deposit, 134, 139, 140, 146, 150, 151
    fillings, 38
    loss, 37, 134, 150, 151; groupings, 37, 80, 85, 88, 134, 146; attitude toward, 113
    stain, 114, 134, 139, 146; groupings, 114
    wear, 80, 88
  diarrhea, 116
  endocrine glands, 119
  eyes, 31, 112–113
    arcus senilis, 133, 144
    blindness, 66, 135, 140, 150; groupings, 81
    cataract, 38, 81, 85
    conjunctivitis, follicular, 112; granular, 112
    crossed, 37
    defective vision, 37, 66, 150
    filmed, 135, 144, 150
    trachoma, 81, 112
  favus, 135
  fractures, 118, 151
  gallstones, 116
  goiter, 146
  hairless, 38
  headache, 81, 85, 113
  hemorrhage, 117
  hemostasis, 117
  influenza, 115
  jaundice, 116
  malaria, 81, 85, 110–112, 135
  metabolism, unbalanced, 119
  miscarriage, 151
  nasal affections, 115
  obesity, 119
  paralysis, hand, 139
  paresis, 116
  plague, 120; bubonic, 31
  pleurisy, 117
  ringworm, 66
  respiratory, 66, 115. *See* influenza, pleurisy, tuberculosis
  scalp infections, 81, 85; scurf, 139. *See* favus
  scars, 38, 66, 114, 151, 152
  skin, 38, 114. *See* "Baghdad boil"
  smallpox, 31, 38, 66, 81, 85, 112, 120, 135, 140, 147, 150, 152
  sprain, 135
  syphilis, 116–117; tertiary, 116
  tuberculosis, 115
  typhus, 31
  vaccination, 112
  venereal diseases, 116–117, 147. *See also* bejel, syphilis, yaws
  ventral disorders, 81, 85, 115–116
  yaws, 116
Peabody Museum (Harvard), 10, 11
Pear trees, 22
Pedersen, Dorothy, 10, 12

INDEX 223

Physical features, 17–18
Pilgrimage, spread of disease resulting from, 31, 120–121
Pipistrellus kuhli, 158
Pomegranate trees, 22
Population, 26
Public health service, 31, 120–121

Qala Sharqat, flora of, 193
Qara-Ghul, 34; tribal list of, 100
Qara-Ghul (Section of the Zoba), 34

Radishes, 22
Rahhaliya, Ar, 101; Negroid element in, 30; population of, 30
Rainfall, 20, 22, 23
Ramadi, 17; date palms at, 30; health inspection at, 120; location of, 30; medical inspection at, 31; population of, 30
Ram-faced types among the Dulaim and the Anaiza, 73–74
Rassam, B. H., 9, 15
Raqqa, Arabs in, 28; Circassians in, 28; population of, 28
Reid, H. C., 9
Religious groups, 26
Reniff, Elizabeth, 10
Reptiles, 24
Rhazes, first account of smallpox by, 112
Rice, 22
Rice, David Talbot, 7
Rickards, A. R. M., 9
Ridhwaniya Canal, description of, 18
Riley, N. W., 15, 163
Riza Shah Pahlavi, 11
Ross, Lillian A., 10
Rowandiz Area, flora of, 193
Royal Geographical Society (London), Permanent Committee on Geographical Names of, 12
Rustam Agricultural Experimental Farm, Hinaidi, 166
Rutba, flora of, 197
Ruwalla, 27, 28, 54; habitat of, 27; importance of, 27; tribal list of, 92

Salt, 24
Samuelsson, Gunnar, 165
Sanborn, Colin C., 15
Sand storms, 22
Saqlawiya Canal, description of, 18
Sbaa (Beduins), 16, 27, 54
Schlimmer, J. L., 114
Schmidt, Karl P., 15
Schroeder, Eric, 110, 118
Scott, Donald, 10
Scully, Theodore, 10
Seltzer, Carl C., 10
Sesame, 22
Shaar, 34; tribal list of, 102
Shaddid, 101
Shamiya, Al, 17
Shammar, 13, 27, 34, 54, 55
Shammar, Southern, 27, 28
Shaw, F. R. S., 9
Shawkat, Shaib, 131
Sheep, 23; oil used as remedy for, 24
Sheikh Adi, flora of, 192
Sheikh Atiyeh, 115
Sheikh *Hajji* Hunta, camp of, 116
Shiahs, 26
Shiti, 34; tribal list of, 101
Showket, S. Y., 8, 83, 116
Shrubs, 23
Shuwartan, 34; tribal list of, 100
Skliros, John, 9
Smeaton, Winifred, see Thomas, Winifred Smeaton
Snow, 20
Spinifex, 23
Standley, Paul C., 16
Subaihat, 34; tribal list of, 101
Subba (Mandeans), 13
Sulaimaniya, classification of land surface of, 106–107; flora of, 194; population of, in 1930, 103, 108, in 1935, registered, 105, unregistered, 104
Sulphur, 28
Sulubba (Sleyb), 13
Sumailat, 34; tribal list of, 101
Sumeria dipotamica, 163, 164
Summerscale, J. P., 24
Sunnis, 26; at Ana, 28; at Kubaisa, 28
Sus attila, 161
Sykes, Mark, 14
Syrian Catholics in Deir-ez-Zor, 28

Tall Afar, flora from west of, 191
Tell Barguthiat, 116
Tell Es Shur, flora of, 190
Temperature, 20
Thomas, Winifred Smeaton (Mrs. Homer), 13, 15
Tobacco, 117
Treatment of disease, fracture, 118; cautery, 116, 117; scarring, 118; venereal disease, 117
Tribal groups, 27
Triticum, 116
Tuch, David, 11
Turkish Petroleum Company, see Iraq Petroleum Company
Turkomans, 13
Turks, age, cephalic indices and head measurements of, from Istanbul, 130, Christian from Turkey, 130, from Van, 130

University of Chicago, Oriental Institute of, 9, 12
Upper Euphrates, historical references to, 14
Ursus arctos, 161

Vulpes persica, 159
Vulpes splendens, 159

Wadi Hauran, flint implements found at source of, 28; Jebel Enaze, source of, 28
Watelin, Louis Charles, 111
Water lift, 22
Water wheel, *see* Noria
Wheat, 22
Wilson, A. T., 103

Wilson, W. C. F., 9
Wiltshire, E. P., 163–164
Wind, 21
Wulud Ali, 27, 54; tribal list of, 93

Yezidis, 13

Zimmerman, Eunice, 10
Zoba, 18; sections of, following the Dulaim, 34; tribal list of, 100–101

GENERAL VIEW OF HADITHA

Field Museum of Natural History Anthropology, Vol. 30, Plate 2

No. 1013 (age 30)
CLASSIC MEDITERRANEAN TYPE

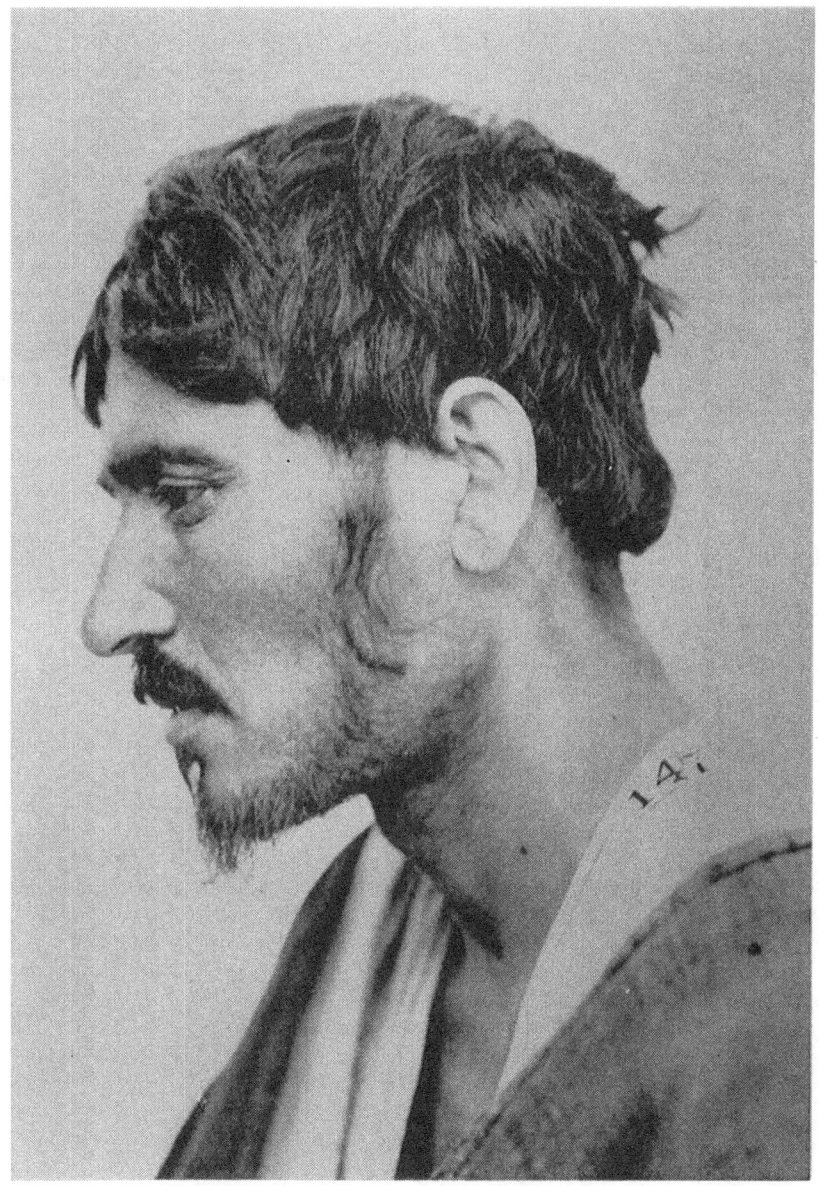

No. 1013 (age 30)
CLASSIC MEDITERRANEAN TYPE

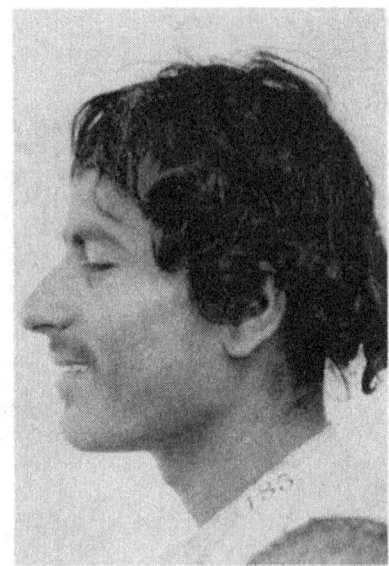

No. 1052 (age 20): Fine Mediterranean type

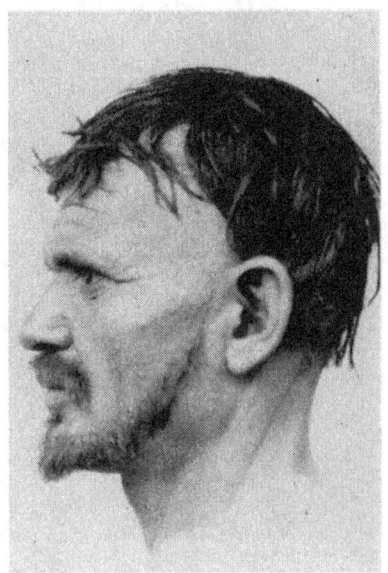

No. 1080 (age 35): Coarse Mediterranean type

MEDITERRANEAN TYPES

No. 1039 (age 27)

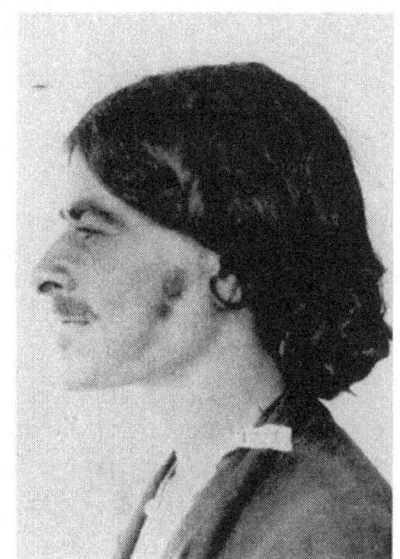

No. 1037 (age 27)

IRAQO-MEDITERRANEAN TYPES

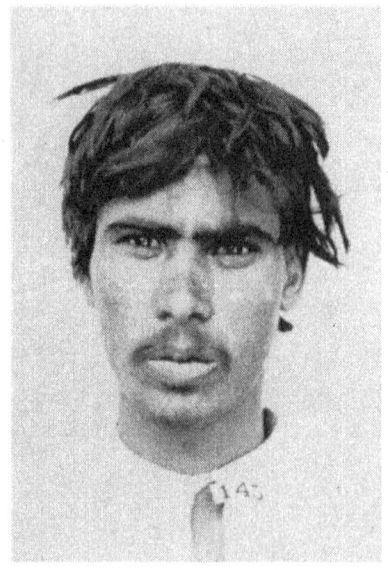 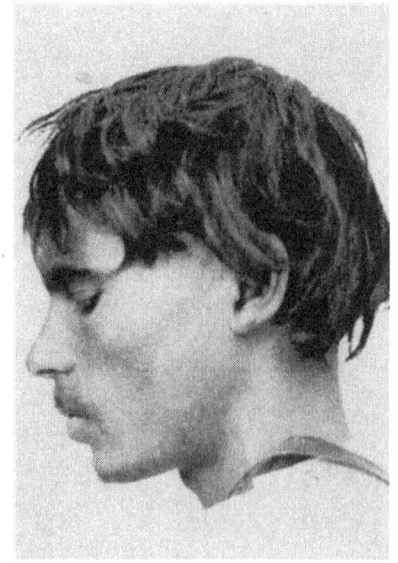

No. 1011 (age 20)

 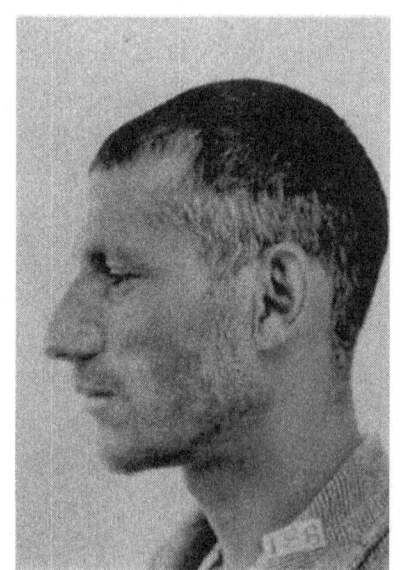

No. 1053 (age 30)

DOLICHOCEPHALS

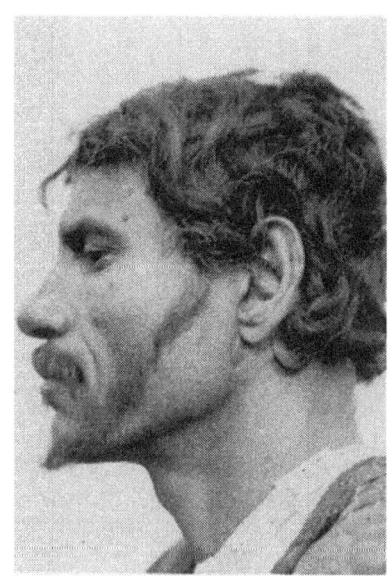

No. 1054 (age 40)

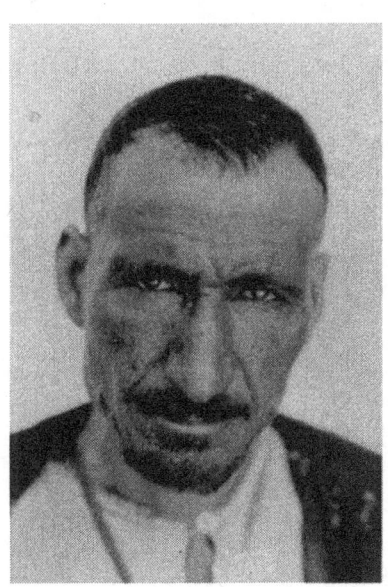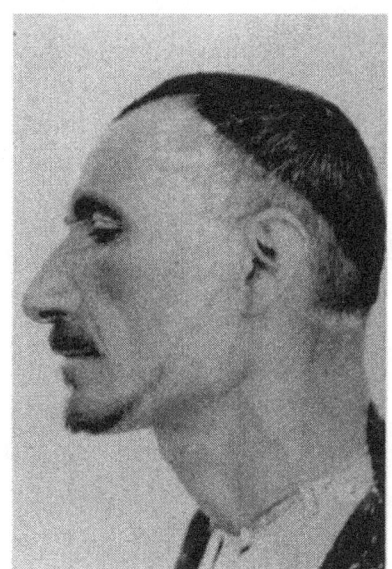

No. 1044 (age 45)

DOLICHOCEPHALS

Field Museum of Natural History   Anthropology, Vol. 30, Plate 8

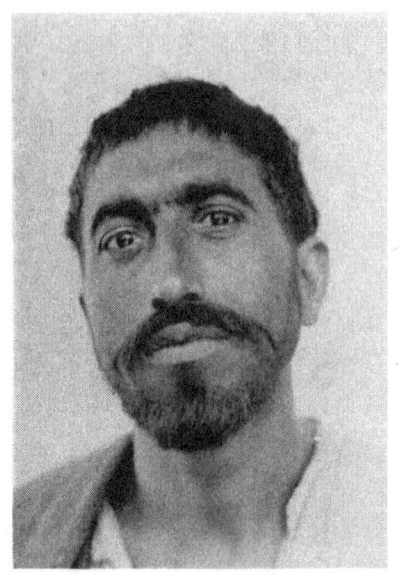
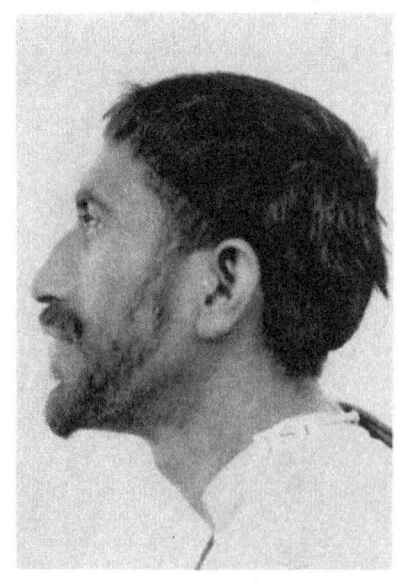

No. 1048 (age 30)

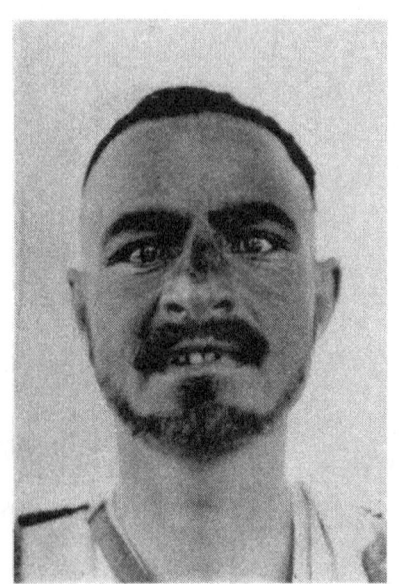
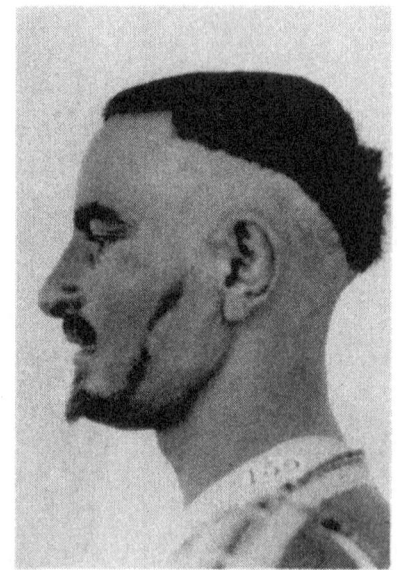

No. 1010 (age 25)

BRACHYCEPHALS

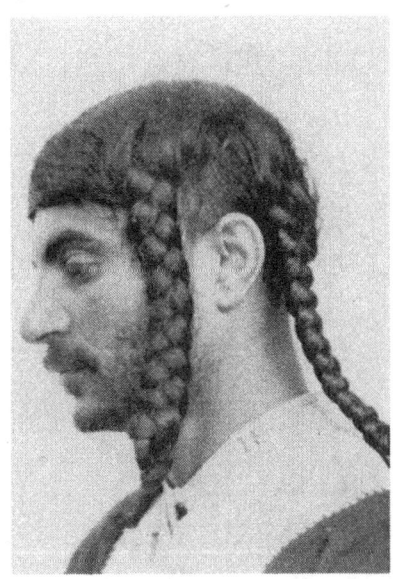

No. 1049 (age 30): Short-faced individual

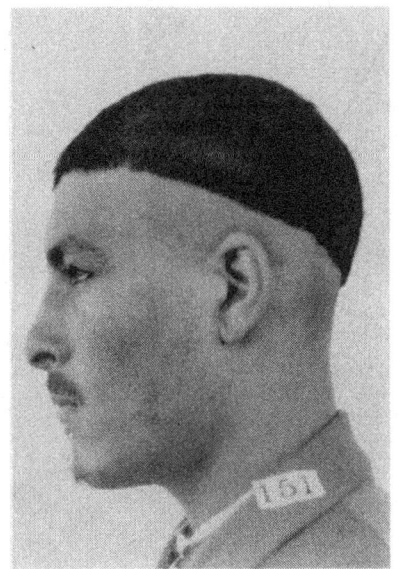

No. 1018 (age 30): Long-faced individual

FACIAL TYPES

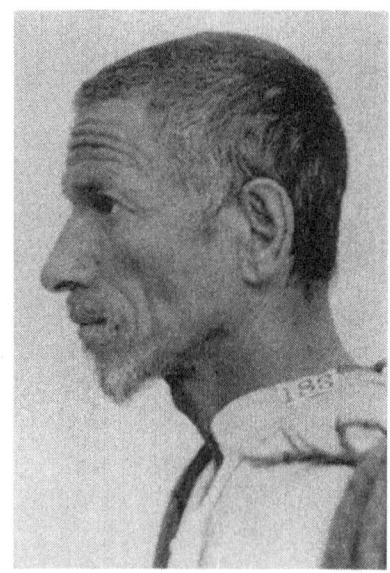

No. 1050 (age 45): Short and narrow-faced type

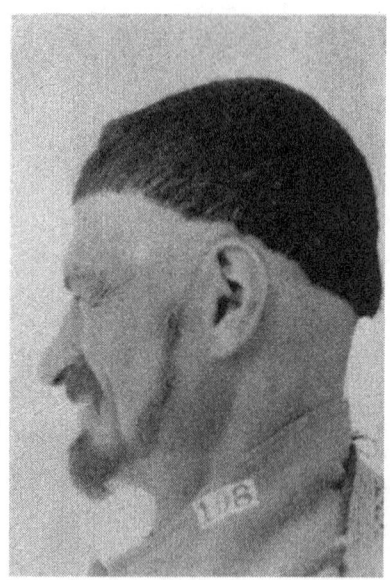

No. 1065 (age 40): Short and broad-faced type

FACIAL TYPES

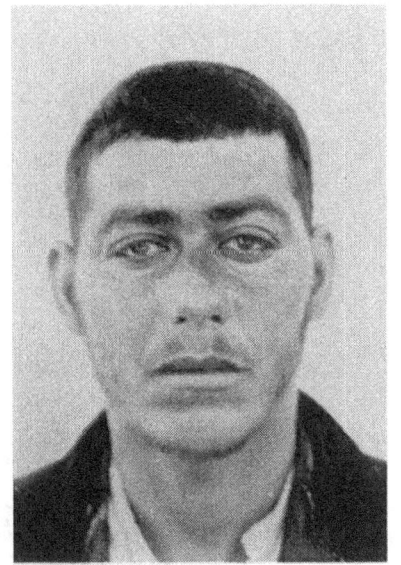
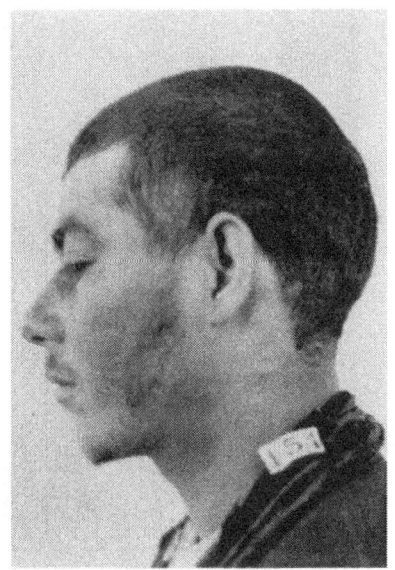

No. 1021 (age 25)

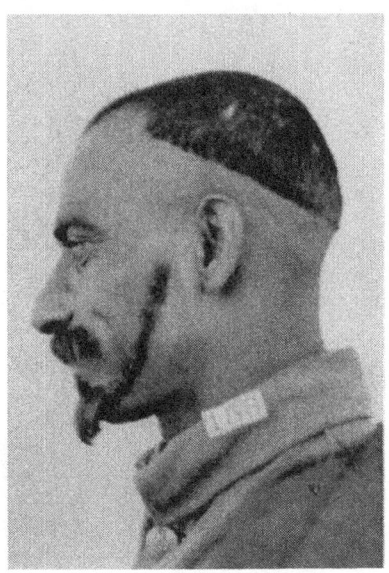

No. 1023 (age 30)

MIXED-EYED INDIVIDUALS

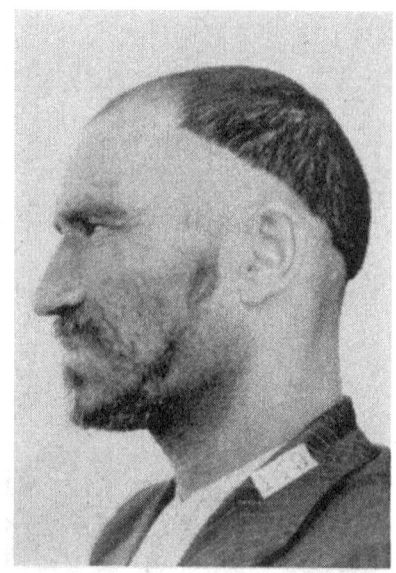

No. 1046 (age 35): Blue-eyed individual

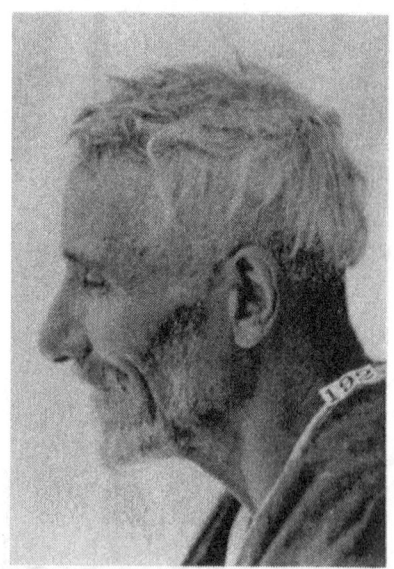

No. 1059 (age 60): Man with green-brown eyes

MIXED-EYED INDIVIDUALS

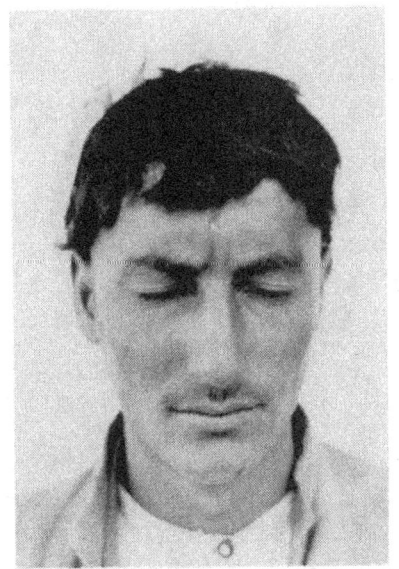
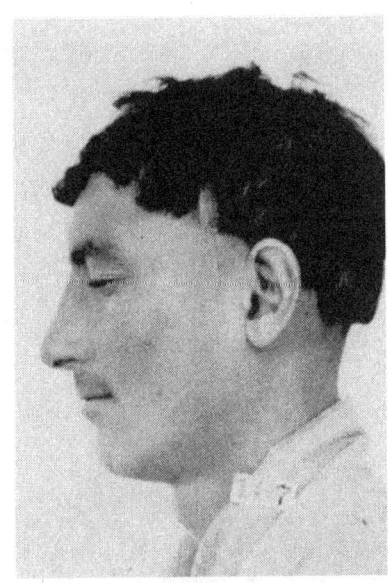

No. 1034 (age 22): Straight-nosed type

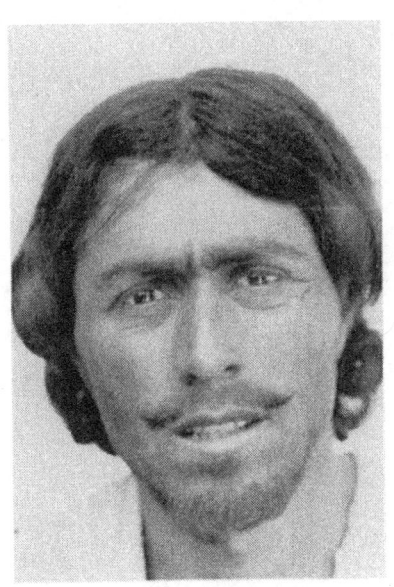
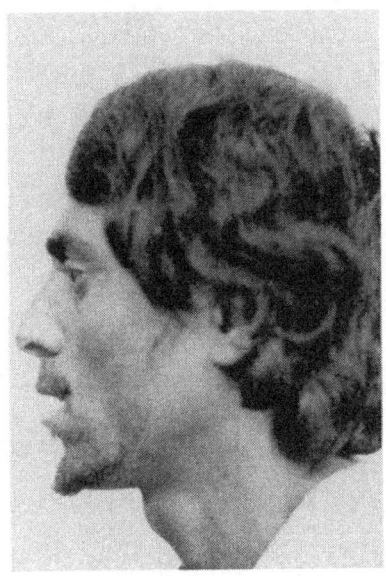

No. 1019 (age 27): Very slightly convex-nosed type

VARIATIONS IN NASAL PROFILE

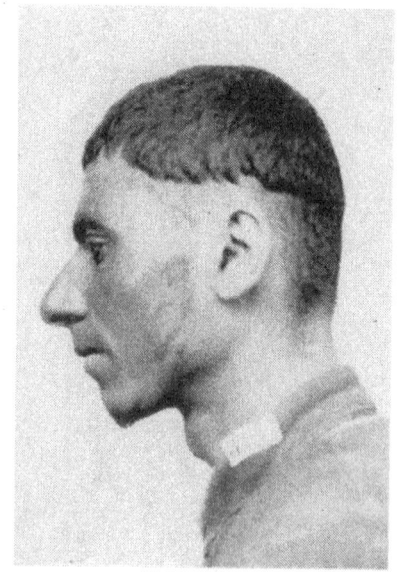

No. 1093 (age 25)

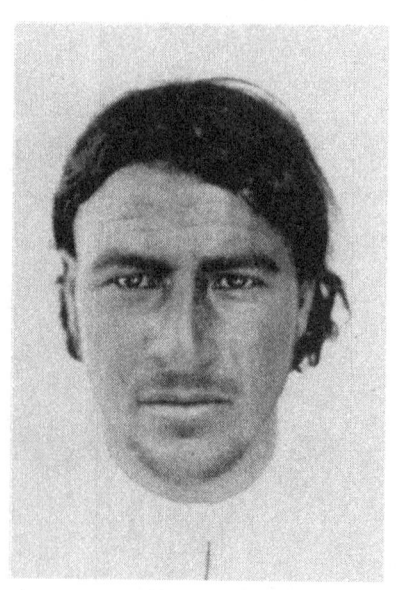
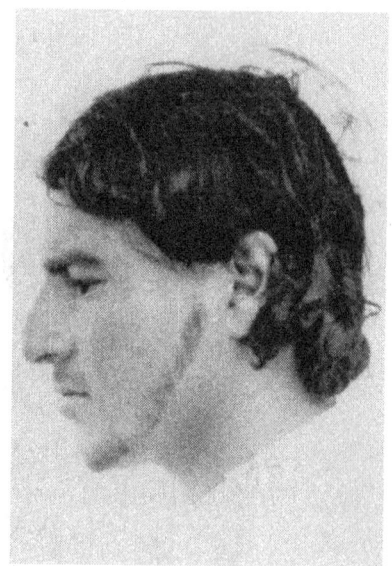

No. 1045 (age 20)

SLIGHTLY CONVEX-NOSED TYPES

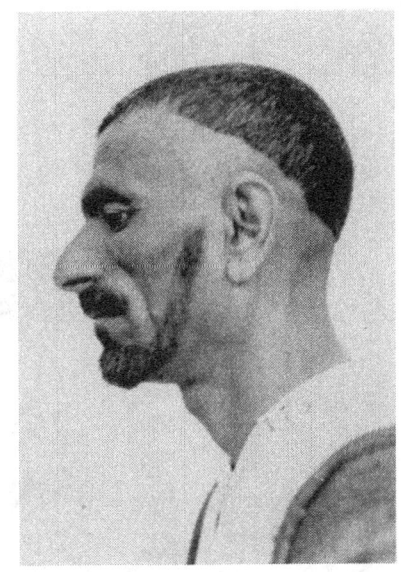

No. 1041 (age 35)

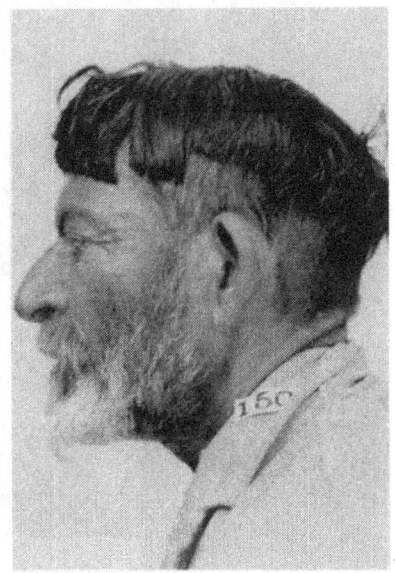

No. 1017 (age 50)

CONVEX-NOSED TYPES

No. 1055 (age 42)
CONVEX-NOSED TYPE

No. 1055 (age 42)

CONVEX-NOSED TYPE

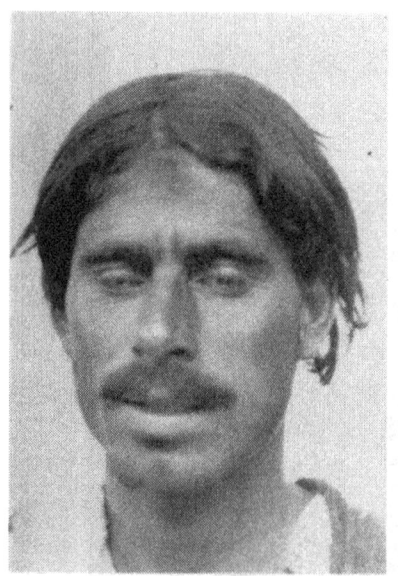 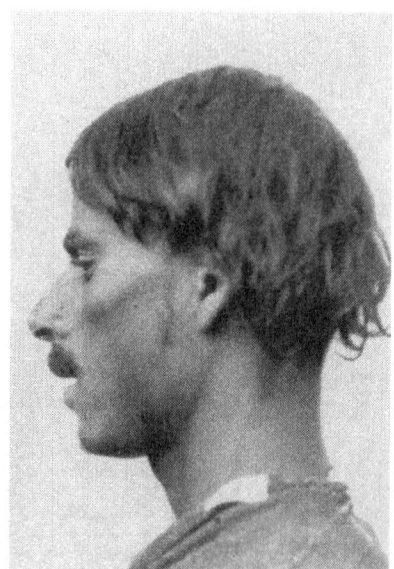

No. 1084 (age 25): Very low wavy hair

 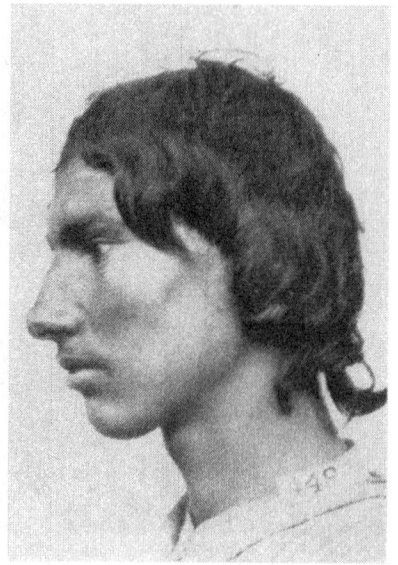

No. 1092 (age 25): Low wavy hair

VARIATIONS IN HAIR FORM

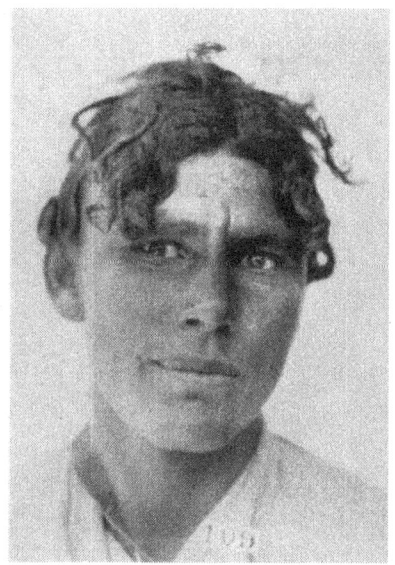
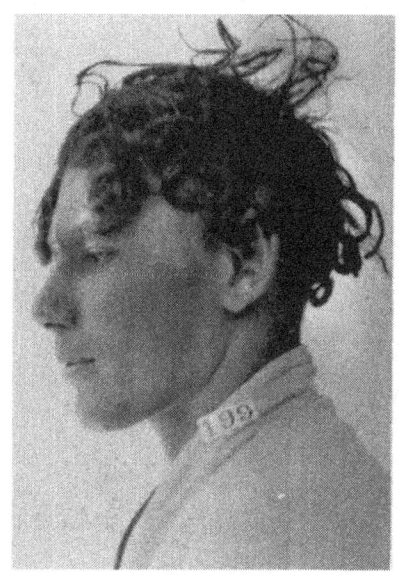

No. 1066 (age 23): Deep wavy hair

No. 1028 (age 40): Very deep wavy hair

VARIATIONS IN HAIR FORM

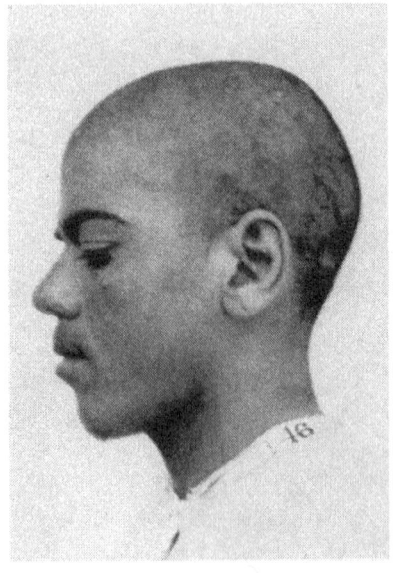

No. 1012 (age 20)

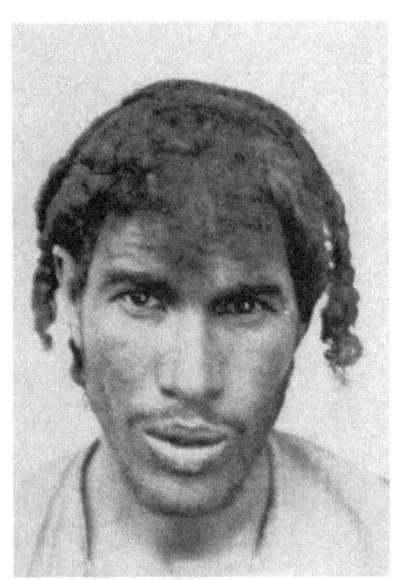
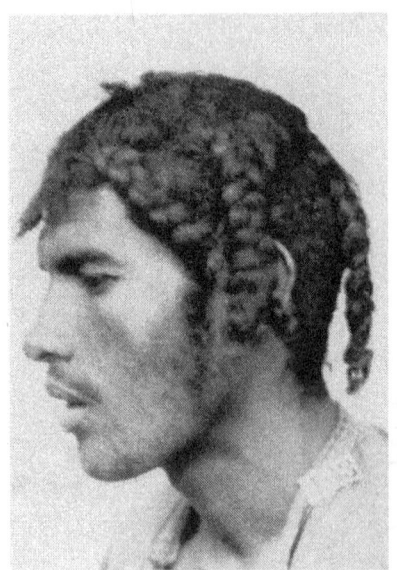

No. 1087 (age 25)

DULAIMIS MEASURED AT HADITHA

 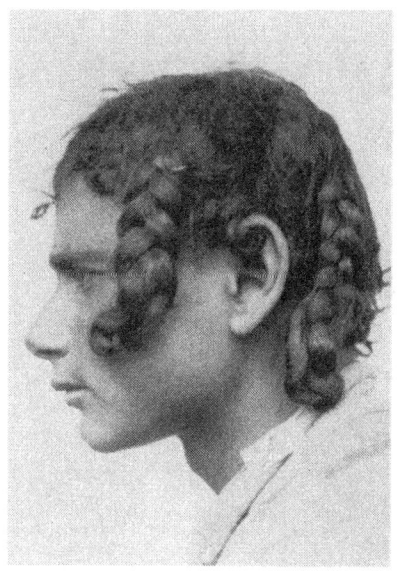

No. 1085 (age 20)

 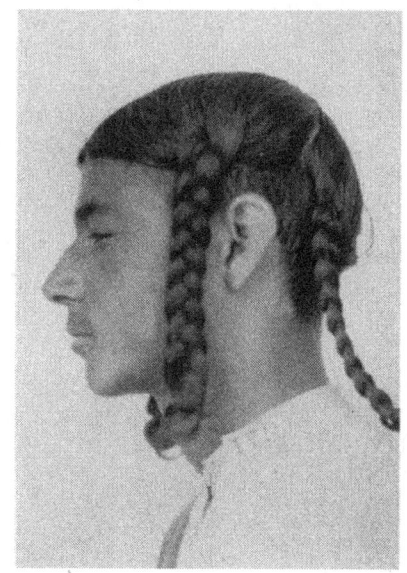

No. 1040 (age 20)

DULAIMIS MEASURED AT HADITHA

No. 1025 (age 20)

 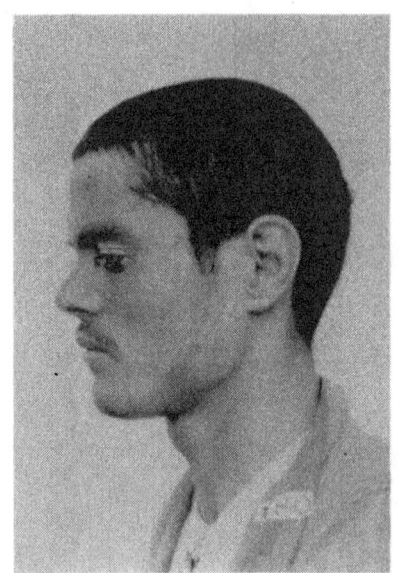

No. 1047 (age 20)

DULAIMIS MEASURED AT HADITHA

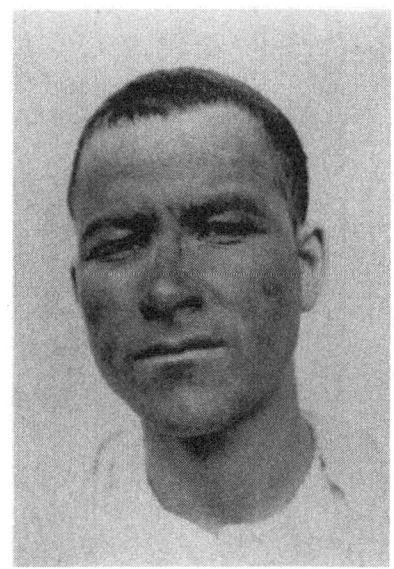 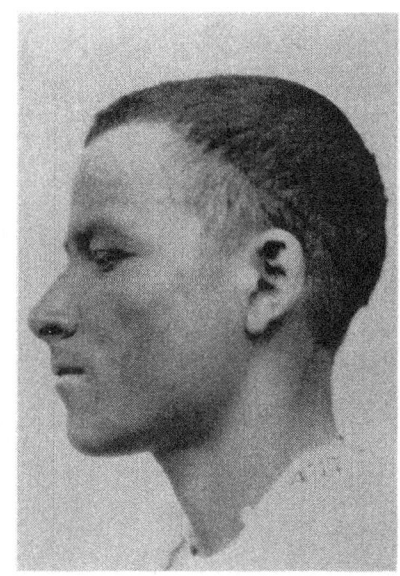

No. 1081 (age 20)

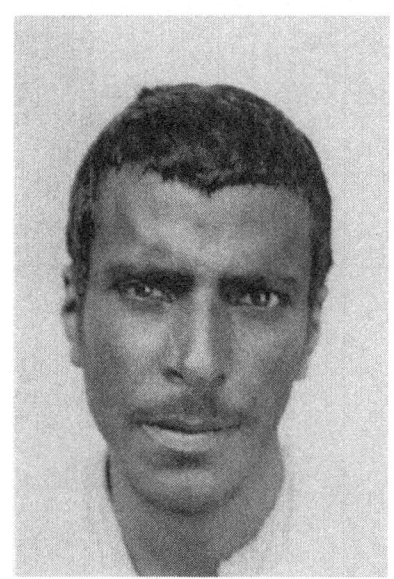 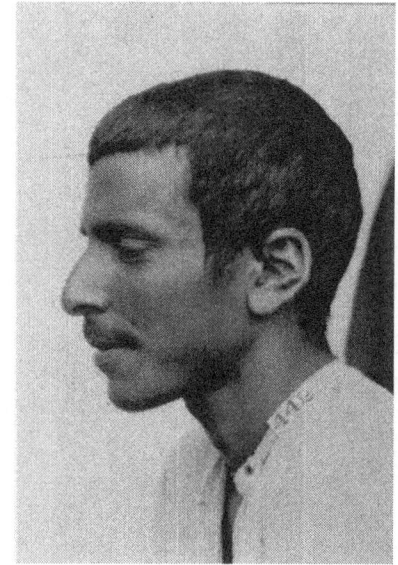

No. 1086 (age 21)

DULAIMIS MEASURED AT HADITHA

 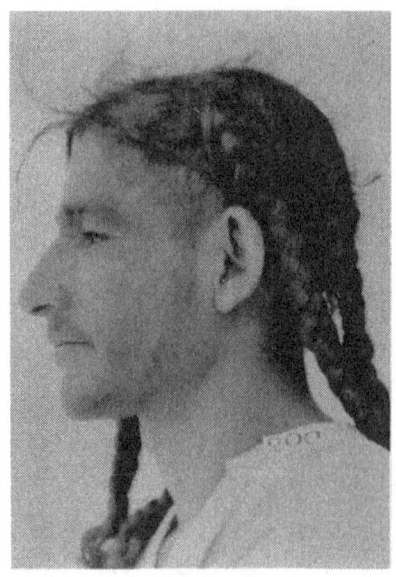

No. 1067 (age 21)

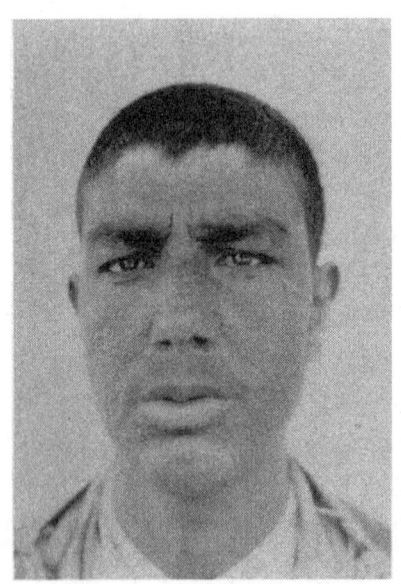 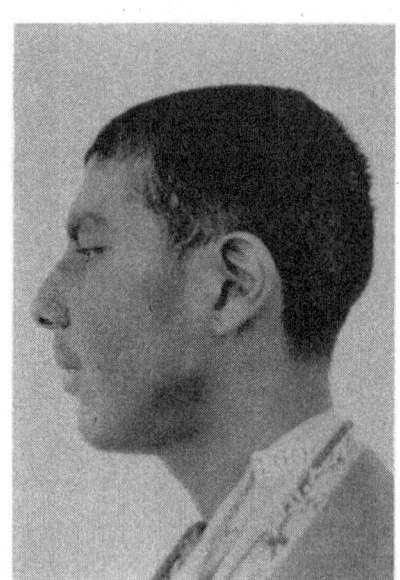

No. 1027 (age 22)

DULAIMIS MEASURED AT HADITHA

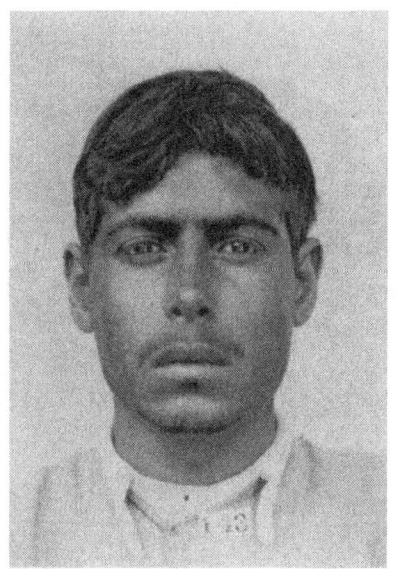
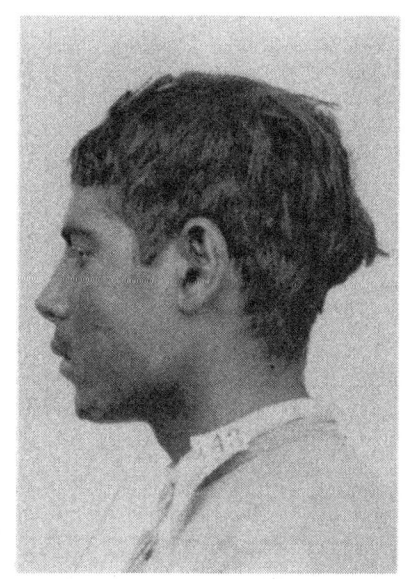

No. 1009 (age 25)

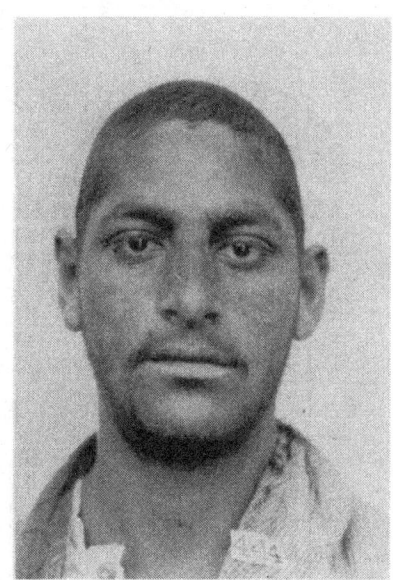
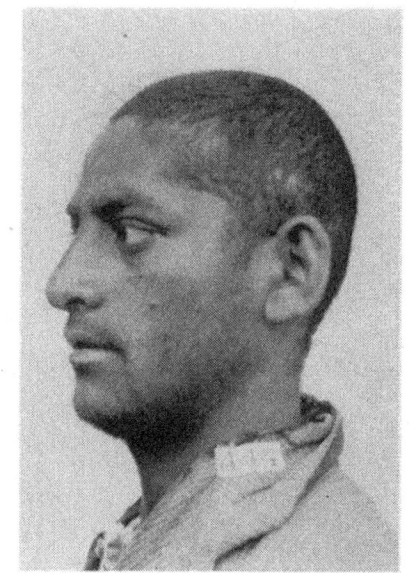

No. 1088 (age 25)

DULAIMIS MEASURED AT HADITHA

Field Museum of Natural History Anthropology, Vol. 30, Plate 26

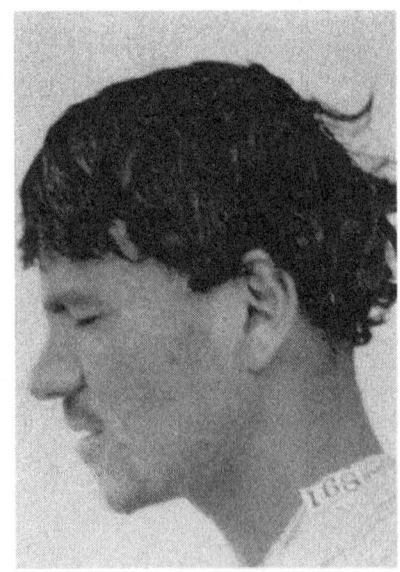

No. 1035 (age 25)

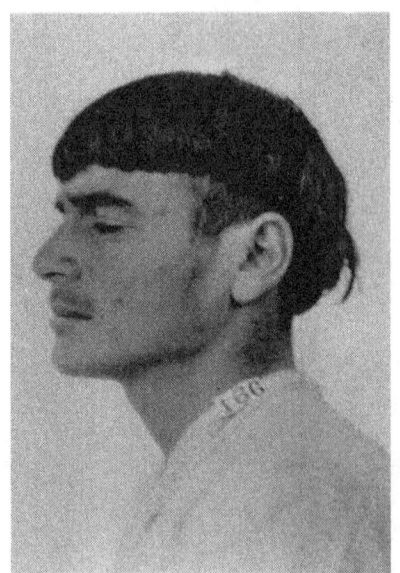

No. 1033 (age 25)

DULAIMIS MEASURED AT HADITHA

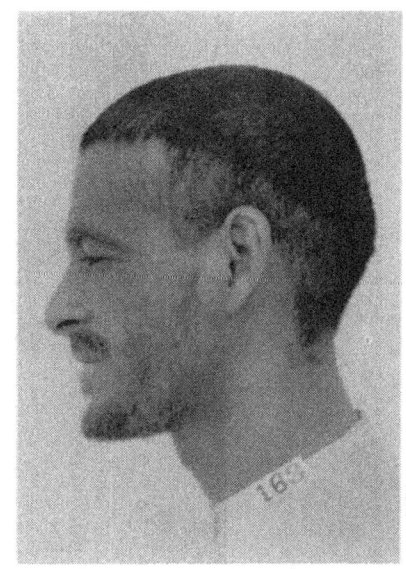

No. 1030 (age 25)

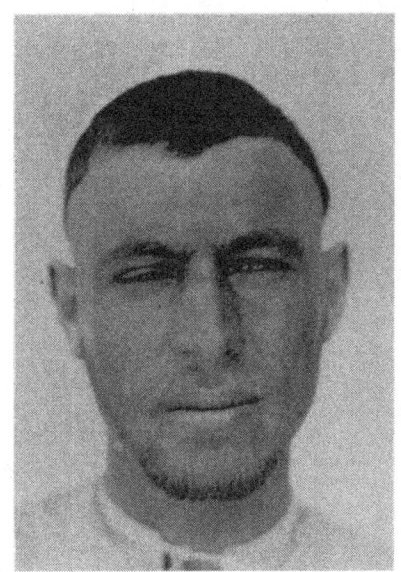
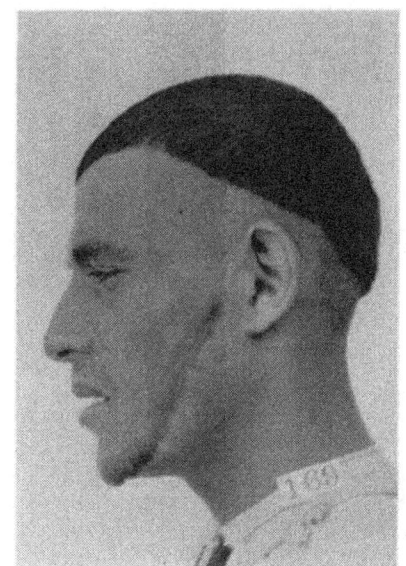

No. 1036 (age 25)

DULAIMIS MEASURED AT HADITHA

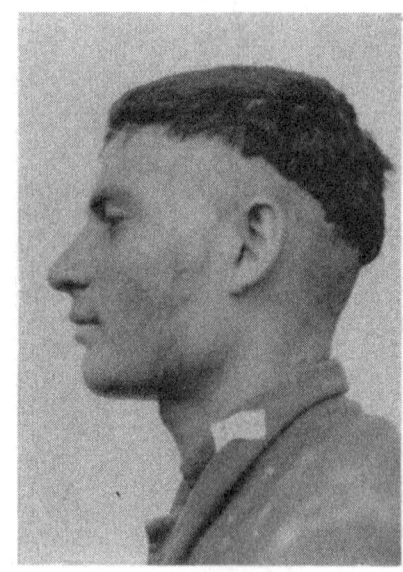

No. 1082 (age 25)

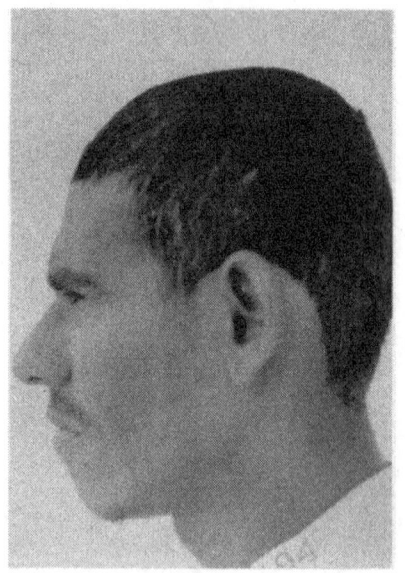

No. 1061 (age 26)

DULAIMIS MEASURED AT HADITHA

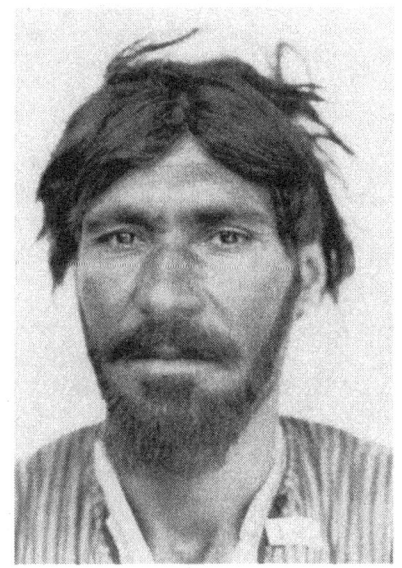
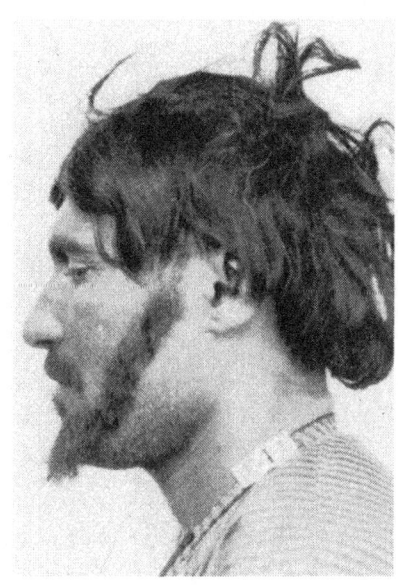

No. 1007 (age 30)

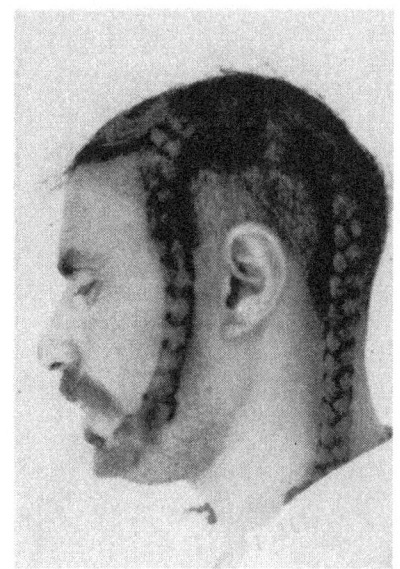

No. 1063 (age 30)

DULAIMIS MEASURED AT HADITHA

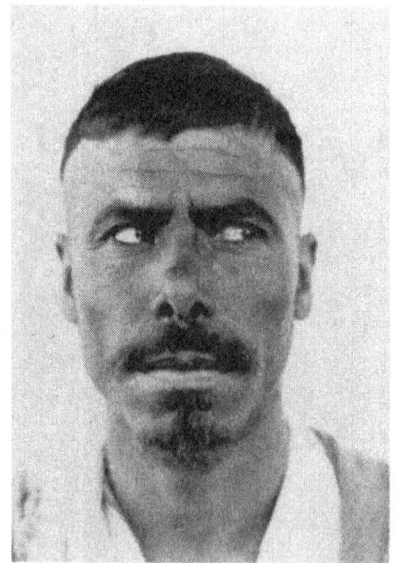
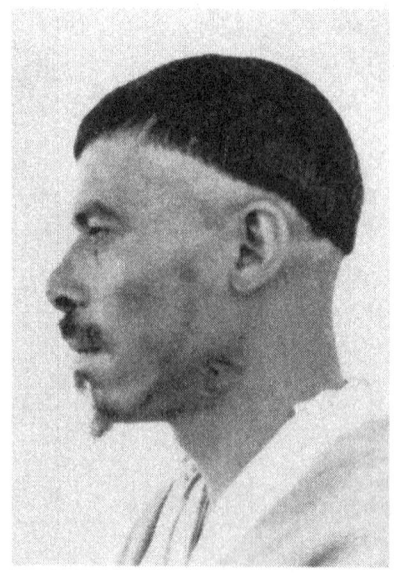

No. 1026 (age 30)

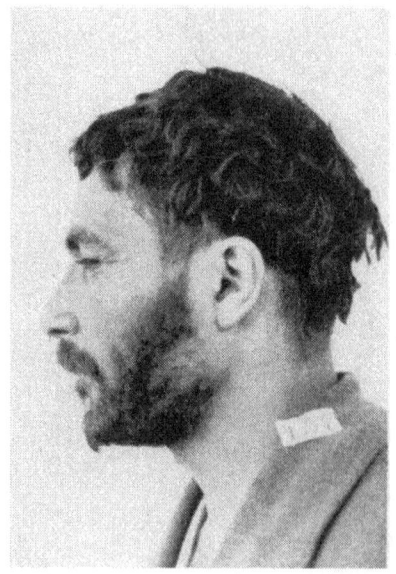

No. 1024 (age 32)

DULAIMIS MEASURED AT HADITHA

 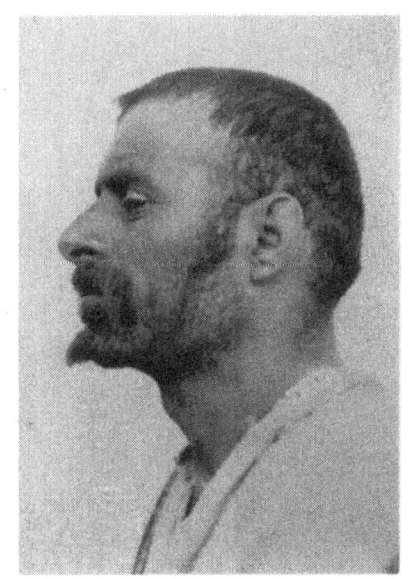

No. 1057 (age 35)

 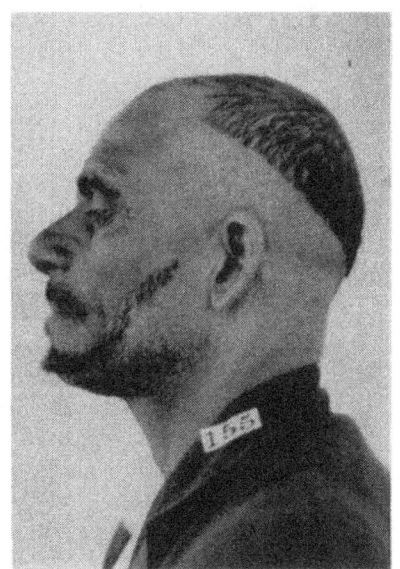

No. 1022 (age 35)

DULAIMIS MEASURED AT HADITHA

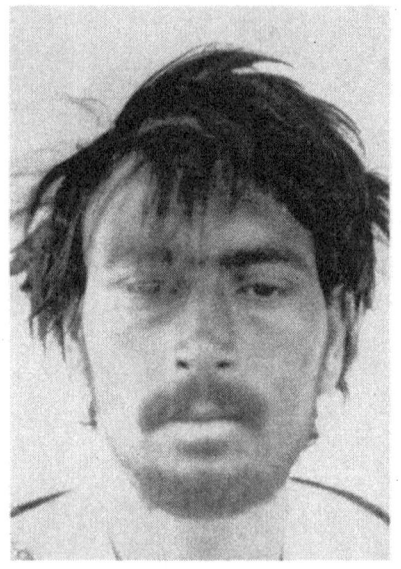 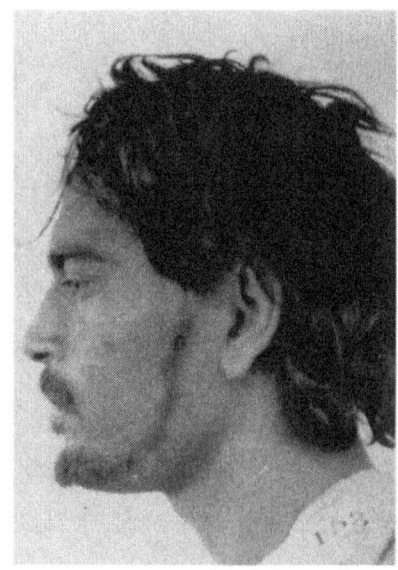

No. 1020 (age 35)

 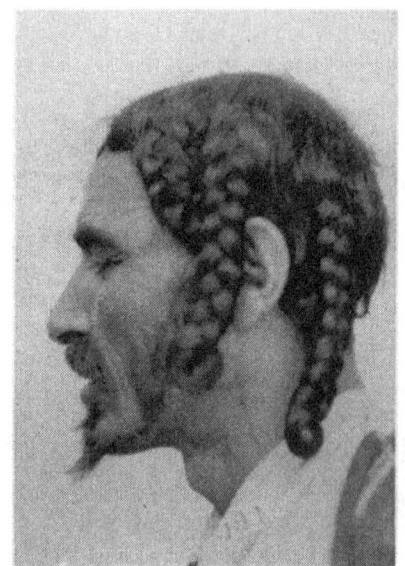

No. 1058 (age 40)

DULAIMIS MEASURED AT HADITHA

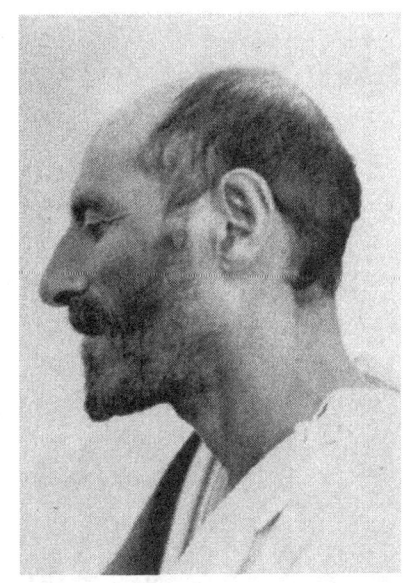

No. 1042 (age 45)

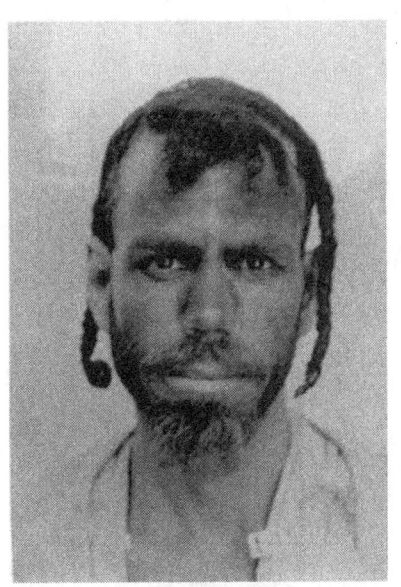
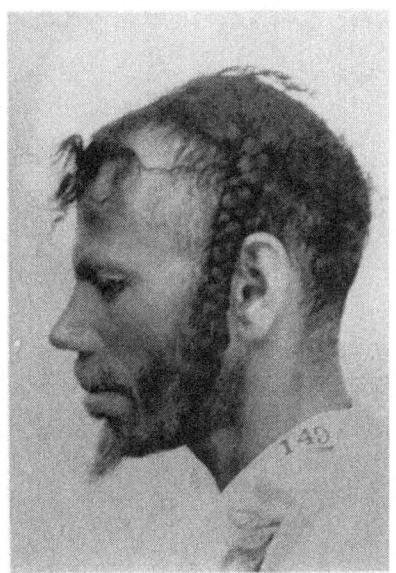

No. 1016 (age 45)

DULAIMIS MEASURED AT HADITHA

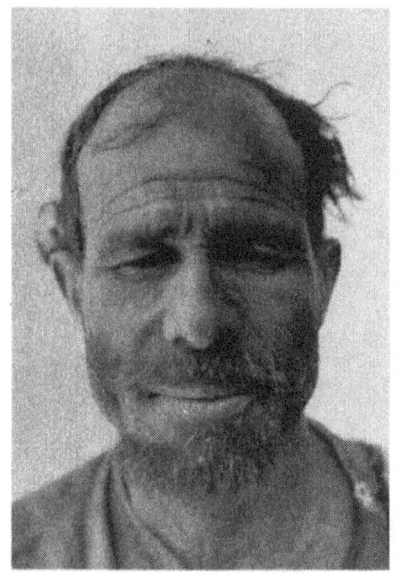 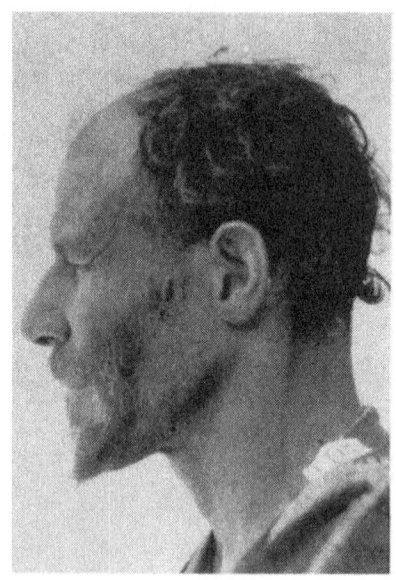

No. 1064 (age 50)

 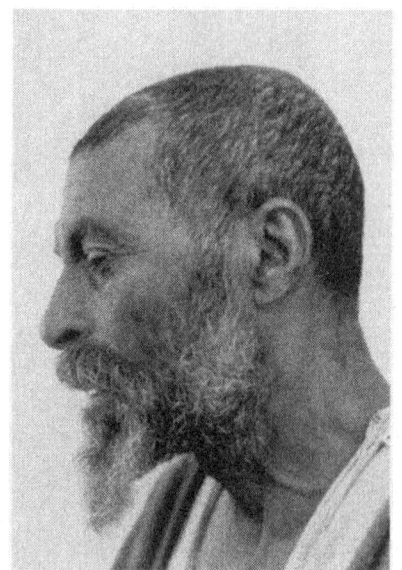

No. 1051 (age 60)

DULAIMIS MEASURED AT HADITHA

 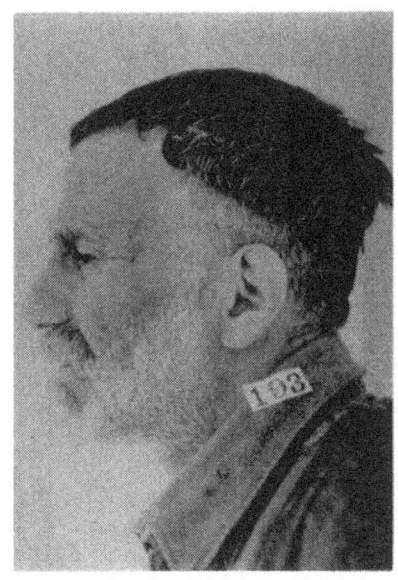

No. 1060 (age 60)

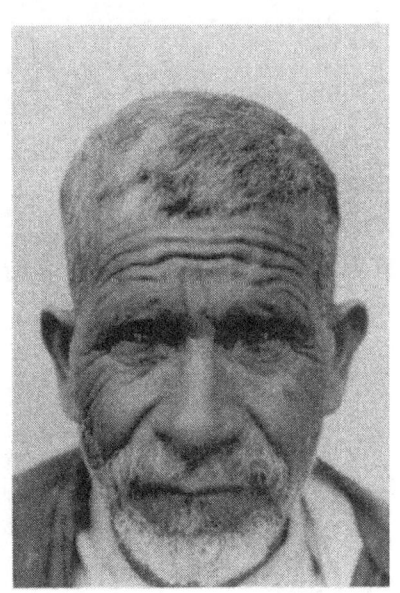 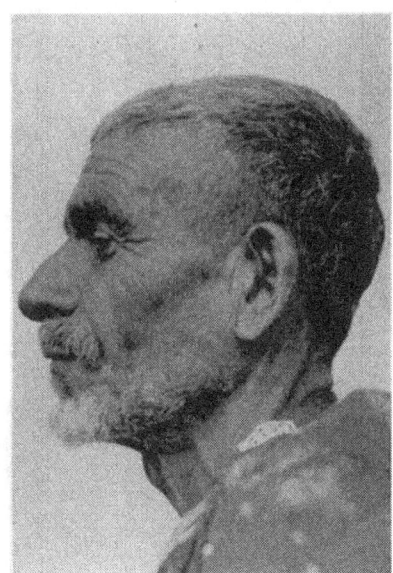

No. 1083 (age 60)

DULAIMIS MEASURED AT HADITHA

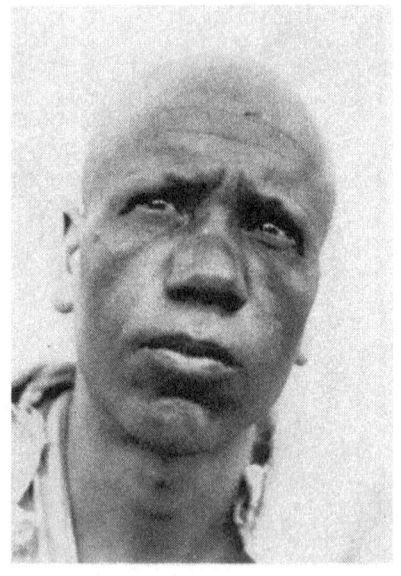
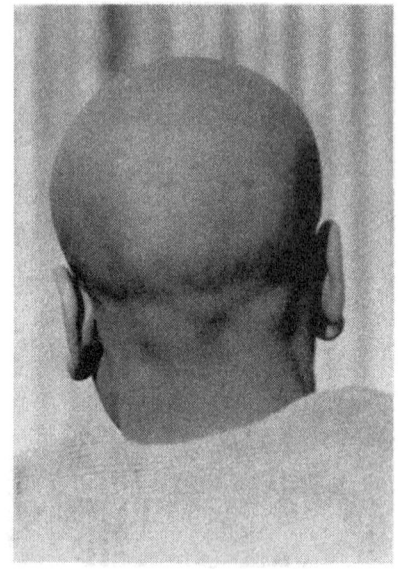
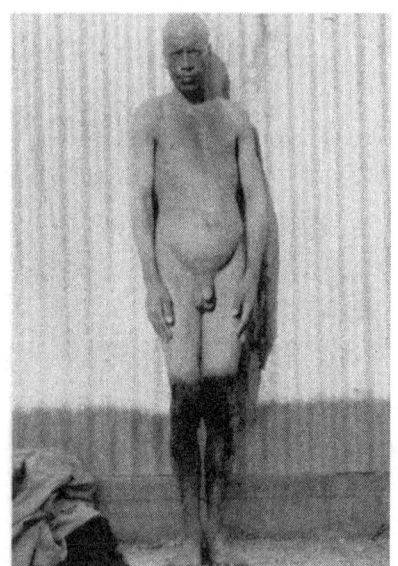
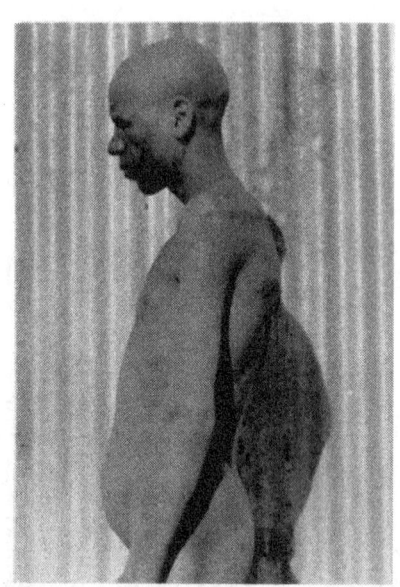

No. 1124 (age 22)
HAIRLESS DULAIMI

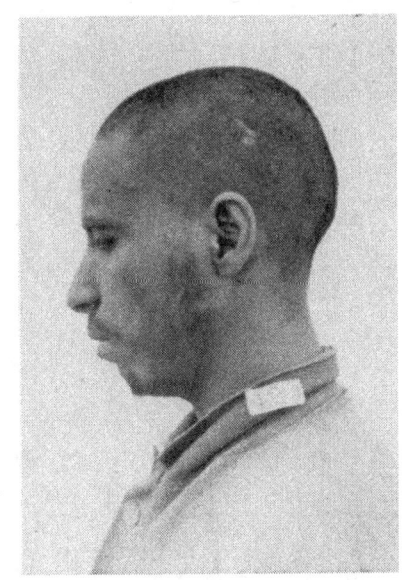

No. 1588 (age 24)

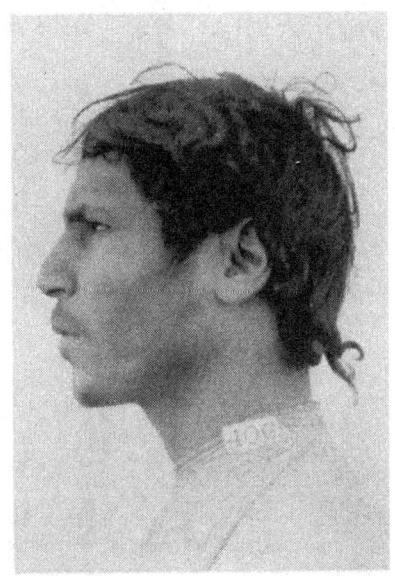

No. 1589 (age 25)

ANAIZA TRIBESMEN

No. 1584 (age 25)

No. 1572 (age 27)

ANAIZA TRIBESMEN

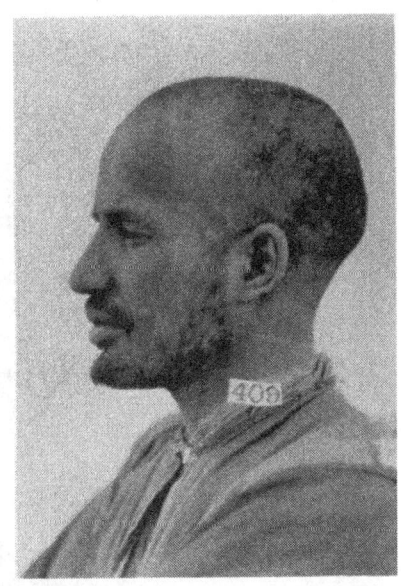

No. 1575 (age 28)

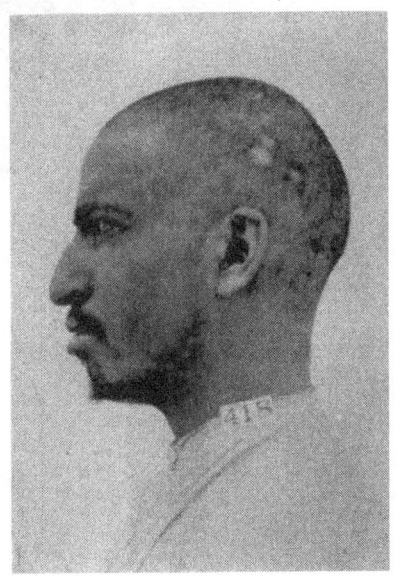

No. 1573 (age 28)

ANAIZA TRIBESMEN

No. 1571 (age 28)
ANAIZA TRIBESMAN

Field Museum of Natural History Anthropology, Vol. 30, Plate 41

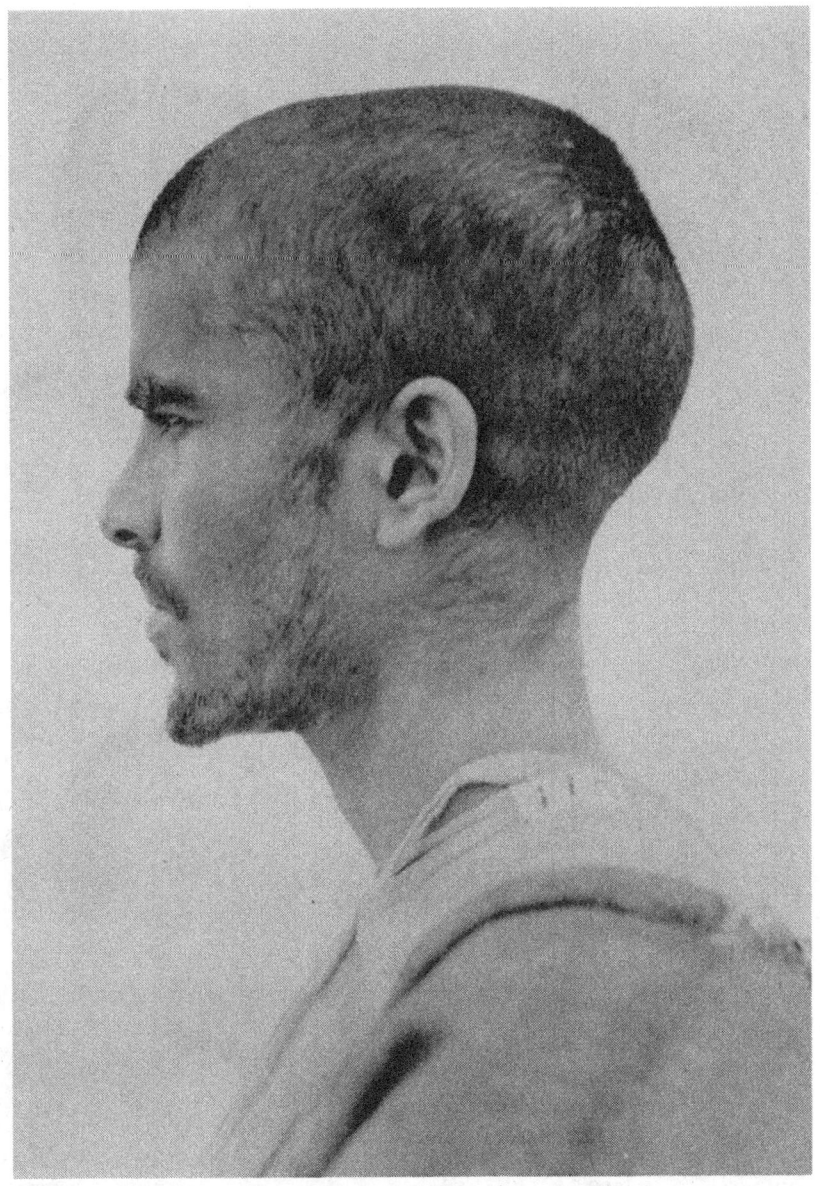

No. 1571 (age 28)
ANAIZA TRIBESMAN

 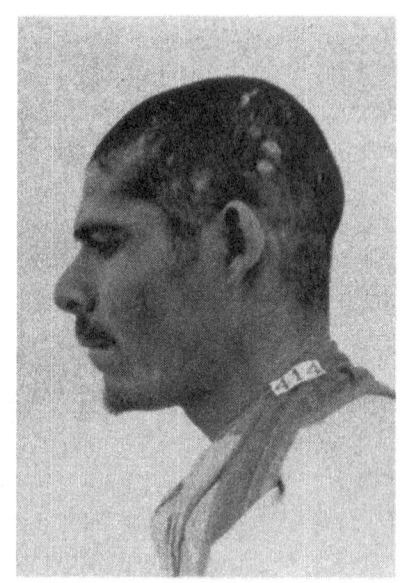

No. 1587 (age 30)

 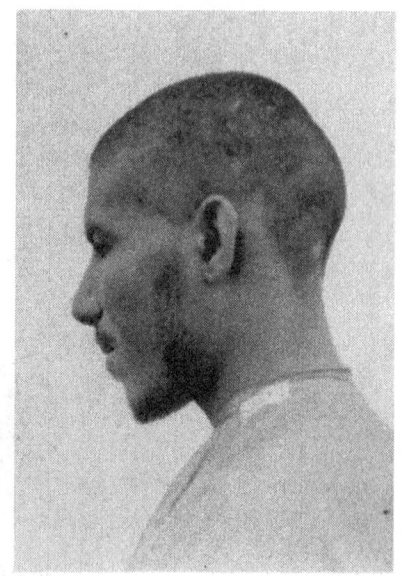

No. 1586 (age 30)

ANAIZA TRIBESMEN

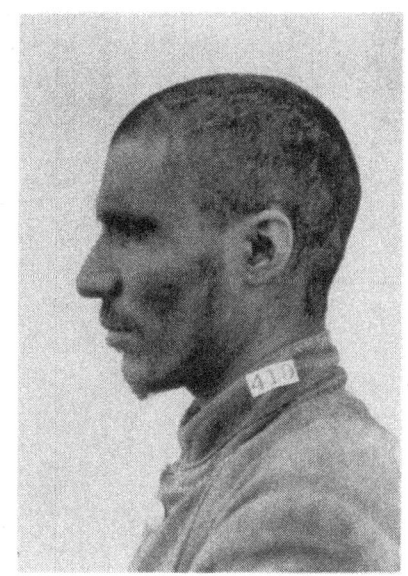

No. 1583 (age 30)

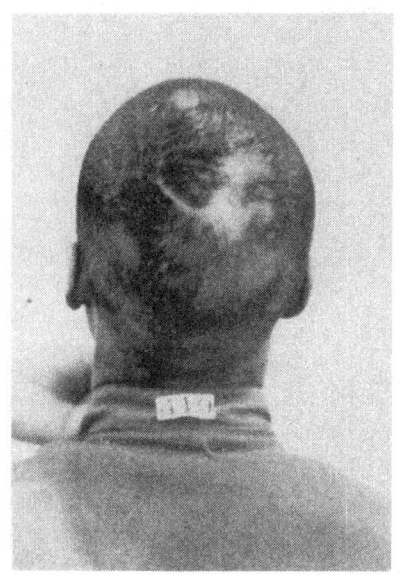

No. 1583 (age 30)　　　　　　No. 1577 (age 38)

ANAIZA TRIBESMEN

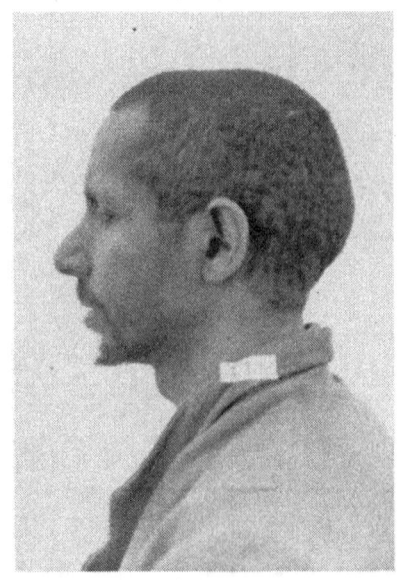

No. 1581 (age 30)

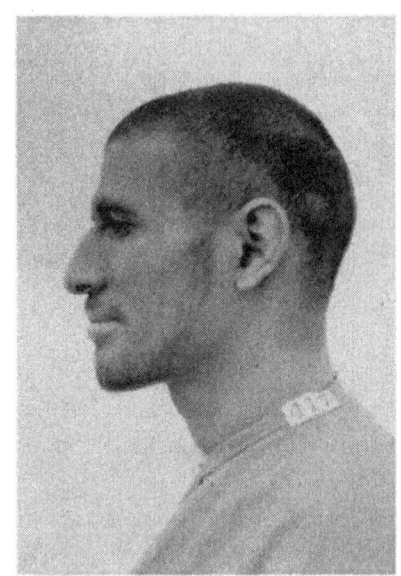

No. 1580 (age 30)

ANAIZA TRIBESMEN

No. 1576 (age 35)

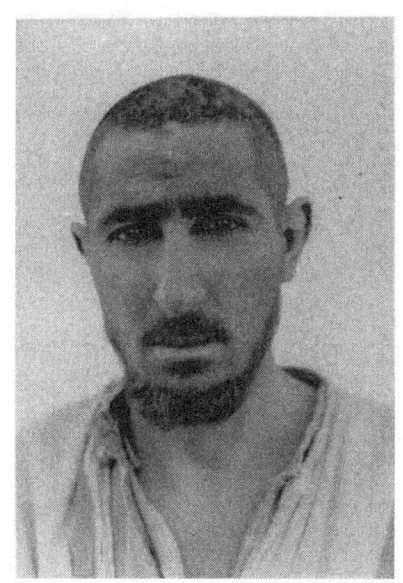
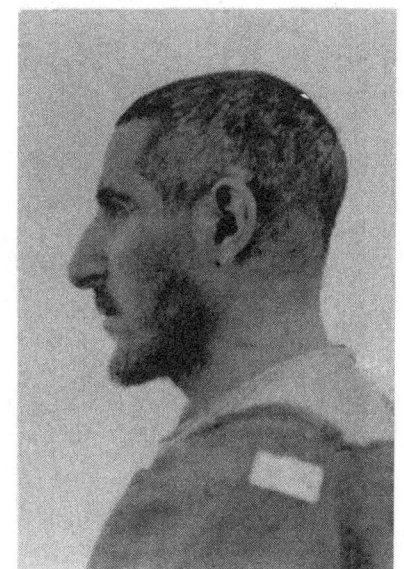

No. 1585 (age 35)

ANAIZA TRIBESMEN

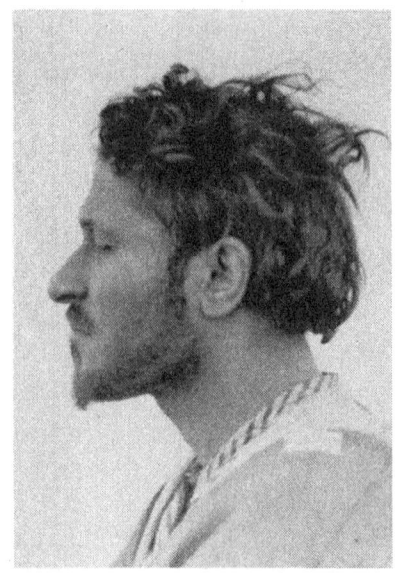

No. 1592 (age 35)

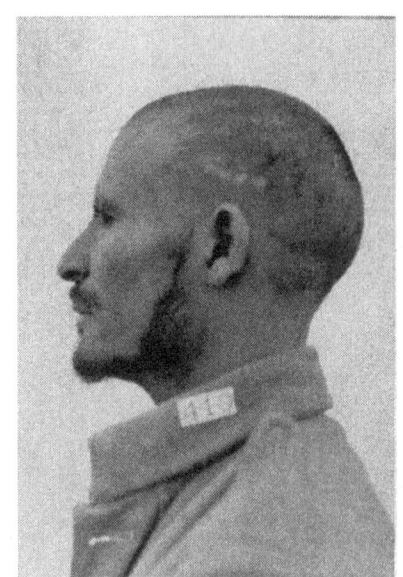

No. 1582 (age 35)

ANAIZA TRIBESMEN

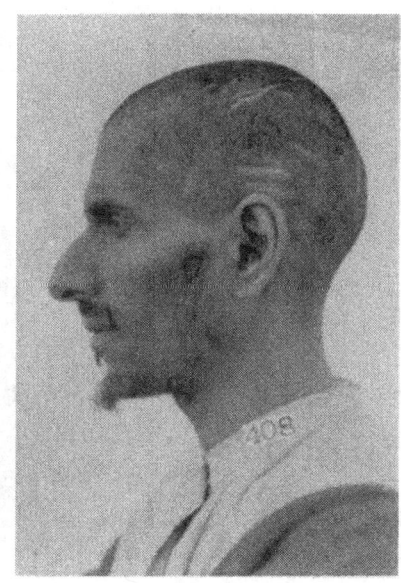

No. 1579 (age 36)

No. 1578 (age 45)

ANAIZA TRIBESMEN

WATER-WHEEL AT HADITHA